Nature-based Solutions for Remediation of Emerging Pollutants in Wastewater

Editors

Vijay Tripathi
Department of Molecular and Cellular Engineering
Jacob Institute of Biotechnology & Bioengineering
Sam Higginbottom University of Agriculture Technology & Sciences
Prayagraj, U.P., India

Pooja Tripathi
Department of Computational Biology & Bioinformatics
Jacob Institute of Biotechnology & Bioengineering
Sam Higginbottom University of Agriculture Technology & Sciences
Prayagraj, U.P., India

Pradeep Kumar Sharma
Department of Environmental Science
Graphic Era Deemed to be University
Clement Town, Dehradun, U.K., India

CRC Press
Taylor & Francis Group
Boca Raton London New York

CRC Press is an imprint of the
Taylor & Francis Group, an **informa** business
A SCIENCE PUBLISHERS BOOK

Cover credit: Image reproduced by kind courtesy of Namratha B, Department of Chemistry, The Yenepoya Institute of Arts, Science, Commerce and Management, Yenepoya (Deemed to be University), Mangaluru - 575002, Karnataka, India.

First edition published 2026
by CRC Press
2385 NW Executive Center Drive, Suite 320, Boca Raton FL 33431

and by CRC Press
4 Park Square, Milton Park, Abingdon, Oxon, OX14 4RN

CRC Press is an imprint of Taylor & Francis Group, LLC

Library of Congress Cataloging-in-Publication Data (applied for)

ISBN: 978-1-032-83392-7 (hbk)
ISBN: 978-1-032-86402-0 (pbk)
ISBN: 978-1-003-52735-0 (ebk)

DOI: 10.1201/9781003527350

Typeset in Times New Roman
by Prime Publishing Services

Preface

In the recent era, human activities are quickly varying ecosystems and contributing to unprecedented environmental challenges, making the need for sustainable, innovative approaches to environmental remediation more urgent. Emerging pollutants—ranging from pharmaceutical residues and microplastics to industrial chemicals—pose a growing threat to environmental health, including to humans. Traditional remediation methods have had varying degrees of success but repeatedly fail to address the full scope of emerging contaminants. In this context, the exploration of nature-based solutions (NBS) offers a promising alternative.

The current book, *Nature-Based Solutions for Remediation of Emerging Pollutants in Wastewater*, comprehensive textbook highlighting the role of nature-based technologies in removing emerging pollutants from wastewater. Complementary chapters focus on the use of different nature-based technologies, including wetlands restoration, phytoremediation, biofiltration, and other forms of ecosystem engineering, providing sustainable, cost-effective, and environmentally harmonious methods for managing pollutants.

The overall objective of this book is to cover the importance of the cutting-edge field of nature-based solutions for the remediation of emerging pollutants. It brings together insights from leading experts, showcasing both the scientific advancements and practical applications of NBS in tackling the growing challenge of environmental contamination. Chapter 1 discusses the different nature-based remediation techniques, including the use of bio-surfactants, adsorption, photocatalysis, nanotechnology, and genetically engineered microorganisms. Combining natural-based materials with specific chemicals can enhance pollutant remediation by achieving complete mineralization and improving the effectiveness of wastewater treatment processes. Chapter 2 delves into case studies and emerging research directions to highlight the efficacy and potential of NBS in addressing the complex challenge of heavy metal pollution in wastewater. Chapter 3 emphasizes the importance of recognizing the urgent need for water recovery and reuse while ensuring the protection and preservation of water resources and natural capital in urban settings. Chapter 4 emphasizes the fabrication/pre-treatment technologies employed for biochar production and addresses its application in removing various pollutants and treating differently sourced wastewater. Research directions on the fabrication of engineered or modified biochar and its application as an adsorbent are discussed. Chapter 5 provides a roadmap for optimizing constructed wetland systems to address emerging contaminants and meet the growing demand for environmentally sustainable wastewater treatment technologies. Chapter 6 explores the ability of microalgae to recover nutrients from wastewater, reduce greenhouse gases, and remove pollutants to improve the sustainability of wastewater treatment systems. Chapter 7 provides future ideas and the main obstacles to the real-scale application of algal-bacterial systems. By turning biomass into a valuable product, the algal biorefinery concept, when combined with wastewater treatment, lowers the total waste component of a wastewater treatment system and creates a sustainable system that supports the circular bioeconomy. Chapter 8 provides a comprehensive detail of biofiltration methods, their working and the efficiency of the process in wastewater treatment. Chapter 9 emphasizes that the widespread use of NBS is promising for enhancing leachate management strategies and protecting ecological integrity for future generations. Chapter 10 concludes by outlining future research directions and emphasizing the critical role of NBS in advancing wastewater treatment. Chapter 11 explores the various strategies to attenuate antibiotic pollution and resistant genes, and the implementation of novel ecofriendly approaches that contribute to mitigating the adverse effects of environmental antibiotic contamination. Chapter 12 focuses on microbial enzymes, enzyme technologies, and computational tools for bioprospecting numerous microbes capable of biodegrading pollutants. It also discusses the use of enzymes in the bioremediation of industrial effluents, pesticides, and pharmaceuticals, as well as the enhancement of enzyme activity.

We hope that this volume will inspire scientists, policymakers, environmental practitioners, and the broader public to embrace nature-based solutions—not as an alternative, but as a critical tool in the fight against emerging pollutants and environmental degradation. Through collaboration, creativity, and a deeper connection with nature, we can move toward a more sustainable and resilient future for generations to come.

Vijay Tripathi
Pooja Tripathi
Pradeep Kumar Sharma

Contents

CHAPTER

1

Perspective of Nature-based Solution in Wastewater Treatment: Challenges and Benefits

Shaziya Siddiqui[1*], Runit Isaac[1], Aparna[3] and Vijay Tripathi[2]

[1] Department of Chemistry, Sam Higginbottom University of Agriculture Technology and Sciences, Prayagraj - 211007, India

[2] Department of Molecular and Cellular Engineering, Jacob Institute of Biotechnology and Bioengineering, Sam Higginbottom University of Agriculture Technology and Sciences, Prayagraj - 211007, India

[3] Department of Microbiology, Graphic Era (Deemed to be) University, Dehradun - 248002, Uttarakhand, India

Introduction

Water contamination is one of the serious problems faced worldwide. About one-third of the population is not able to access clean water. Water scarcity has been continuously increasing due to larger populations, industrialization, urbanization, improper waste disposal, climate change, etc., leading to human and environmental health disturbances. Urban water management systems should decide the land use that affects the water supply. Industrialization and urbanization are the key sources of water pollution due to toxic contaminants like heavy metals, dyes, pharmaceutical products, sewage discharge, oil spills, etc. Water pollution is also caused by various non-anthropogenic mechanisms like natural disasters such as floods, droughts, and earthquakes, as well as atmospheric deposition. Agricultural practices, including the use of pesticides, fungicides, manure, crop residue, and improper disposal of contaminants, can also disseminate emerging pollutants into the environment. Radioactive wastes and microplastics also constitute pollution and are discharged into water bodies without being treated properly (Koul et al. 2022). Wastewater reuse can be important in reducing the clean water problem.

*Corresponding author: shaziya.shoaib@gmail.com

DOI: 10.1201/9781003527350-1

NBS are described as actions that protect, sustainably manage, restore, and modify the ecosystem to manage societal challenges. NBS are methods that employ natural processes in urban landscapes required for waste management with less energy and fewer chemicals. Most of the wastewater treatment technologies like advanced oxidation processes (Otter et al. 2020), adsorption (Isaac et al. 2023), biological membranes (Wang et al. 2023), constructed wetlands (Brix 1994), and disinfection methods (Alfiya et al. 2017) are employed to produce clean water. These greener technologies are highly efficient in removing contaminants with lesser waste discharge. Nowadays, green and grey technologies are widely implemented to deal with urban usage, for example, constructed wetlands and ozonation for the removal of pharmaceutical contaminants in domestic wastewater and industrial waste (Lancheros et al. 2019), and mobile vertical flow coupled with membrane filtration for treating greywater for potable use (Lakho et al. 2020).

This chapter deals with the role of NBS in treating greywater and converting it into potable water. It discusses the role, types, and economic and cost-effectiveness of NBS using various approaches.

Nature-Based Solution for Wastewater Treatment

Our climate is changing drastically, the population is expanding, and resources, including water supply, are getting scarce. This calls for a sustainable plan of utilization, which led to the development of the Sustainable Development Goals (SDGs) by the United Nations (UN). Out of the 17 SDGs, SDG 6 deals with achieving a hygienic water supply, water reuse, water management, and sanitation by 2030 (Vanham et al. 2018). One of the main challenges in achieving the SDG of clean water and water recycling is the presence of pharmaceutical compounds, heavy metals, and pathogens in wastewater that are difficult to remove using conventional technologies (Evaristo et al. 2023).

NBSs are a group of “green techniques” involving the utilization of natural alternatives such as microbial ecology, plants, organic wastes, etc., for the bioremediation of pollutants with simultaneous benefits of biodiversity conservation, sustainability, and economics. NBS works in line with nature, co-existing together rather than working against the flow. Another advantage of NBS over conventional methods is that it is cost-effective, less energy-intensive, and carbon neutral, meaning they leave little or no carbon footprint in the environment (Boano et al. 2020). Various types of NBS are discussed as follows:

Constructed Wetlands

Constructed wetlands mimic natural wetland systems but in a controlled, low-cost, and easy-to-maintain environment. They involve the simultaneous use of physical, chemical, biological, and phytoremediation processes for the eco-remediation of pollutants. CWs can be categorized into surface flow type and subsurface flow type. The surface flow type involves the free flow of water above a surface (e.g. gravel bed). In contrast, the subsurface type involves water flow through a porous surface while undergoing microbial degradation. The flow can be horizontal or vertical

(Justino et al. 2023). A study done by Huong et al. built a hybrid CW with multiple layers of substrates where plants and earthworms were also employed. The study achieved over 70% nutrient removal, 71.5% copper (Cu) removal, and 67.8% lead (Pb) removal; removal efficiency increased with the number of substrate layers (Huong et al. 2020). CWs utilize various substrates, mainly porous adsorbents such as quartz sand, zeolites, or biochar, which support microbial growth or biogeochemical cycling by acting as electron acceptors and enhance removal efficiency manyfold (Bai et al. 2022).

Apart from being a stable, cheap, and eco-friendly solution, CWs also prevent flooding, enhance soil quality, promote biodiversity, sequester carbon, and remove an array of pollutants. Constructed wetlands require a large surface area and optimum plant species to achieve good efficiency. Their performance may vary according to seasonal changes, such as temperature or sunlight variations, which affect biodegradation and photodegradation activities. Also, regular monitoring and removal of sediments are required to prevent clogging.

Biochar

Biochar is a porous solid adsorbent with a high ratio of fixed and stabilized carbon, produced by the pyrolysis of organic biomass. This organic biomass can be municipal solid waste, agricultural wastes, fruit and vegetable peels, wood or fallen leaf biomass, activated sludge, or algae. Being highly porous, biochar acts as an excellent adsorbent material that captures various micropollutants. Functional groups present in biochar, such as carboxyl, hydroxyl, and amines, form covalent bonds with the micropollutant molecules.

Biochar pores provide an ideal environment and a large surface area for the growth of degradative bacteria. It increases pollutant availability to the degrading bacteria but at the same time also protects the bacteria from possible toxicity due to the accumulated concentration of toxic compounds (Ajala et al. 2023). Biochar can be modified by physical, chemical, or biological means to increase its adsorption capacity. Physical methods include steam activation, crushing, and ball milling, which increase the surface area of biochar. Chemical activation involves corrosive acid/base modifications (e.g. HCl/KOH) or mineral etching (e.g. nitrogen doping), which enhances pore volume and number and provides oxygen or hydroxyl functional groups; iron modification imparts magnetic properties to biochar, making it easy to recover from wastewater after use. Biological means involve the attachment of biodegrading bacteria onto the surface of biochar (Wang et al. 2020, Sen et al. 2023). Biochar can be used for soil amendments or road/building construction, provided all toxicants have been sufficiently removed (Muoghalu et al. 2023).

Biological Consortium

Biological species such as gram-positive or gram-negative bacterial species, microalgae, fungi, and worms act as single entities or in combination for bioremediation. Microbial cells have a high surface-area-to-volume ratio, allowing them to convert significant amounts of toxic compounds into less harmful final products (Chettri et al. 2024). The concentration of the xenobiotic compound governs the degree of biotransformation. For example, a study reported about 90% degradation of four dyes found in textile wastewater using a bacterial consortium of *Proteus mirabilis, Morganella morganii* and *Enterobacter cloacae* (Madhushika et al. 2019).

A synergistic microbial consortium provides a survival advantage over a single strain as it cooperatively distributes pollutant and environmental load, stabilizing bacteria for longer durations, which in turn maximizes nutrient removal from wastewater (Congestri et al. 2020). A fungal-bacterial biological consortium was used successfully to treat sewage water. The consortium consisted of *Aspergillus*, *Lactobacillus*, *Pseudomonas*, and *Saccharomyces*, of which *Lactobacillus* acted on the lignocellulosic compounds, *Aspergillus* decomposed organic matter, and *Pseudomonas* removed metal ions (Sharma et al. 2023).

Algae have also been increasingly explored for wastewater treatment. Algae carry several advantages, such as the ability to utilize both organic and inorganic pollutants, having fewer nutritional requirements, and sequestering carbon. Algae can be used in algae-algae, algae-bacterium, or algae-fungal associations, as they often already co-exist in nature in a symbiotic manner—for example, lichen. Also, algae-bacterium synergism can be simultaneously used to treat pollutants and produce biofuel (Yao et al. 2019). Some examples of algae used in wastewater treatment include *Chlorella*, *Scenedesmus*, *Halochlorella rubescens*, and *Anabaena* (Gururani et al. 2022). External factors such as pH, temperature, oxygen availability, and nutritional requirements also affect the growth of biological organisms. Of these, temperature and pH primarily affect bacterial enzyme activity and the degradation of pollutants. Oxygen is important in the case of aerobes, and such cultures require agitation. Also, high amounts of salt in wastewater may inhibit bacterial growth. It is also very important to choose biological species that exhibit cooperative rather than competitive growth.

Phytoremediation

Phytoremediation comes from two words: *phyto*, meaning plants, and *remediation*, meaning clean. It is an emerging technology that utilizes tolerant plants for the remediation of targeted pollutants such as heavy metals, dyes, pesticides, pharmaceuticals, and hydrocarbons. The roots are primarily involved in taking up toxicants because of their relatively high surface area. Plants are selected based on their tolerance levels, ability to grow in polluted sites, and growth time. The mechanisms that govern phytoremediation include phytostabilization, phytodegradation, phytofiltration, and phytovolatilization (Khan et al. 2022). In phytostabilization, the pollutants accumulate in the roots and remain stationary. In phytodegradation, the toxic compounds are transformed into non-active byproducts by the plant's metabolic enzymes. Phytofiltration involves the uptake or adsorption of toxins from the surroundings. Phytovolatilization refers to the conversion of pollutants into less harmful substitutes and the release of these compounds through transpiration. The process of phytoremediation can be enhanced by the simultaneous use of rhizosphere bacteria, which work in sync with the roots to improve uptake and degradation. The removal process is controlled by factors like duration of treatment, efficiency of the root system, pH, and temperature (Diaconu et al. 2023).

For wastewater treatment, three of the most notable species—*Salvinia molesta*, *Pistia stratiotes*, and *Eichhornia crassipes*—are preferred. *Salvinia molesta* (Giant Salvinia), *Eichhornia crassipes* (Water Hyacinth), and *Pistia stratiotes* (Water Lettuce) have a high absorption capacity, exhibit a high growth rate, and show good degradation efficiency up to a temperature of 30 °C (Mustafa and Hayder 2020). Wibowo et al. used *Pistia stratiotes* and *Eichhornia crassipes* for wastewater treatment at room temperature and obtained 89% iron (Fe) removal, 98% reduction in BOD and COD, and over 70% nutrient reduction. Bioaccumulation was the main mechanism involved (Wibowo et al. 2023). Plants may also be engineered to increase their tolerance to toxic compounds, survival, and remediation capacity (Hu et al. 2020). The best approach is to combine phytoremediation with other methods of wastewater treatment for a synergistic effect and better results.

Types of Pollutants

Organic Pollutants

The main elemental composition of organic pollutants includes carbon, hydrogen, and oxygen. Organic pollutants are released into the environment through various petrochemicals, cosmetics, pharmaceuticals, paint, paper, textile industries, agricultural sources, and domestic households. By 2050, about 2.5 billion people will be affected by rivers polluted due to organic toxicants received from partially treated or untreated wastewater (Wen et al. 2017). The first primary pollutant included in this category is chlorophenols, which are used in pesticides, herbicides, antiseptics, wood preservation, and as a paint additive. Chlorophenols are divided into five groups and listed among the priority hazardous chemicals by the U.S. Agency for Toxic Substances and Disease Registry. Chlorophenol accumulates in tissues and breaks down to release reactive oxygen species (ROS), classifying it as a possible carcinogen (Group 2B) by the International Agency for Research on Cancer (IARC) (Yadav et al. 2023). Chlorophenols have been detected in drinking water, causing gastrointestinal infections, respiratory illness, hypotension, and nausea (Lampi et al. 2000). Since they exhibit quick absorption, they can also damage cell membranes, enzymes, DNA, and organs such as the liver, lungs, and kidneys. They can also disrupt the mitochondrial membrane and cause uncoupling during oxidative phosphorylation, thus affecting ATP synthesis.

Another organic contaminant detected in wastewater is Bisphenol A (BPA), an endocrine disruptor commonly used in the manufacturing of plastics and soda cans. It may leach out when exposed to heat. Its poor solubility in water and lipophilic nature cause it to accumulate in body tissues over time (Cimmino et al. 2020). BPA mimics estrogen and disrupts its signaling pathway by competitively binding to estrogen receptors. It has been reported to induce cancer as well as resistance to chemotherapeutic agents at low doses (Khan et al. 2021) and negatively affect thyroid hormone synthesis and signaling pathways (Kim and Park 2019). Another important organic pollutant is Persistent Organic Pollutants (POPs) in the environment. The use of POPs started during the Industrial Revolution post-World War II, as they were comparatively cheaper than the previous chemicals used in pesticides, agriculture, and various industrial processes. POPs persist in the environment due to their resistance to degradation. Their semi-volatile nature allows them to disperse over long distances, and being lipophilic makes them bioaccumulate in different species across continents (Titchou et al. 2021). These compounds are highly carcinogenic, can cause neurological and reproductive disorders, and weaken the immune system. They also pose a significant threat to aquatic animals and bioaccumulate in fish.

Another organic pollutant used in the textile and food industries is dyes. They are soluble organic compounds classified as reactive, direct, basic, and acidic. They exhibit high solubility in water, making them difficult to remove by conventional methods. They impart color to substances due to the chromophoric groups in their molecular structure. They are toxic, carcinogenic, and mutagenic agents. They persist as environmental pollutants and spread through the food chain, resulting in biomagnification (Lellis et al. 2019).

Inorganic Pollutants

Inorganic pollutants are heavy metals discharged by industries into wastewater. These pollutants enter water bodies through various sources, including textile, petroleum, metallurgy, chemical and paper mill industries, agricultural runoff, mining, electroplating, and burning fossil fuels. Among them, heavy metals are the most hazardous, as they persist in the environment for extended periods. For example, five heavy metals—arsenic (As), cadmium (Cd), chromium (Cr), lead (Pb), and mercury (Hg)—are classified as Group 2B (probable) carcinogens by the United States Environmental Protection Agency (USEPA) (Kumar et al. 2023). Heavy metals can bind to proteins or enzymes in the human body, disrupting their function by causing misfolding and impairing normal biological processes. Additionally, heavy metals generate reactive oxygen species (ROS), inducing oxidative stress in cells and causing cell membrane damage (Isaac et al. 2023). They have also been reported to cause hepatic and renal damage, neurological disorders, high blood pressure, vomiting, electrolyte imbalance, birth defects, and cancer.

Biological Pollutants

Excessive nutrients (eutrophication) and organic pollutants in wastewater create an ideal environment for microbial growth. Moreover, discharge from hospitals, poultry farms, and domestic fecal waste adds to the pathogenic load, which includes bacteria, viruses, fungi, protozoa, and worms. For example, *Ascaris lumbricoides*, a commonly found parasitic roundworm, is the causative agent of ascariasis (Yesilay et al. 2023). If wastewater is not efficiently treated, these pathogens may spread among the human population and cause various diseases or outbreaks. Among them, the most common diseases include typhoid, cholera, dysentery, hepatitis, salmonellosis, stomach ulcers, and skin infections. The overuse of antibiotics has further worsened the issue by promoting the development of antibiotic-resistant strains, making infections difficult to treat. Pathogens can also be transmitted to plants through irrigation practices using partially treated wastewater and subsequently to humans.

Emerging Pollutants

Emerging pollutants are a group of new contaminants that are presently not under any environmental regulation, and their toxicological profiles are not yet well understood. They can pose a significant ecological threat. This category includes Pharmaceutical and Personal Care Products (PPCPs), newly launched drugs, antiseptics, antibiotics, hormones, previously undocumented microbes, cosmetics, microbeads, and microplastics. Between 2011 and 2021, there has been an exponential rise in studies citing emerging contaminants. Their main sources include domestic households,

Table 1: Type of pollutants and their health effects

Type of Pollutant	Source	Health Effects	Reference
Arsenic	Industrial waste, fertilizers	Carcinogenic, skin irritation, cardiovascular illness, diarrhea & vomiting	Kumar et al. 2021
Lead	Mining, pesticides, paint industries	Anemia, kidney stones, hypertension, neurological disorders	Collin et al. 2022
Mercury	Waste incineration, mining, hospitals, cement industries, chlor-alkali industry	Neurotoxic-depression, hallucination & tremors affect the thyroid gland and gastrointestinal tract.	Verma et al. 2018
Bacteria	*Salmonella typhi*	Typhoid	Chahal et al. 2016
	E. coli	Urinary tract infection, gastroenteritis	
	Vibrio cholerae	Cholera	
Virus	Adenovirus	Flu, respiratory infections	Chahal et al. 2016
	Hepatitis A virus	Hepatitis	
	Rotavirus	Diarrhea	
Diclofenac Sodium	Pharmaceutical industries, household sewage discharge	Liver and kidney cytotoxicity, Gastrointestinal damage, malfunction and human ovulation	Higgins et al. 2022
Ciprofloxacin	Pharmaceutical industries, household sewage discharge	Liver and kidney cytotoxicity, Gastrointestinal damage, malfunction and human ovulation	Higgins et al. 2022
Crystal Violet	Textile, printing and paint industry	Skin irritation, intestinal tract infection, kidney failure, and cancer	Paul et al. 2024
Methylene Blue	Textile, printing and paint industry	Skin irritation, intestinal tract infection, kidney failure, and cancer	Siddiqui 2018

laundry services, animal farms, chemical industries, hospitals, and pharmaceutical industries (Krishnakumar et al. 2022, Wang et al. 2024).

Microplastics are particles generated through the crushing or degradation of plastics, resulting in particle sizes smaller than 5 mm. Data suggests that about 65 million microplastic particles are released into freshwater bodies daily. In humans, they can cross the blood-brain barrier and exert cytotoxic and neurotoxic effects (Li et al. 2023). Similarly, triclosan—an antibacterial compound found in skin-care cosmetics, toothpaste, and soaps—has been shown to negatively affect fish, algae, and phytoplankton. PPCPs are another large group of emerging organic contaminants. They include all kinds of drugs—prescription or over-the-counter—hormones, antibiotics, painkillers, and daily-use products such as body lotions, fragrances, sunscreens, and shampoos. They have been reported to bioaccumulate in fish and plants (Osuoha et al. 2023). PPCPs are frequently released from households, hospitals, and pharmaceutical industries. The presence of antibiotics in wastewater has exacerbated the issue of antibiotic resistance. Wastewater serves as a significant reservoir for the development of antibiotic-resistant bacteria, as it harbors numerous pathogenic microbes and receives different classes of antibiotics discharged from homes and hospitals (Wydro et al. 2024).

Remediation Technique for Nature-based Wastewater Treatment

Natural phenomena and systems are applied in nature-based wastewater treatment techniques. Many of these techniques regularly require filtration, biological degradation, and plant uptake to remove pollution; thus, they often require fewer mechanical and chemical inputs to enhance water quality. The following sub-section highlights several remediation techniques that utilize natural resources, including biosurfactants, adsorption, photocatalytic oxidation, nanotechnology, and genetically engineered microorganisms.

Biosurfactant

Biosurfactants are bioactive molecules that originate from natural settings and are described as efficient enough to remove mean wastewater. These compounds reduce the interfacial tension that is present between gases, liquids, and solids, and this makes it possible to wipe out and disintegrate hydrophobic pollutants, including heavy metals, oil, and organic compounds. Recent studies mention the advantages of biosurfactants, and therefore, their application is possible for the enhancement of wastewater treatment processes because of their non-toxic, biodegradable, and effective features in diverse environmental conditions (Miller et al. 2023). They are highly soluble and enhance foams and emulsion stability.

Biosurfactants are unique and versatile and can be classified into many categories, including glycolipids, polysaccharide proteins, and lipopeptides. The glycolipids, which are synthesized by bacteria and yeasts, such as rhamnolipids and sophorolipids, are known to be very effective in the emulsification of hydrophobic compounds and enhancing the biodegradation of these compounds (Mulligan et al.

2022). Their effectiveness has been evidenced in the treatment of industrial effluents and petroleum hydrocarbons, as shown by a recent study (Vijayakumar et al. 2021). High surfactant abilities are declared(change into processed) by lipopeptides, for example: surfactin produced by *Bacillus* species, which are used to treat wastewater contaminated with heavy metals (Kumar et al. 2023).

Adsorption

It is noteworthy to point out that pollutants in wastewater may be easily removed from the liquid phase using the process of adsorption if they adhere to the surface of the adsorbent material. This technique is used because it is simple to apply, effective, and can be applied to treat all types of pollutants. Many adsorbents, such as biochar, zeolites, and activated carbon, have been well studied and even used in numerous approaches to tackle the problem of water pollution (Khan et al. 2023). The major phenomenon associated with the adsorption process is the physical and chemical adsorption between the adsorbent and the pollutants. Chemisorption, which involves covalent or ionic forces and is often irreversible, differs from physisorption, which requires weak van der Waals forces and is generally reversible (Ahmad et al. 2022). The degree of adsorption is regulated by a string of factors, such as the surface area of the adsorbent, the type of adsorbate, and conditions including the temperature and the pH (Zhu et al. 2023). Activated carbon remains one of the most popular and effective adsorbents because of its large surface area and porous structure. As per Namasivayam et al. (2023), it is highly successful in removing organic pollutants such as dyes and volatile organic compounds (VOCs) from wastewater. The preparation and regeneration processes have largely improved the performance and affordability of activated carbon (Namasivayam et al. 2023).

Zeolites

Natural and synthetic zeolites have been used to remove ammonia and heavy metals from wastewater due to their high ion exchange capacity and well-defined pore structures. According to Yoon et al. (2023), it has been suggested that modified zeolites can be employed effectively for better contaminant removal efficiency with subsequent recovery (Yoon et al. 2023).

Charcoal

Pyrolysis of organic molecules produces chars, a new class of adsorbent materials potentially applicable in wastewater treatment. They are advantageous because they are inexpensive and environmentally friendly. According to recent research (Lehmann et al. 2022), charcoal biochar can be used to eliminate organic contaminants, heavy metals, or even pharmaceuticals from wastewater.

Photocatalysis

Photocatalysis is the activity of a photocatalyst material utilized for decomposing contaminants in wastewater when exposed to light and generating reactive species. Photons from ultraviolet (UV) or visible light sustain the process by exciting the photocatalyst to produce reactive oxygen species (ROS), which include superoxide anions and hydroxyl radicals. This makes photocatalysis a powerful technology for cleaning up wastewater due to its efficiency in degrading different organic and inorganic pollutants (Liu et al. 2023). Several important stages are involved in the photocatalytic process: Initially, photons are captured by the photocatalyst, thereby energizing electrons from the valence band into the conduction band. Consequently, electron-hole pairs that migrate onto the surface of the photocatalyst are formed. ROS comes about when electron-hole pairs combine with water and oxygen molecules. These ROS break down any pollutants in the wastewater, either completely mineralizing them into CO_2 and H_2O or degrading them into less harmful substances (Li et al. 2023). Various photocatalysts that are stable, non-toxic and have strong photo-catalytic activity include:

Titanium Dioxide

TiO_2 is widely used as a photocatalyst. It has high potential to decompose many types of organic matter, such as dyes, insecticides, and drugs. Considerable progress has been made in this area, including the development of nanomaterials and TiO_2-based composites. These advances have increased its light-driven catalytic reactivity and extended it to visible light (Huang et al. 2023).

Zinc Oxide

ZnO is a well-recognized, highly active, and cost-effective photocatalyst. It can break down organic pollutants and remove heavy metals from wastewater. Recent studies have focused on improving the stability and photocatalytic activity of ZnO by doping and forming composites (Zhang et al. 2022).

Graphene-based Photocatalysts

Advanced catalysis based on graphene and its corresponding products, such as reduced graphene oxide or GQDs, has been developed. These materials exhibit a large surface area, excellent electrical conductivity, and the ability to absorb visible light. Frequently, these materials are combined with traditional catalysts to boost performance (Singh et al. 2023).

Nanotechnology

Nanotechnology is the engineered design and application of materials at a scale ranging from 1 to 100 nanometers. The high surface area, reactivity, and catalytic behavior of nanomaterials have resulted in applications in wastewater treatment (Feng et al. 2023). Nanomaterials can interact with pollutants through adsorption, catalytic degradation, and photocatalysis. Their large surface area increases adsorption capacity, and their catalytic properties offer additional advantages for contaminant degradation. For example, nanoparticles such as zero-valent iron (ZVI) and titanium dioxide (TiO_2) are used to catalyze the decomposition of organic contaminants and reduce metal ions in groundwater (Zhang et al. 2023). In addition, photocatalytic nanomaterials can generate reactive oxygen species from light-driven reactions, which support the photomineralization of pollutants or their degradation into minerals (Liu et al. 2023).

Metal Nanoparticles

Metal nanoparticles, including silver (Ag), gold (Au), and iron (Fe), are used in a range of applications owing to their antibacterial qualities and capacity to catalyze redox processes. Zero-valent iron nanoparticles often reduce chlorinated organic compounds and heavy metal ions in wastewater (Zhang et al. 2023). Silver nanoparticles have shown potent antibacterial properties, making them valuable for eliminating pathogens (Rao et al. 2023).

Metal Oxide Nanoparticles

Titanium dioxide, zinc oxide, and cerium dioxide have high photocatalytic activity under either ultraviolet or visible light. Degradation by nanoparticles is potentially implemented to degrade various organic contaminants and disinfect wastewater. Among the family of semiconductors, titanium dioxide (TiO_2) has attracted much attention as a photocatalyst due to its ability to generate hydroxyl radicals that can efficiently degrade many organic pollutants (Liu et al. 2023).

Carbon-Based Nanomaterials

Nanoscale additives can adsorb due to their vast surface area. Materials like graphene oxide (GO) and carbon nanotubes have demonstrated their efficiencies in removing organic dyes and heavy metals from aqueous solutions through adsorption or electrochemical methods. Current studies have focused on enhancing the specificity of these materials through chemical modifications (Chen et al. 2023).

Genetic Engineering in Wastewater Treatment

Genetic engineering alters an organism's genetic material to enable certain characteristics. It is employed in wastewater treatment to increase an organism's metabolic capacities, strengthen microorganisms' resistance to detrimental compounds, or help produce useful by-products. Genetically engineered microorganisms (GEMs) are being developed to improve the efficiency and precision of wastewater treatment and to overcome some of the drawbacks of existing practices (Zhou et al. 2023).

Mechanisms of Genetic Engineering in Microorganisms

Genetically modified microbes are obtained by adding, deleting, or modifying one or more genes, thus improving their metabolic pathways or physiological characteristics. Standard approaches include:

- Improved degradation pathway: Adding genes encoding enzymes that decompose more complex contaminants (e.g. xenobiotics or medicines) into less complex and less harmful compounds.
- Improved heavy-metal resistance: Changing the genetic material to improve

the microorganism's ability to withstand and neutralise heavy metals, thereby increasing its capacity to remove these heavy metals from contaminated wastewater.

- More specific contaminant bioremediation involves biological manipulation to equip microbes with novel receptors or a transport system, which allows them to target and break down hydrocarbons such as benzene derivatives specifically.
- Engineering of enzymes: A common approach involves inserting genes encoding certain enzymes into microbes. These enzymes can degrade complex organic matter like hydrocarbons, insecticides, or pigments. For example, *Pseudomonas putida* engineered with genes for hydrocarbon breakdown shows enhanced bioremediation of petroleum-contaminated soil and effluent (Gao et al. 2023).
- Metal resistance engineering: Genetic modifications increase the ability of microbes to withstand and neutralise heavy metals. *Cupriavidus metallidurans* has been genetically modified to increase the ability of metal resistance species to tolerate toxic metals such as mercury and cadmium. It is, therefore, very effective in cleaning industrial wastewater heavily contaminated with heavy metals (Wang et al. 2023).

Recent progress in synthetic biology has resulted in the creation of genetically edited microorganisms (GEMs) that possess modified pathways capable of breaking down certain contaminants. This involves the development of microbes engineered to possess specific biosynthetic pathways capable of breaking down newly identified pollutants, such as medicines and personal care items (Li et al. 2022).

Challenges and Drawbacks in Treatment Techniques

Nature-based solutions (NBS) in wastewater treatment involve using natural processes and ecosystems to enhance treatment sustainably. Nevertheless, despite their considerable promise in terms of cost-effectiveness, environmental restoration, and improved aesthetics of urban areas, the widespread implementation and success of NBS encounter significant obstacles. An important obstacle is in guaranteeing the efficacy and consistency of NBS, notwithstanding their susceptibility to external influences such as climate, seasonal variations, and local site circumstances.

One instance of NBS, artificial wetlands, may experience reduced treatment effectiveness in colder climes due to decreased microbial activity inside the system (Brix 1994, Vymazal 2014). To tackle this issue, it is necessary to implement thorough monitoring and adaptive management strategies to guarantee the effectiveness of the system and its adherence to treatment requirements (Kadlec and Knight 1996, Miller et al. 2018). One significant obstacle is the land and space needs of NBS, such as built wetlands and vegetated swales. These NBS often need large areas, which may be problematic in densely populated metropolitan areas (Zhai et al. 2018). Creating huge artificial wetlands in densely populated urban areas with limited space is unfeasible (Ghosh et al. 2018). Integrating NBS into urban planning involves incorporating multi-functional spaces, such as green roofs or parks with integrated treatment elements, to optimize land use (Falkenmark and Rockström 2006, Li et al. 2020). One of the major drawbacks of NBS is the complexity associated with design and engineering. Achieving efficient execution of this task requires a profound comprehension of ecological processes, which may be challenging due to the diverse environmental circumstances and the need for precise system specifications (Hussey et al. 2020).

Furthermore, NBS may be likened to a framework that requires ongoing upkeep and administration to ensure its long-term functionality. Continuous management is necessary to prevent system deterioration caused by invasive species, siltation, and plant diseases (Tanner et al. 1995, Wang et al. 2016). By including explicit maintenance instructions and actively engaging local communities, there are further prospects to enhance the sustainability of these systems. Additional significant barriers include legislative and institutional factors (Barton et al. 2016, Wang and Li 2019). Existing regulations and technology for water treatment seldom, if ever, fully consider or acknowledge NBS (Davis and Slavin 2017). This obstacle has hindered the full integration of NBS into traditional treatment frameworks, making it challenging to deploy NBS effectively (Katsigiannis et al. 2017). It is crucial to emphasize that the promotion and implementation of revised legislation and standards should be backed by evidence of their effectiveness, shown via case studies, in overcoming these hurdles and gaining broader acceptance of the notion of NBS (Jenkins et al. 2017).

Undoubtedly, the way the general public perceives and accepts something may make the process of adopting NBS much more complex. Doubts may be raised about its efficacy compared to conventional technologies, which might hinder its adoption (Niemelä et al. 2011). Public educational campaigns and outreach programs will be crucial in ensuring the success of NBS by effectively showing the positive impact on environmental quality and ecosystem services, hence gaining and maintaining public support (Barton and Lindhjem 2015, Zhai et al. 2018). In addition, economic factors may also have a significant impact, since implementing NBS generally involves substantial upfront expenses and may have uncertain long-term economic advantages (Costanza et al. 2014, Zhang et al. 2019). The financial viability of NBS is supported by thorough cost-benefit evaluations and the possibility of significant long-term savings resulting from reduced operating and maintenance expenses (Jenkins et al. 2017). Additionally, providing funding possibilities and incentives would alleviate early financial difficulties and facilitate the implementation of NBS applications (Zhao et al. 2018).

Scaling up NBS poses a significant issue due to the discrepancy between their successful performance at small scales and the difficulties encountered when attempting to implement them on a larger scale (Grove et al. 2018). Conducting fundamental research on scalability, implementing pilot projects, and adopting modular design strategies that enable gradual extension may help address scalability challenges and make NBS feasible on larger scales (Li et al. 2020).

References

Ahmad, M., Wani, A.H. and Kumar, P. 2022. Recent advances in adsorption technology for wastewater treatment: A review. J. of Environ Mangtt, 319: 115639.

Ajala, O.A., Akinnawo, S.O., Bamisaye, A. et al. 2023. Adsorptive removal of antibiotic pollutants from wastewater using biomass/biochar-based adsorbents. RSC Advances, 13: 4678.

Alfiya, Y., Friedler, E., Westphal, J., Olsson, O. and Dubowski, Y. 2017. Photodegradation of micropollutants using V-UV/UV-C processes; Triclosan as a model compound. Sci of Total Environ, 601-602: 397-404.

Barton, H. and Lindhjem, C. 2015. Urban green infrastructure and its role in improving human health. Urban Forestry & Urban Green, 14: 510-518.

Bai, S., Wang, X. and Zhang, Y. 2022. Constructed wetlands as nature-based solutions for the removal of antibiotics: Performance, microbial response, and emergence of antimicrobial resistance (AMR). Sustainability, 14: 14989.

Barton, H., Grant, M. and Mitcham, C. 2016. The role of public perception in the acceptance of urban green infrastructure. Landscape and Urban Plann, 151: 78-86.

Boano, F., Caruso, A., Costamagna, E. et al. 2020. A review of nature-based solutions for greywater treatment: Applications, hydraulic design, and environmental benefits. Sci Total Environ, 711: 134731.

Brix, H. 1994. Functions of constructed wetlands. Water Sci and Technol, 29(4): 77-88.

Chahal, C., Van den, A.B., Young, F., Franco, C. et al. 2016. Pathogen and particle associations in wastewater: significance and implications for treatment and disinfection processes. Advan Applied Microbiol, 97: 63-119.

Chen, X., Zhang, Q. and Li, X. 2023. Carbon-based nanomaterials for wastewater treatment: Recent advances and future directions. Environm Sci: Nano, 10: 763-781.

Chettri, D., Verma, A.K. and Verma, A.K. 2024. Bioaugmentation: An approach to biological treatment of pollutants. Biodegradation, 35: 117-135.

Cimmino, I., Fiory, F., Perruolo, G., Miele, C. et al. 2020. Potential mechanisms of bisphenol A (BPA) contributing to human disease. Intern J. Molecul Sci, 21: 5761.

Collin, M.S., Venkatraman, S.K., Vijayakumar, N. et al. 2022. Bioaccumulation of lead (Pb) and its effects on human: A review. J. of Hazard Mater Adv, 7: 100094.

Congestri, R., Savio, S., Farrotti, S. et al. 2020. Developing a microbial consortium for removing nutrients in dishwasher wastewater: Towards a biofilter for its up-cycling. Wat Sci and Technol, 82: 1142-1154.

Costanza, R., Groot R. de, Sutton, P., Ploeg, S.V. der, Anderson, S.J., Kubiszewski, I. et al. 2014. Changes in the global value of ecosystem services. Global Environ Change, 26: 132-144.

Davis, A.P. and Slavin, S.M. 2017. The role of policy in advancing the adoption of green infrastructure. J. Environ Managt, 203: 178-187.

Diaconu, L.I., Covaliu-Mierlă, C.I., Păunescu, O. et al. 2023. Phytoremediation of wastewater containing lead and manganese ions using algae. Biology, 12: 773.

Evaristo, J., Jameel, Y. and Tortajada, C. et al. 2023. Water woes: The institutional challenges in achieving SDG 6. Sustain Earth Rev. doi.org/10.1186/s42055-023-00067-2.

Falkenmark, M. and Rockström, J. 2006. The new green revolution: Water scarcity and the role of urban planning. Agricul Water Managt, 86(1-2): 1-15.

Feng, Y., Liu, X. and Wang, H. 2023. Nanotechnology in wastewater treatment: A review of recent advances and emerging applications. J. Environ Managt, 328: 115634.

Gao, J., Wang, H. and Liu, X. 2023. Engineering pseudomonas for enhanced bioremediation of petroleum hydrocarbons: Recent advances and applications. J. of Hazard Mater, 454: 131218.

Gaur, A., Prasad, V. and Singh, R. 2022. Recent advances in biosurfactant production and application for wastewater treatment: A review. Biotechnol Rep, 33: 00550.

Ghosh, S., Vale, R. and Vale, B. 2018. Integrating nature-based solutions in urban environments: Opportunities and challenges. Sustain Cities and Soc, 39: 520-528.

Grove, J.M., Locke, D.H., O'Neil-Dunne, J.P.M., Nowak, D.J. and Walton, J.T. 2018. The role of green infrastructure in urban sustainability. Urban Ecosystems, 21(3): 491-504.

Gururani, P., Bhatnagar, P., Kumar, V. et al. 2022. Algal consortiums: A novel and integrated approach for wastewater treatment. Water, 14: 3784.

Hein, L., Gray, P., Vulturius, G., Bakkenes, M. and Dunford, R. 2018. Evaluating the ecosystem services provided by nature-based solutions. Ecosystem Services, 34: 1-10.

Higgins, P., Siddiqui, S.H. and Kumar, R. 2022. Design of novel graphene oxide/halloysite nanotube@polyaniline nanohybrid for the removal of diclofenac sodium from aqueous solution. Environm Nanotechnol Monitor & Managt, 17: 100628

Hu, H., Li, X., Wu, S. et al. 2020. Sustainable livestock wastewater treatment via phytoremediation: Current status and future perspectives. Biores Technol, 315: 123809.

Huang, H., Zhang, X. and Yang, Y. 2023. Advancements in titanium dioxide photocatalysts for wastewater treatment: A review. Environ Sci & Technol, 57: 852-870.

Huong, M., Costa, D.T., and Hoi, B.V. 2020. Enhanced removal of nutrients and heavy metals from domestic-industrial wastewater in an academic campus of Hanoi using modified hybrid constructed wetlands. Water Sci & Technol, 82: 1995-2006.

Hussey, K., Pittock, J., McKay, J. and Chan, T. 2020. The challenges and opportunities of nature-based solutions for water management. Water Research, 173: 115481.

Isaac, R., Siddiqui, S., Aldosari, A. and Kashifuddin, M. 2023. Magnetic biochar derived from *Juglans regia* for the adsorption of Cu^{2+} and Ni^{2+}: Characterization, modeling, optimization, and cost analysis. J. Saudi Chem Soc, 27: 101749.

Jenkins, J.M., Andrews, B.R. and Patel, S. 2017. The economic implications of implementing nature-based solutions in wastewater treatment. Environ Econ and Policy Studies, 19: 359-379.

Justino, S., Calheiros, C.S.C., Castro, P.M.L. and Gonçalves, D. 2023. Constructed wetlands as nature-based solutions for wastewater treatment in the hospitality industry: A review. Hydrology, 10(7): 153.

Kadlec, R.H. and Knight, R.L. 1996. *Treatment Wetlands*. CRC Press.

Katsigiannis, Y., Papageorgiou, M. and Tsihrintzis, V.A. 2017. Policy frameworks for integrating nature-based solutions in wastewater treatment. Environm Sci & Policy, 69: 95-103.

Khan, S.A., Bibi, H. and Khan, M.N. 2023. Activated carbon and its applications in wastewater treatment: A review of recent developments. Environm Technol & Innovat, 26: 103885.

Khan, A.U., Khan, A.N., Waris, A. et al. 2022. Phytoremediation of pollutants from wastewater: A concise review. Open Life Sciences, 17: 488-496.

Khan, N.G., Correia, J., Adiga, D., Rai, P.S. et al. 2021. A comprehensive review on the carcinogenic potential of bisphenol A: Clues and evidence. Environmental Science and Pollution Research International, 28: 19643-19663.

Kim, M.J. and Park, Y.J. 2019. Bisphenols and thyroid hormone. Endocrinol and Metabol, 34: 340-348.

Koul, B., Bhat, N., Abubakar, M., Mishra, M., Arukha, A.P. and Yadav, D. 2022. Application of natural coagulants in water treatment: a sustainable alternative to chemicals. Water, 14: 3751.

Krishnakumar, S., Singh, D.S.H., Godson, P.S. et al. 2022. Emerging pollutants: Impact on environment, management, and challenges. Environm Sci Pollut Res, 29: 72309-72311.

Kumar, A., Ali, M., Kumar, R. et al. 2021. Arsenic exposure in Indo Gangetic plains of Bihar causing increased cancer risk. Scientific Report, 11: 2376.

Kumar, A., Singh, P. and Sharma, A. 2023. Lipopeptide biosurfactants for heavy metal removal and organic pollutant degradation: Current trends and future prospects. J. Environm Managt, 310: 114742.

Lakho, F.H., Le, H.Q., Van Kerkhove, F., Igodt, W., Depuydt, V., Desloover et al. 2020. Water treatment and re-use at temporary events using a mobile constructed wetland and drinking water production system. Sci of Total Environ, 737: 139630.

Lampi, P., Vohlonen, I., Tuomisto, J. and Heinonen, O.P. 2000. Increase of specific symptoms after long-term use of chlorophenol polluted drinking water in a community. Europ J of Epidemiol, 16: 245-251.

Lancheros, J.C., Madera-parra, C.A., Caselles-osorio, A., Torres-lópez, W.A., Vargas-ramírez, X.M. 2019. Ibuprofen and naproxen removal from domestic wastewater using a horizontal subsurface flow constructed wetland coupled to ozonation. Ecolog Engg, 135: 89-97.

Langerhans, R.B., Meyer, J.L. and Fulton, S. 2017. The role of nature-based solutions in managing urban water systems. J Urban Water Managtt, 5: 67-79.

Lehmann, J., Joseph, S. and Pan, G. 2022. Biochar as a sustainable material for wastewater treatment: Recent advances and future prospects. Environ Sci & Technol, 56: 1864-1882.

Lellis, B., Polanio, C.Z.F., Pampile, J.A. and Polanio, J.C. 2019. Effect of textile dyes on health and the environment and bioremediation potential of living organisms. Biotechnol Res and Innov, 3: 275-290.

Li, M., Zhang, Y., Chen, H. and Liu, X. 2020. Modular approaches to scaling up nature-based solutions. Water, 12(7): 1896.

Li, Y., Tao, L., Wang, Q. et al. 2023. Potential health impact of microplastics: A review of environmental distribution, human exposure, and toxic effects. Environm & Health, 1: 249-257.

Li, X., Chen, Y. and Zhang, Q. 2022. Synthetic biology approaches for engineering microorganisms to degrade emerging contaminants in wastewater. Environm Sci & Technol, 56: 3123-3135.

Li, X., Wang, Y. and Wang, S. 2023. Recent developments in photocatalysis for wastewater treatment: Mechanisms, materials, and applications. Chem Engg J, 454: 140463.

Liu, J., Zhang, Z. and Xu, J. 2023a. Recent advances in genetic engineering of microorganisms for wastewater treatment: Mechanisms, applications, and challenges. Chem Engg J, 459: 140267.

Liu, X., Wang, L. and Zhang, Y. 2023b. Photocatalytic nanomaterials for water treatment: Recent advances and challenges. Chem Engg J, 460: 141924.

Madhushika, H.G., Ariyadasa, T.U. and Gunawardena, S.H.P. 2019. Biological decolourization of textile industry wastewater by a developed bacterial consortium. Water Sci and Technol, 80: 1910-1918.

Miller, J., Thompson, L. and Green, R. 2018. Assessing the performance of constructed wetlands under varying environmental conditions. Ecolog Engg, 117: 43-52.

Miller, R., Johnson, M. and Chen, L. 2023. Biosurfactants in wastewater treatment: Recent advances and emerging applications. J of Hazard Mater, 454: 131362.

Mulligan, C.N., Yong, R.N. and Gibbs, B.F. 2022. Environmental applications for biosurfactants: Enhanced pollutant removal and wastewater treatment. Current Opinion in Environ Sci & Health, 24: 100316.

Muoghalu, C.C., Owusu, P.A., Lebu, S. et al. 2023. Biochar as a novel technology for treatment of onsite domestic wastewater: A critical review. Front of Environm Sci, 11: 1095920.

Mustafa, H.M. and Hayder, G. 2020. Recent studies on applications of aquatic weed plants in phytoremediation of wastewater: A review article. Ain Shams Engg J, 12: 355-365.

Namasivayam, C., Kadirvelu, K. and Rajarathnam, S. 2023. Application of activated carbon for removal of organic pollutants from wastewater: A review. J. Clean Product, 393: 136460.

Niemelä, J., Saarela, S.-R., Söderman, T., Kopperoinen, L., Yli-Pelkonen, V., Väre, S. et al. 2011. The role of urban green infrastructure in promoting public health. Health & Place, 17(1): 761-768.

Osuoha, J.O., Anyanwu, B.O. and Ejileugha, C. 2023. Pharmaceuticals and personal care products as emerging contaminants: Need for combined treatment strategy. J. of Hazard Materi Advanc, 9: 100206.

Otter, P., Mette, K., Wesch, R., Gerhardt, T., Krüger, F., Goldmaier, A. et al. 2020. Oxidation of selected trace organic compounds through the combination of inline electro-chlorination with UV Radiation (UV/ECl2) as alternative AOP for decentralized drinking water treatment. Water, 12: 1-19.

Paul, A.S., Khan, S.H., Hassan, I. and Siddqiui, S. 2024. Preparation and performance of novel bio-nanocomposite ZnMg@PH for the removal of hazardous brilliant green, crystal violet, and diclofenac sodium. Biomass Conversion and Biorefinery, 10.1007/s13399-024-05801-0.

Rao, P., Kumar, A. and Singh, A. 2023. Application of metal nanoparticles in wastewater treatment: A review. J. of Hazard Mater, 450: 131580.

Sen, U., Esteves, B., Aguiar, T. and Pereira, H. 2023. Removal of antibiotics by biochars: A critical review. Applied Science, 13, 11963.

Sharma M., Agarwal, S., Malik, R.A. et al. 2023. Recent advances in microbial engineering approaches for wastewater treatment: A review. Bioengineered, 14(1): 2184518.

Siddiqui, S.H. 2018. The removal of Cu^{2+}, Ni^{2+} and Methylene Blue (MB) from aqueous solution using Luffa Actangula Carbon: Kinetics, thermodynamic and isotherm and response methodology. Groundwat for Sustain Develop, 6: 141-149.

Singh, P., Sharma, A. and Yadav, S. 2023. Graphene-based photocatalysts for environmental remediation: Recent advances and future prospects. Mater Today Advanc, 15: 100320.

Tanner, C., Kadlec, R.H. and Headley, T.R. 1995. Constructed wetlands for wastewater treatment: An overview. Water Sci and Technol, 32(3): 127-136.

Tanner, C. and Clayton, T. 2008. Design and management of nature-based solutions for wastewater treatment. J. of Environ Managt, 87: 128-139.

Titchou, F.E., Zazou, H., Afanga, H., El Gaayda, J. et al. 2021. Removal of Persistent Organic Pollutants (POPs) from water and wastewater by adsorption and electrocoagulation process. Groundwat for Sustain Develop, 13: 100575.

Vanham, D., Hoekstra, A.Y. and Wada, Y. 2018. Physical water scarcity metrics for monitoring progress towards SDG target 6.4: An evaluation of indicator 6.4.2 Level of water stress. Sci of the Total Environ, 614: 218-232.

Verma, R.K., Sankhla, M.S. and Kumar, R. 2018. Mercury contamination in water and its impact on public health. Intern J Foren Sci, 1(2): 72-78.

Vijayakumar, S., Lee, J. and An, S. 2021. Sophorolipids for petroleum hydrocarbon bioremediation: Recent advances and applications. J. of Clean Product, 316: 128211.

Vymazal, J. 2014. Constructed wetlands for wastewater treatment: A review. Environ Sci & Technol, 48: 9178-9193.

Wang, R. and Li, Z. 2019. Case studies of successful nature-based solutions in urban wastewater management. Urb Wat J, 16: 356-367.

Wang, R., Tang, H., Yang, R. and Zhang, J. 2024. Emerging contaminants in water environments: Progress, evolution, and prospects. Wat Sci and Technol, 89: 2763-2782.

Wang, S., Zhang, Y. and Liu, X. 2023. Genetically modified *Cupriavidus metallidurans* for enhanced heavy metal removal from industrial wastewater. Environ Technol & Innovat, 25: 103885.

Wang, X., Guo, Z., Hu, Z. and Zhang, J. 2020. Recent advances in biochar application for water and wastewater treatment: A review. PeerJ, 8: e9164.

Wang, Z., Brown, G. and Liu, Y. 2016. Community involvement in the management of urban green infrastructure. Landscap and Urb Plann, 150: 90-99.

Wen, Y., Schoups, G. and van de Giesen, N. 2017. Organic pollution of rivers: Combined threats of urbanization, livestock farming and global climate change. Sci Rep, 7: 43289.

Wibowo, Y.G., Nugraha, A.T. and Rohman, A. 2023. Phytoremediation of several wastewater sources using *Pistia stratiotes* and *Eichhornia crassipes* in Indonesia. Environ Nanotechnol Monitor and Managtt, 20: 100781.

Wydro, U., Wołejko, E., Luarasi, L. et al. 2024. A review on pharmaceuticals and personal care products residues in the aquatic environment and possibilities for their remediation. Sustainability, 16(1): 169.

Yadav, S., Kumar, S. and Haritash, A.K. 2023. A comprehensive review of chlorophenols: Fate, toxicology and its treatment. J. of Environ Managt, 342: 118254.

Yao, S., Lyu, S., An, Y. et al. 2019. Microalgae-bacteria symbiosis in microalgal growth and biofuel production: A review. J Appl Microbiol, 126: 359-368.

Yesilay, G., dos Santos, O.A.L., Roger, A.B. and Hazeem, L.J. 2023. Impact of pathogenic bacterial communities present in wastewater on aquatic organisms: Application of nanomaterials for the removal of these pathogens. Aquatic Toxicol, 261: 100620.

Yoon, Y., Lee, J. and Kim, J. 2023. Enhanced heavy metal removal using modified zeolites: Recent advances and applications. Chemosphere, 297: 134232.

Zhai, Y., Chen, W. and Xu, H. 2018. Advancements in nature-based solutions for sustainable wastewater management. Sustain Wat Resour Managt, 4: 1-15.

Zhang, W., Liu, J. and Zhang, Y. 2022. Zinc oxide photocatalysts for water treatment: A comprehensive review of recent progress and future challenges. J. Hazard Mater, 438: 129547.

Zhang, W., Yang, M. and Zhang, H. 2023. Zero-valent iron nanoparticles for wastewater treatment: Synthesis, characterization, and applications. J Clean Product, 392: 136425.

Zhang, X., Li, H. and Zhou, Y. 2019. Funding opportunities and financial mechanisms for nature-based solutions. Environmental Finan, 22: 11-21.

Zhao, X., Liu, Y., Wang, Y., & Huang, J. 2018. Evaluating land use for nature-based wastewater treatment systems. Landscap Ecol, 34: 1559-1572.

Zhou, J., Zhang, L. and Li, X. 2023. CRISPR/Cas9-mediated genetic engineering in microorganisms for wastewater treatment: Recent advances and future directions. Biotechnol Advan, 57: 107974.

Zhu, L., Wang, L. and Zhang, Y. 2023. Nanomaterials and composite adsorbents in wastewater treatment: A review of recent advances. Environ Res Lett, 18: 025001.

CHAPTER
2

Mitigation of Heavy Metal Pollutants in Wastewater through Nature-based Solutions

Nitesh Kumar[1*], Varun Kumar[2] and Priyanka Verma[3]

[1] Amity Institute for Advanced Research and Studies (Materials & Devices) Amity University, Noida - 201313, India

[2] Department of Infection and Immunology, Translational Health Science and Technology Institute, NCR Biotech Science Cluster, 3rd Milestone, Faridabad-Gurugram Expressway, Faridabad - 121001

[3] Department of Chemistry, Biochemistry, and Forensic Science, Amity School of Applied Sciences, Amity University Haryana - 122413

Introduction to Heavy Metal Pollution in Wastewater

Water is an essential resource for the survival of all living organisms globally. However, with the increasing population and various human activities like urbanization, industrialization, recreation, and agriculture, the demand for water has surged, leading to reduced access to water resources. These activities significantly elevate water consumption and demand, straining diminishing water supplies. The extensive use of water in these sectors often results in high levels of pollution, including organic and inorganic compounds and heavy metals, which are discharged into the environment through inappropriate wastewater disposal (Odumbe et al. 2023).

Overview of Heavy Metals and their Sources in Wastewater

With the ongoing decline in freshwater availability, there is a growing push to promote water reuse as a solution to meet the increasing demand for clean water. However, this approach can intensify contamination problems, as certain pollutants inevitably re-enter the environment, posing risks to ecosystems and living organisms. For example, using treated wastewater for irrigation has been linked to a

*Corresponding author: nkumar10@amity.edu

DOI: 10.1201/9781003527350-2

higher incidence of health-related infections. Heavy metals, with an atomic density exceeding 5 g/cm^3 and relative atomic masses between 63.5 and 200.6, are among the most significant contaminants in freshwater due to their harmful effects on both human health and ecosystems (Akpor et al. 2014, Ince and Ince 2017, Odumbe et al. 2023). Being non-biodegradable, these metals persist in wastewater, contributing to extensive environmental pollution.

There are various sources of heavy metal pollution in the environment like industrial discharge, municipal waste, agricultural waste, and urban runoff, as shown in Figure 1. Among them, industrial and municipal effluents are primary sources of heavy metal pollution. The rise in industrial activities results in the deposition of large quantities of waste and wastewater polluted with heavy metals (Varma et al. 2013). The discharge of these heavy metal ions into the environment due to industrial development and urbanization poses significant global challenges, making heavy metal ion pollution a major environmental concern (Odumbe et al. 2023). Consequently, there has been an increased focus on the disposal of heavy metals due to their severe impacts on human health. The removal of metal ions from wastewater effluents is crucial due to their toxicity, non-biodegradable nature, and associated health risks for all living organisms (Shakir et al. 2017).

Environmental and Human Health Impacts of Heavy Metal Pollution

Heavy metal contamination in wastewater poses severe risks to the environment and human health. These metals can persist in the environment, bioaccumulate in the food chain, and cause toxic effects at various trophic levels (Ali et al. 2019). The key impacts of heavy metal pollution are briefly described below:

Environmental Impact

- *Aquatic Life:* Heavy metals can be toxic to aquatic organisms, leading to decreased biodiversity and disrupted ecosystems. For example, cadmium and mercury can cause mortality and reproductive issues in fish and invertebrates (Khayatzadeh and Abbasi 2010, Sharma et al. 2024).
- *Soil Contamination:* Irrigation with contaminated water can lead to the accumulation of heavy metals such as chromium, lead, arsenic, mercury, etc. (Kumar et al. 2019, Kumar N et al. 2020). In soil, this affects plant growth and crop quality. High levels of chromium and lead can inhibit seed germination and root development (Kumar N et al. 2020, Madhu and Sadagopan 2020).
- *Bioaccumulation:* Heavy metals can bioaccumulate in organisms, increasing in concentration up the food chain. This can affect top predators, including humans, as they consume contaminated fish and other wildlife (Ali and Khan 2018).

Human Health Impact

- *Neurological Damage:* Metals like lead and mercury are neurotoxic, particularly to children. Lead exposure can result in cognitive deficits, developmental delays, and behavioral issues (Kaur et al. 2021).
- *Carcinogenic Effects:* Arsenic and cadmium are classified as carcinogens. Long-term exposure to these metals can increase the risk of cancers, including skin, lung, and bladder cancer (Zhu and Costa 2020).

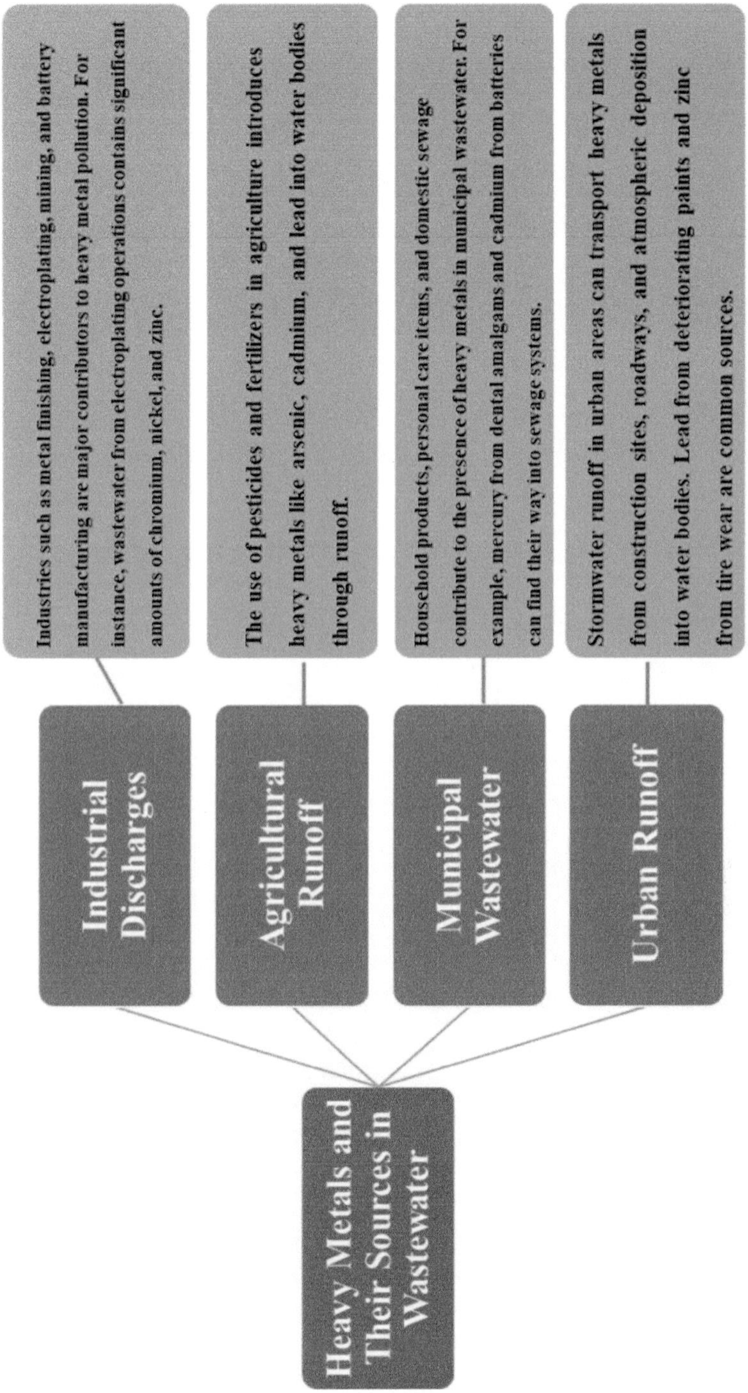

Figure 1: Various key sources available to discharge the heavy metals.

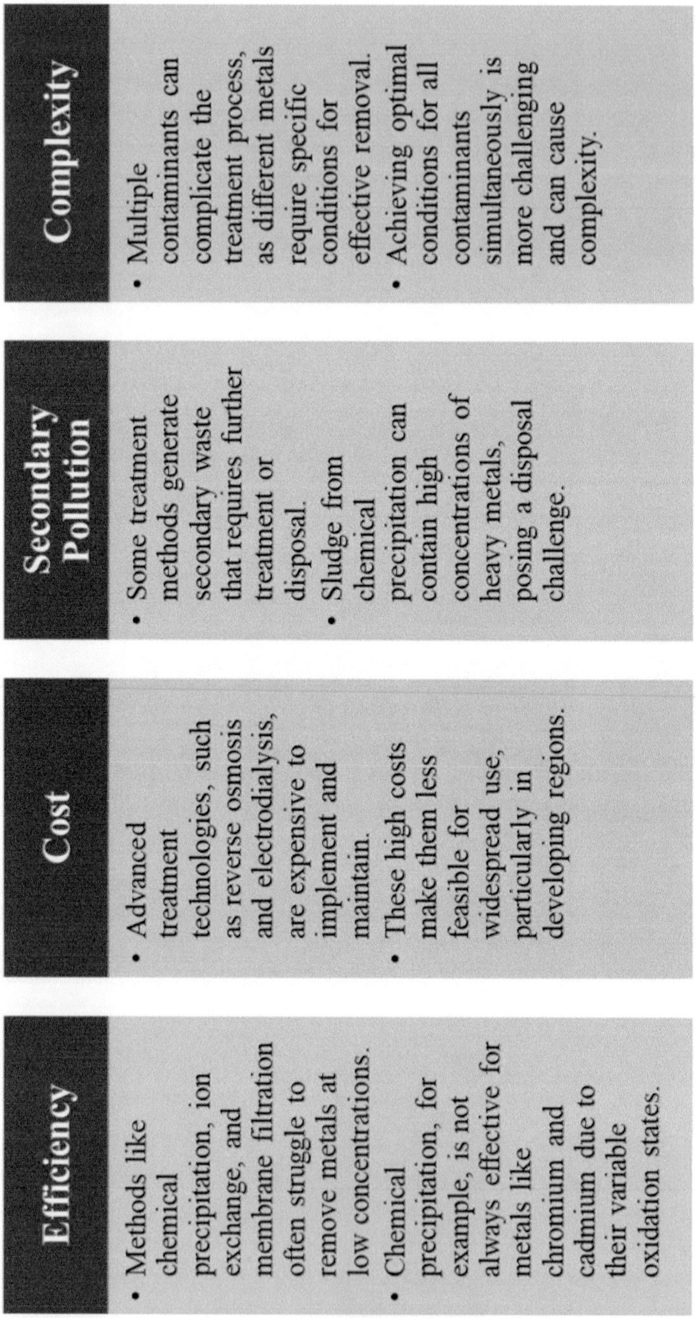

Figure 2: Challenges associated with traditional wastewater treatment methods for removing heavy metals.

- *Organ Damage:* Chronic exposure to heavy metals can cause damage to vital organs. For example, cadmium can damage kidneys, while mercury exposure can impair liver and lung function (Balali-Mood et al. 2021).

Challenges Associated with Conventional Wastewater Treatment Methods

Conventional wastewater treatment methods face several challenges in effectively removing heavy metals. Methods like chemical precipitation, ion exchange, and membrane filtration often struggle to remove metals at low concentrations. Chemical precipitation, for example, is not always effective for metals like chromium and cadmium due to their variable oxidation states. Moreover, advanced treatment technologies, such as reverse osmosis and electrodialysis, are expensive to implement and maintain, making them less feasible for widespread use, particularly in developing regions. Additionally, some treatment methods generate secondary waste that requires further treatment or disposal. For instance, sludge from chemical precipitation can contain high concentrations of heavy metals, posing a disposal challenge. Furthermore, the presence of multiple contaminants can complicate the treatment process, as different metals require specific conditions for effective removal. Achieving optimal conditions for all contaminants simultaneously is challenging (Vidu et al. 2020). These challenges can be categorized into further points such as efficiency, cost, secondary pollution, and complexity, as illustrated in Figure 2.

Nature-based Solutions for Heavy Metal Remediation

Concept and Principles of Nature-Based Solutions (NBS)

Nature-based solutions involve the sustainable management and use of natural processes and ecosystems to address environmental challenges, including the remediation of heavy metal pollutants in wastewater. NBS relies on the inherent capabilities of plants, microbes, and natural materials to absorb, accumulate, transform, or immobilize heavy metals, thus reducing their bioavailability and toxicity (Song et al. 2019).

Nature-based solutions for heavy metal remediation hinge on several core principles. The utilization of natural processes involves leveraging natural cycles, biota, and materials to remove or stabilize contaminants effectively. This approach ensures sustainability by maintaining long-term effectiveness and minimizing the environmental footprint. By enhancing or restoring natural ecosystems, these solutions improve water quality through integration with existing environmental systems. Furthermore, nature-based solutions are cost-effective, utilizing low-cost and locally available resources for both implementation and maintenance. This combination of natural processes, sustainability, ecosystem integration, and cost-effectiveness makes nature-based solutions an attractive and viable option for managing heavy metal pollution in wastewater (Wang et al. 2023a).

Advantages of NBS over Conventional Treatment Methods

NBS offers several benefits over traditional wastewater treatment methods, making them an attractive alternative for managing heavy metal pollution:

Cost-Effectiveness

NBS often incurs lower capital and operational expenditures compared to conventional methods. For instance, constructed wetlands and phytoremediation systems utilize natural vegetation and microbial communities, which can be established and maintained at relatively low costs.

Sustainability

NBS is based on renewable natural processes that do not rely on synthetic chemicals or energy-intensive operations. This minimizes the environmental impact and carbon footprint associated with treatment.

Biodiversity Enhancement

Implementing NBS can enhance local biodiversity by creating or restoring habitats for various plant and animal species. For example, constructed wetlands provide habitats for birds, amphibians, and invertebrates.

Community Engagement

NBS projects can engage local communities in environmental stewardship, raising awareness and fostering a sense of ownership. Community involvement in maintaining green infrastructure, such as rain gardens and green roofs, can enhance social cohesion and environmental awareness.

Classification of NBS Techniques for Heavy Metal Removal

Nature-based solutions for heavy metal remediation can be classified into several categories based on their mechanisms and applications:

Phytoremediation

Plants are used to remove, stabilize, or transform heavy metals from soil and water. Phytoremediation techniques include:

- *Phytoextraction:* Plants absorb heavy metals from the soil and accumulate them within their biomass. Hyperaccumulator plants such as *Thlaspi caerulescens* and *Sedum alfredii* have been used to extract metal ions like cadmium and zinc from contaminated soils (Gill et al. 2012).
- *Phytostabilization:* Plants immobilize metals in the soil, reducing their mobility and bioavailability. For example, species like *Vetiveria zizanioides* are effective in stabilizing lead and arsenic in contaminated soils (Yu et al. 2020).
- *Rhizofiltration:* Plant roots absorb and concentrate metals from aqueous solutions. This technique is commonly used in constructed wetlands to treat wastewater containing heavy metals (Sukumaran 2013).

Bioremediation

The use of microorganisms to break down or transform contaminants into less toxic forms is known as bioremediation. Bioremediation techniques include:

- *Bioaccumulation:* Microorganisms accumulate metals intracellularly. For instance, certain bacteria and fungi can accumulate significant amounts of heavy metal ions such as cadmium and mercury (Girdhar et al. 2022).
- *Biosorption:* Microbial cell walls adsorb metals from wastewater. The use of biosorbents derived from microbial biomass, such as *Saccharomyces cerevisiae* (yeast), has shown promise in removing metals like lead and copper from aqueous solutions (Ghosh et al. 2021).

Green Infrastructure

It refers to the incorporation of natural elements in urban settings to manage and treat stormwater. Green infrastructure techniques include:

- *Green Roofs:* Roofs planted with vegetation that absorb rainfall, reduce runoff, and improve water quality. Green roofs have been shown to reduce concentrations of heavy metals like lead and zinc in stormwater runoff (Vijayaraghavan et al. 2019).
- *Rain Gardens:* Shallow, vegetated basins that capture and treat stormwater on-site. Rain gardens can effectively remove metals such as cadmium and copper through plant uptake and soil filtration (Sounthararajah 2016).

Constructed Wetlands

Constructed wetlands are engineered systems designed to replicate natural wetlands for wastewater treatment. These systems employ physical, chemical, and biological processes to remove contaminants. Types of constructed wetlands include:

- *Surface Flow Wetlands:* Shallow basins planted with aquatic vegetation that treat water as it flows over the surface. These systems are effective in removing metal ions such as zinc and copper through sedimentation and plant uptake.
- *Subsurface Flow Wetlands:* Water flows through a porous medium planted with vegetation, allowing for microbial and plant-based treatment. Subsurface flow wetlands are particularly effective in removing metals like chromium and nickel through adsorption and microbial activity.

Constructed Wetlands (CW) for Heavy Metal Removal

Heavy metal and metalloid (HMM) contamination in wastewater is becoming a global concern. HMMs are non-biodegradable and come from both natural and human sources. As a result, they accumulate throughout the environment (Yu et al. 2022). Waterbodies contain HMMs that are hazardous to the human population directly through ingestion and indirectly through enrichment in the food chain (Yu et al. 2021). Due to different geographical regions, climates, and degrees of development, the types of heavy metals and metalloids (HMMs) prevalent in contaminated water bodies and the emphasis on heavy metal and metalloid remediation studies varies globally. Heavy metals and metalloids (HMM) remediation by constructed wetlands (CWs) is quickly emerging as the most effective and eco-friendly technology worldwide because of its operational simplicity and being less expensive as compared to conventional technologies used in the remediation of HMM contamination from wastewater (Davis 1995, Ang et al. 2023).

A constructed wetland is a natural system for the treatment of wastewater that utilizes biological and geochemical processes of a wetland ecosystem. This approach relies on the intricate interactions between water, soil or gravel media, plants, microorganisms, and the atmosphere. The remediation process is powered by natural energy, primarily derived from solar energy (Davis 1995). These engineered wetlands are capable of treating waste from various sources, including landfill leachate, sludge dewatering processes, mine drainage, stormwater, domestic or municipal wastewater, industrial effluents, and agricultural operations. Mine drainages and industrial effluent are the typical sources to find heavy metal contamination (Yu et al. 2021). According to Thullen et al. (2005), the primary heavy metals found in wastewater generated by mining and industrial activities are lead, copper, cadmium, chromium, mercury, iron, and zinc (Thullen et al. 2005). Numerous techniques, such as sedimentation, filtration, adsorption, plant uptake, and geochemical precipitation, are employed to eliminate heavy metals from constructed wetland systems (Stottmeister et al. 2003). It has been stated that artificial wetlands can remove up to 100% of heavy metals (Camacho et al. 2017). Sheoran and Sheoran (2006) reported that constructed wetlands (CWs) exhibit significant potential for heavy metal removal, with efficiencies of 7-99% for cadmium, 26% for lead, 76% for silver, and 67% for zinc. Furthermore, the CWs achieved removal rates of 75-80% for COD, BOD, and TSS.

Design, Principle, and Components of Constructed Wetlands

General Designing of the Constructed Wetlands

Though a great deal of study and material has been published, the best way to develop constructed wetlands for different uses has not yet been established. A large number of wetland systems (constructed) have not been sufficiently observed or have not been in operation long enough to yield meaningful data for analysis (Davis 1995, Hua and Haynes 2016). Performance has varied among the systems that have been observed, and it has been challenging to quantify the effects of the many different aspects that influence performance, including preferred location, the kind of wastewater or runoff, the architecture of the wetlands, climate, disturbance, weather, and daily or seasonal variability. Wetland designs generally aim to replicate the overall architecture of natural wetlands while promoting the wetland processes that are believed to have the most positive impact on improving the quality of the water (Kumar P et al. 2020, Yu et al. 2022). According to (Mitsch 1992), the following rules can be followed to create successful wetlands (Mitsch 1992):

1. Keep the design simple for the constructed wetland. Complex technological methods lead to failure.
2. Make a design that requires minimal maintenance.

3. Design the system to use innate energies, like gravity flow.
4. Instead of designing for ordinary weather and climate, design for extremes. Droughts, floods, and storms should not be feared, but rather anticipated and prepared for.
5. Design the constructed wetland to work with, not against, the surrounding terrain. Complement the site's natural topography with the design.
6. Refrain from overly complicating the design with inflexible channels, rectangular basins, and regular morphology. Mimic natural systems.
7. Allow time for the system. Wetlands may take several years to reach optimal performance levels; they do not always become operational quickly. Attempts to overmanage or circumvent the system development process frequently backfire.
8. Focus on designing the system for functionality rather than appearance. For example, if the initial plantings do not succeed but the overall performance of the wetland aligns with its intended purpose, the system should still be considered successful.

Planning

A conceptual planning stage is essential. Wetlands can be created in various configurations and system types to deal with specific wastewater needs, alternative sites are often available, and a range of plant species native to the area can be chosen (Davis 1995, Hua and Haynes 2016). Since each site is different, a constructed wetland system's design will be tailored to the particular location. The planning stage comprises characterizing the quality and quantity of the wastewater to be treated, evaluating the discharge standards to be fulfilled, selecting the site, deciding configuration and system type, and outlining the designing specifications to be met by the complete engineering plans. Economic concerns include the land area requirements, the type of water confinement, vegetation, and the management and transport of water through the system. Establishing and prioritizing the goals of the wetland system is essential to its development. The features of a native natural wetland should serve as a model for the constructed wetland, adapted to meet the requirements of the project and the particulars of the constructed wetland site. A site-sensitive wetland design that integrates existing characteristics of the site reduces the amount of earthmoving needed and enhances the visual beauty of the site (Davis 1995).

Site Selection

Making a wise choice while selecting an appropriate site can result in significant cost savings. When choosing a location, one should take into account the following factors: topography, soils, land availability, land usage and access, surrounding environmental resources, and potential effects on neighbors. In order to allow water to flow through the system by gravity, the site location should be as close to the wastewater source as feasible, preferably down gravity. Even though a wetland can be installed on practically any site, if costly liners or significant earthmoving are needed, the expense of construction may become unaffordable. An appropriate location for a constructed wetland is one that (Davis 1995, Qasaimeh et al. 2015):

1. Is conveniently situated near wastewater source
2. Is gradually sloping, to allow gravity to carry water through the system
3. Contains soils that can be compacted enough to reduce seepage of water into the earth
4. Is above the water level
5. Does not include any endangered or vulnerable species

Land Usage and Access

Access is a crucial consideration; the location of the wetland should allow for gravity-driven water flow. The wetland should be located as far away from dwellings as feasible if the smells or insects could be an issue, as with some agricultural wastewater. Access to the site is necessary for workers, delivery trucks, and construction and maintenance equipment. The landowner needs to be picked with caution. Cooperativeness and a thorough understanding of the risks and limits inherent in a new technique, like constructed wetland treatment, are crucial for the landowner (Qin and Chen 2016). A significant buffer zone should exist between the next property and the constructed wetland. The wetland shouldn't be situated close to the property's edge (Raza et al. 2023).

Land Availability

The retention time of wastewater in a constructed wetland determines how well it treats stormwater or wastewater. Therefore, the size of the constructed wetland is required for sufficient retention time to determine the performance of the created wetland. The chosen location should have sufficient space for both current needs and any future expansion (Si et al. 2014).

Topography

Landform designs include optimum shape, size, and alignment with the prevailing winds. Even though a constructed wetland may be developed practically anywhere, choosing a location with easily adjustable, gradual slopes to collect and hold water makes the design and construction process easier and less expensive (Davis 1995). Previously drained sites for wetlands involving prior converted (PC) agricultural areas may be ideal for wetland construction since the topography is typically favorable to gravity flow. The ideal site for a constructed wetland is a low, plain area where water flows by gravity. It is crucial to make sure the area isn't already a wetland because not all wetlands have standing water year-round (Hua and Haynes 2016).

Environmental Resources

The presence or absence of substantial environmental resources must be determined to circumvent damaging valuable resources on the site. The sources of such information that can help select an appropriate site include the US Geological Survey Topographic Quadrangle maps, National High Altitude Photography Program (NHAPP) photographs, and National Aerial Photography Program (NAPP); Geographical Information System (GIS) maps are also additionally available. It is advisable to search the County Soil Survey, which includes a list of county hydric soils, and the National Wetlands Inventory (NWI) maps for potential locations of existing wetlands (Davis 1995, Kumar P et al. 2020).

Permits and Regulations

The appropriate regulatory bodies must be contacted to determine the regulatory requirements related to the planned constructed wetland and its discharge. A permit is needed to perform work in a natural wetland or waterway. Permission is also needed for wetland discharges into natural waters. Zoning approval could be necessary in some zoned communities. Local and state stormwater regulations must be complied with by any stormwater plan. Stormwater regulations have been included in certain local ordinances, and they must be followed. Before creating a stormwater management strategy, local stormwater regulations should be taken into consideration as they differ from place to place. Before choosing a site layout, it is necessary to review the regulatory status of a planned stormwater wetland and its link to surrounding natural wetlands and streams with the state or federal wetland permitting agency (Sunding and Zilberman 2002).

Components of Constructed Wetland

A constructed wetland is a system made up of various components that cooperate to guarantee the accomplishment of its intended goal. A well-planned and built basin that contains water, a substrate that offers filtration pathways, habitat/growth media for the required organisms, and communities of microorganisms and aquatic invertebrates—which, for the most part, develop naturally—are all components of engineered wetland systems. Constructed wetlands have three principal components:

1. Impervious clay layer
2. A layer of gravel-based substrate
3. Ground vegetation zone

Typically, a constructed wetland consists of three main components. There is an impervious layer (usually clay) in the constructed wetland. Additionally, it has a layer of gravel that serves as a substrate for the root zone, providing nutrients and support. A vegetative zone above the surface is also present (Davis 1995, Qin and Chen 2016). The constructed wetland system's impermeable barrier keeps wastes from seeping into subterranean aquifers. The middle gravel layer and root zone consist of a layer where water travels and facilitates bioremediation and denitrification processes. The vegetative layer above the ground consists of well-adapted plant material. Within constructed wetlands, both anaerobic and aerobic processes take place, and these processes can be divided into distinct cells. Pumping or gravity can be used to force groundwater through the constructed wetland naturally. However, in wetlands designed to remediate water contaminated by explosives, specific plant species are employed in a procedure known as phytoremediation to speed up the process of degradation (Qasaimeh et al. 2015).

Water

Wetlands are designed on substrates that are partially or completely submerged in water, where surface water cannot penetrate the ground due to a relatively impermeable subsurface layer. These settings can be made with minimal modifications to build a constructed wetland. By sculpting the land's surface to collect surface water and sealing the basin to hold onto the water, one can create a constructed wetland in practically any place in the terrain (Davis 1995, Hua and Haynes 2016). Since hydrology is the main element determining whether most constructed wetlands succeed or fail, it is the most significant design consideration for the construction of wetlands. Hydrology also improves the connection of all the functions in a wetland system. Therefore, to guarantee that all hydrological needs and conditions are met, the planning and construction of wetlands require the assistance of a certified hydrologist (Davis 1995).

Substrate

Litter or silt can be used as substrates for constructed wetlands. Wetlands are built using a variety of substrates, including soil, sand, rock, gravel, and organic materials like compost (Qasaimeh et al. 2015). Sediment and litter accumulation happens within wetlands because of their high productivity and low water velocities. Numerous processes carried out by the substrates, sediments, and litter improve the functional efficacy of constructed wetlands. They offer growth support to many of the living organisms in wetlands, and the permeability of the substrate also influences the water flow through the wetland and facilitates a variety of biological and chemical processes, many of which are microbial and also enhance the degradation/transformation of contaminants within the substrates.

Vegetation

Both vascular (higher plants) and nonvascular (algae) plants can coexist in constructed wetlands. Algae photosynthesis raises the concentration of dissolved oxygen in water, which in turn influences the availability of nutrients and metals (Davis 1995). Large animals that can consume pollutants, such as birds, are drawn to constructed wetlands. They also create surfaces on which other protozoans and other microorganisms, like phytoplankton, zooplankton, and bacterioplankton, can attach themselves, helping to remove impurities and pollutants (Zaghloul 2020).

Microorganisms

A primary characteristic of constructed wetlands is that their operational capacity is principally regulated by the presence of microorganisms and their metabolism. Protozoa, fungi, yeasts, bacteria, and rind algae are examples of microorganisms used in constructed wetlands. Microbial biomass is a significant sink for several nutrients and organic carbon. Many organic and inorganic compounds are transformed by microbial activity into harmless or insoluble molecules (Stottmeister et al. 2003).

Mechanism of Heavy Metal Removal using Constructed Wetlands

A constructed wetland is made up of water, plants, substrate, invertebrates, and a variety of microorganisms, as was previously indicated. The methods used in constructed wetlands to raise the water's quality can differ and are frequently interrelated (Sheoran and Sheoran 2006, Sahu 2014). The following is a series of events that occurs in constructed wetlands:

1. Settlement of the suspended matter

2. The wastewater coming in contact with the substrate undergoes filtration, chemical precipitation, and chemical transformation by denitrification and bioremediation processes
3. The wastewater then undergoes absorption and ion exchange on the surface of vegetation, substrate, and sediment.
4. The microorganisms and plants within the constructed wetland bring down the transformation of the pollutants
5. The microorganisms consume the nutrients
6. The predation and natural die-off of the pathogens

Physical Removal Mechanisms

One of the main methods for removing heavy metals from CWs is frequent sedimentation with suspended solids (Pachaiappan et al. 2022). In other words, heavy metal-associated suspended particles are filtered and retained within the constructed wetland. According to Johnson (1995), the wetland's sedimentation rate is directly related to the wastewater's residence period and particle settling velocity (Martin and Johnson 1995). After heavy metals have accumulated into sufficiently large particles to settle through other means, a physical process takes place. Prior to sedimentation, the particles may aggregate together through chemical reactions (e.g. adsorption, ion exchange, precipitation, co-precipitation, and complexation with organic matter). Therefore, the entire process is physicochemical as opposed to simply physical, and this leads to the removal of heavy metals from the wastewater and their retention in the wetland sediment. For particles that are less dense than water, the process of floc formation is crucial for them to settle (Sheoran and Sheoran 2006). Physicochemical variables like pH changes and oxidation-reduction effects will also have a major impact on the availability and solubility of metals in sediments (Davis 1995).

Adsorption

Metals can be adsorbed in wetland substrates by chemisorption or cation exchange, which is a type of physical adsorption. Compared to cation exchange, chemisorption is a more powerful and long-lasting method of binding. The adsorption process is crucial to the metal removal mechanism, particularly in the early stages of a constructed wetland's operational lifetime when sorption sites are not saturated (Yu et al. 2021). The carboxyl and phenolic hydroxyl residues of molecules (such as humic acids) generated or exposed during plant breakdown act as a medium for the adsorption of heavy metals onto the organic materials. Further breakdown of this detrital material could result in chelate complexes that are more stable. It has been demonstrated that copper (Cu) and nickel (Ni) have a special affinity for organic materials and are adsorbed into wetlands, where they are retained. It has been noted that, especially in acidic environments, metals may bind to the functional groups of organic materials adsorbed on Fe-hydro (oxides) (Song 2003). The pH affects adsorption because it causes the functional groups that bind dissolved metals to become deprotonated (Kumar P et al. 2020). Certain synthetic or natural sorbents can improve the adsorption process.

Filtration and Sedimentation

When metals and metalloids are in a particle or colloidal form, sedimentation and filtration are especially effective methods (Sheoran and Sheoran 2006). The use of fine filtration media or systems with free water surfaces, low flow velocities, and preferably dense vegetation to capture metallic particulates are ways to improve these processes. Precipitates that originate from external sources and those that occur in coastal waters are eliminated by sedimentation and filtration. For instance, the bed media of CWs may contain residual metal hydroxides from processed industrial effluent. Additionally, the efficiency of the filtration and sedimentation procedures affects the removal of metals in the form of sulfides or (oxy)hydroxides to some degree (Raza et al. 2023).

Precipitation as Sulfides

The biological reduction of sulfate to metal sulfides has been the main focus of much of the research on the removal of heavy metals in CWs. Low effluent metal concentrations are the result of the intractable nature of most metal sulfides, which hold steady under reduced conditions (Sochacki et al. 2018).

Chemical Removal Mechanisms

Association with Metal Oxides and Hydroxides

Based on the simplified processes, iron (Fe), manganese (Mn), and aluminum (Al) can create insoluble compounds in CWs through hydrolysis or oxidation (abiotic or bacterially induced). Precipitation as hydroxide is the most frequent solution for Al removal. Al may also precipitate as hydroxysulfate if dissolved sulfate is present. In an oxygenated environment, ferric iron (Fe^{3+}) is typically found as insoluble (oxy)hydroxides, represented as FeOOH. In a reduced condition, the presence of Fe^{2+} causes a reduction in the precipitation of oxidized manganese (e.g. MnO_2 and MnOOH) (Sobolewski 1999). Bacteria can significantly speed up the oxidation of Mn, which happens considerably more slowly than it does for Fe (Sheoran and Sheoran 2006). Cu, Ni, Pb, Zn, and other metals are thought to be effectively scavenged by iron and manganese oxides and (oxy)hydroxides (Kadlec 2009). Adsorption or co-precipitation may associate these Fe or Mn oxyhydroxides with other metals (Sobolewski 1999). For instance, in the top layer of marsh sediments, 97% of Zn was found to be bound to the hydroxide phase (Kadlec 2009).

Microbial Removal Processes

Heavy metal removal in CWs is impacted by some microbiological activities, either directly or indirectly (Stottmeister et al. 2003). These activities can be categorized into four groups: (i) biosorption; (ii) methylation; (iii) microbial oxidation-reduction of heavy metals; and (iv) heavy metal removal supported by the biogeochemical cycle. Biosorption, which occurs frequently on both living and dead microbial biomass, is the passive uptake of heavy metals (Stottmeister et al. 2003, Yu et al. 2021). If the metal-loaded biomass is not harvested, sorption sites become saturated, making this a short-term approach (Kosolapov et al. 2004). Many functional groups, such as carboxyl, phosphate, hydroxyl, sulfonate, and amino moieties are involved in biosorption (Yu et al. 2022). Metals are consumed by anaerobic metal-reducing bacteria as terminal electron acceptors during the process of respiration; the reduced metals' solubility varies based on the particular elemental chemistry. One example of this process is the microbial conversion of Cr (VI) to Cr (III) (Kosolapov et al. 2004). Certain microbial oxidation reactions, such as the oxidation of dissolved iron (II) and manganese (II), occur in non-reduced zones of CWs and precipitate these metals as metal (oxy)hydroxide. Aerobic and anaerobic bacteria of CWs can biomethylate some metals, like Hg, Se, and As, producing volatile derivatives such as trimethylarsine, dimethylmercury, and dimethyl selenide (Kosolapov et al. 2004). Due to their lipophilic nature, these volatile derivatives are typically more hazardous than their inorganic counterparts even though the metalloid is eliminated from water (Kosolapov et al. 2004).

Plant Uptake of Heavy Metals in Constructed Wetlands

One important factor in removing heavy metals from CWs may be plant absorption. While certain plants accumulate metals, the majority of plants do not (Sochacki et al. 2018). Metals are often converted by accumulator plants into a non-reactive state that can be safely retained in their tissues. Plants have an effective system in place to absorb micronutrients from their surroundings, and metals could be among them. Plant roots can solubilize and absorb micronutrients from nearly insoluble precipitates found in the soil, as well as from extremely low concentration levels, thanks to chelating agents produced by the plants, pH changes induced by the plants, and redox processes (Zaghloul 2020, Yu et al. 2022).

Other Processes

The production of carbonates, metal hydrolysis (catalyzed by bacteria in an acidic environment), reduction to non-mobile forms (chemically or biologically), and biological methylation (and subsequent volatilization, particularly for mercury) are additional processes that may contribute to the removal of metals (Hua and Haynes 2016). Metals may be complexed by microbiological sources, organic ligands from plants, or organics in wastewater. However, normal wetland conditions are not conducive to these phenomena (Kosolapov et al. 2004). Lead (Pb) is one example of very complicated metal chemistry found in natural systems; in addition to sulfides, hydroxides, carbonates, and oxides, Pb can also form sulfate ($PbSO_4$), oxide (PbO), biplumbite ions ($HPbO_2^-$), charged hydroxides ($PbOH^+$), and pyromorphite ($Pb_5(PO_4)_3Cl$) if phosphate is present (Reddy and DeLaune 2022). Metals like lead (Pb) in wetlands are complex, but their removal may be facilitated by various abiotic and biological processes. The interactions between different biotic and abiotic processes are still a topic for research, despite being examined as individual processes to some extent (Kosolapov et al. 2004).

Limitations and Concerns Associated with the Application of Constructed Wetlands

Even though constructed wetlands offer a useful substitute technique for removing heavy metals from the environment, they also have certain downfalls (Davis 1995, Hua and Haynes 2016):

1. The long-term efficacy of constructed wetlands to contain or treat some contaminants is not well documented.
2. Wetland aging may contribute to a decline in the removal rate of heavy metals over time.
3. Like other biological techniques, constructed wetlands are constrained by the biota's capacity to tolerate exposure to their surroundings. In order to be successful, natural systems need to establish themselves.
4. Events related to weather, wildlife, and pollution concentrations could pose challenges when setting up the systems. In one demonstration, for instance, a tadpole infestation severely defoliated the plants within two months of planting; a hailstorm destroyed one of the few plants that managed to re-establish itself; and the photodegradation of explosives in the contaminated groundwater caused the water to turn a dark red, inhibiting photosynthesis (Kumar P et al. 2020).
5. In the case where heavy metals are the primary contaminant, constructed wetlands do not destroy the metals; they limit their mobility through the process of sorption.
6. Metals that have accumulated in plants may have a negative impact on wildlife throughout the operation of the constructed wetlands (Mitsch 1992).
7. The ecology of constructed wetlands shifts when the pumping of contaminated water stops. This could severely impact the wetland's supporting plant and animal life and result in waste contaminated with heavy metals and some other pollutants. It could be necessary to dispose of or cap this residue or sludge (Camacho et al. 2017).
8. It is important to keep a close eye on the wetland's outlet. It's also necessary to keep an eye on the underlying aquifers to make sure the impermeable base hasn't spilled.
9. It is best to avoid using exotic and invasive species when constructing wetlands, and to have a strategy in place for dealing with them should they arise (Thullen et al. 2005).

Phytoremediation: Harnessing Plants for Heavy Metals and Metalloids Clean-up

The environment and natural resources are under extreme stress due to the world's population growth and the increasing need for food, energy, water, and other ecosystem services (Zaghloul 2020). Human activities such as agricultural intensification, quick urbanization, and rapid industrialization are examples of those that are gravely contaminating the environment with heavy metals (HMs) and metalloids. To tackle the issues of an increasing spectrum of heavy metal contamination in wastewater, phytoremediation using plant species can be a potential solution (Stottmeister et al. 2003).

Plant Selection Criteria for the Phytoremediation

The characteristics of ideal plant species that can employed in the process of heavy metal remediation using plants are discussed here (Stottmeister et al. 2003, Zaghloul 2020):

1. Plant should be naturally hard,
2. Should produce a lot of biomasses,
3. Should withstand the toxic impact of heavy metals and contaminants,
4. Should be easily cultivated,
5. Should have a high capability of absorption,
6. Should not be appealing to herbivores.

Phytoremediation Mechanisms for the Decontamination of Heavy Metals

Plants can use a variety of methods to remove both organic and inorganic contaminants from soil and water, depending on the type of plant and its environment. The removal of pollutants is mostly reliant on the interactions between plants and microbes in the water, air, and soil (Zaghloul 2020). The type of vegetation, the soil, and the pollutant all affect the efficacy of the clean-up process. Numerous elements, including soil pH, electrical conductivity, the concentration of organic matter, microbial activity, and other soil additives, affect the efficacy of remediation. Of these, plant biomass and their metabolism to remove heavy metals are crucial (Zaghloul 2020).

Phytoextraction

The process known as phytoextraction, sometimes called phytoaccumulation, entails the uptake of hazardous heavy metals by plant roots, which are then transferred to shoots and deposited at cell walls, vacuoles, cell membranes, and other areas of the plant tissues that are not metabolically active (Stottmeister et al. 2003). The plants called hyperaccumulators, known for the remediation of HMs, retain a higher concentration of metals in their shoot and root tissues. The absorption of metal cations is one of the general pathways during the accumulation of toxic heavy metals. This is followed by the formation of metal-ligand complexes or metal-phytochelatin complexes (M-PC) inside the plant cell. These complex molecules are translocated to the plant vacuole and kept thereafter, forming the M-PC (Yan and Lo 2012).

Plant biomass and the quantity of heavy metals in plant tissues (aboveground) are the primary factors influencing a species' capacity for extraction (Zaghloul 2020). As a result, the ideal species for phytoremediation must not only be able to withstand and efficiently absorb HMs but also develop quickly, produce a large amount of biomass, and offer financial advantages (Stottmeister et al. 2003). Some plant species are better adapted to use the phytoextraction method in the phytoremediation of particular metals, like *Sedum alfredii* for Cd (Zhou and Qiu 2005), *Pteris vittata* for As (Salido et al. 2003), *Sesbania drummondii* for Pb (Salido et al. 2003), *Commelina communis* for Cu (Wang and Liu 2014), etc. The Bioconcentration Factor (BCF) and Translocation Factor (TF) are determined to assess the phytoremediation potential of plants. The Bioconcentration Factor is defined as the ratio of pollutant concentration in the plant parts to that in the media, and the Translocation Factor (TF) is defined as the ratio of elemental accumulation in the plant's shoot compared to plant root (Zaghloul 2020).

Phytostabilization/Phytoimmobilization

Toxins or contaminants inside the roots or rhizosphere might be rendered inactive or immobile to accomplish remediation. The stabilizing action of plant roots reduces the mobility and bioavailability of contaminants, which lessens their harmful effects. Certain plants accumulate harmful pollutants that are bonded to their residue and either remain inactive in their toxic form or are unable to be released from the solid matrix. According to Li et al. (2002), the herbicide trifluralin, for instance, can be absorbed by the ryegrass plant and converted into bound residue (Li et al. 2002). Likewise, to convert them into non-toxic compounds, HMs and some organic pollutants can form complexes in the rhizosphere or conjugate with derivatives of sugar, protein, and amino acids. For instance, arsenic (As) was rendered immobile by the epidermis of plant roots, which bound it to ferric sulfate in vacuoles and formed the trivalent complex As-tris-thiolate in the plant rhizosphere (Hammond et al. 2018). The stabilization efficiency of contaminants can be increased by using a variety of bioaccumulation enhancers.

Rhizofiltration

In order to remove aqueous pollutants from wastewater, rhizofiltration entails using the roots of plants in a saturated zone to remediate toxic substances via processes like adsorption, concentration, and precipitation onto roots or other submerged organs of metal-tolerant aquatic plants (Zaghloul 2020). Based on root-mediated mechanisms, rhizofiltration of HMs-contaminated water sources may cause precipitation and fixation inside the root that is intractable. For instance, Pb-phosphate precipitates in the roots of plants like sunflower and mustard, which can filter Pb-contaminated water (Zaghloul 2020). In addition, some of the endophytic bacteria that reside on the surface of roots support the rhizofiltration mechanism. Several bacterial strains such as *Pseudomonas* and *Ochrobactrum* can convert hexavalent Cr (Cr-VI) into non-toxic trivalent Cr (Cr-III). Once inside the plant root, the reduced Cr-III readily precipitates, purifying the water (Dimitroula et al. 2015).

Phytovolatilization

There are several phases involved in phytovolatilization. Initially, pollutants are absorbed by plants from the soil, whereupon they change from less volatile to more volatile forms before volatilizing and releasing the pollutants into the atmosphere. When the volatilized pollutants have reduced harmful effects upon partitioning from soil to the atmosphere, this approach is effective. For organic contaminants, the volatilization approach is more effective (Zaghloul 2020). Certain compounds are capable of undergoing volatilization directly from stems and leaves, whereas other compounds can volatilize with the aid of interaction between the soil and roots. The hydrophobic organic molecules can leave the stem and leaves of the plant through hydrophobic barriers like cuticles, the epidermis, suberin, and other dermal layers.

Furthermore, certain compounds may be lost to evaporation during transpiration as they move upward in the plant system via the transpiration stream. According to Li et al. (2002), ryegrass transpires the trifluralin compounds from its leaves, which are said to be residue-bound. Through the process of volatilization, several plants can release harmful organic pollutants into the surroundings in a harmless form (Li et al. 2002).

Phytodegradation and Rhizodegradation

Phytodegradation is the process by which contaminants are broken down into their non-toxic form and removed from the tissues of plants. The process of phytodegradation may be significantly impacted by plant components either directly or indirectly. In the direct removal process, the pollutants are absorbed by the different components of the plant, which are further metabolized into smaller, less toxic chemicals to be distributed throughout the plant tissue. Since plants don't have active transporters, these organic contaminants are absorbed passively. The breakdown of contaminants that occurs specifically in the rhizosphere is referred to as rhizodegradation (Stottmeister et al. 2003, Zaghloul 2020).

Key Factors that Limit the Usage of Phytoremediation in the Remediation of Heavy Metals

Despite offering an efficient alternative to other methods of pollutant removal from the environment, phytoremediation has several limitations and disadvantages.

1. For starters, most research is conducted in a controlled environment for a brief period. This might not produce good results that are truly indicative of those obtained over an extended time frame in the field. To fully understand the potential of phytoremediation, many on-site studies based on extended time frames are therefore essential (Zaghloul 2020).
2. The success of phytoremediation is also based on the fast and successful growth of the plant species employed. Because every site has a different soil composition and temperature, a particular phytoremediation process used for one plant species in one location might not be applicable in another scenario (Stottmeister et al. 2003).
3. The presence of contaminants such as toxic metals, organic pollutants, radionuclides, or some antibiotics, combined with insects, pests, and pathogens, may increase the susceptibility of plants to disease and imperil phytoremediation efforts. According to Huang et al. (2000), plants that are more sensitive to greater levels of pollutants may develop more slowly, which could have an impact on their ability to perform phytoremediation. For instance, when remediating highly concentrated metallic pollutants like Pb, As, and U, it was demonstrated that metal hyperaccumulators grew and developed very slowly and had low biomass (Huang et al. 2000).
4. Toxic substances are also able to reduce the biomass and microbial diversity that help the phytoremediation process, which can further reduce the effectiveness of phytoremediation.

5. In addition, the full effects of phytoremediation may not be seen before three to five years because it is a slower process than conventional approaches. Handling and appropriately disposing of plants that may have absorbed harmful substances (metals) is another hurdle with technology usage. Animals and other living things may be impacted if they consume such plants because they contain high concentrations of translocated pollutants (Huang et al. 2000).

Biofiltration Systems for Heavy Metal Removal

Background of the Biofiltration Technique

Biofiltration is the biologically based method used to clean polluted water and air (Srivastava and Majumder 2008). Biological filters, or biofilters, are used to conduct the biofiltration process. The filter media used in biofilters support the growth of microorganisms by aiding them in adhering to and colonizing the filter membrane. These microorganisms are responsible for oxidizing and getting rid of the available pollutants in the water and air due to their propensity to absorb and break down both organic and inorganic contaminants found in the air and water (Pachaiappan et al. 2022). The routine application of biofiltration technology began after the 1950s when the management of hazardous volatile and toxic chemicals from the industrial sector became a major environmental concern in countries like the US and Europe. A biofiltration system with the proper filter beds and microorganisms was created for the treatment (Swanson and Loehr 1997). In countries like Japan and Europe, over 500 biofilters have demonstrated viable success with optimal functional proficiency (Bohn 1992, Wani et al. 1997). Numerous research investigations have been documented on the utilization of biofiltration techniques to clean up contaminated water and/or air. It has been stated that biofiltration is more effective than other current technologies at controlling volatile organic compounds, with an effectiveness rate of approximately 90% (Pachaiappan et al. 2022).

Biofiltration Mechanism in the Remediation of Heavy Metal Contaminants

The remediation of pollutants using biofiltration methods from air or water is accomplished through a sequence of processes. Initially, the biofilters receive the contaminated air or water as feed, and the pollutants are absorbed into the biofilm or cellular membrane of the biofilter bed. In the aqueous phase, the contaminants are transferred to the bed medium (Srivastava and Majumder 2008, Pachaiappan et al. 2022). As a source of carbon, contaminants function as feed for the microorganism. As a result of their metabolism, the microbes proliferate, colonize, and degrade or transform contaminants into their less toxic state. Eventually, treated water or air containing carbon dioxide, water, and biomass is released. The biofiltration method can be simplified to the following equation (Srivastava and Majumder 2008):

Contaminants (Heavy metals) + Microorganism + Oxygen → Biomass + Water + Carbon dioxide

The two primary mechanisms or processes in biofiltration are sorption and biodegradation (Srivastava and Majumder 2008). The water or air that was contaminated is fed into the biofilter bed and is transitioned into a solid or an aqueous phase in the biofilter media. Microbial agents perform phase transformation and constant degradation of contaminants in the below mechanisms (Pachaiappan et al. 2022).

1. A pollutant is adsorbed onto organic media (a biofilter bed), followed by biotransformation with the aid of microbial activities.
2. Direct adsorption of pollutants onto the biofilm membrane and biodegradation/ transformation.
3. Dissolution of pollutants in aqueous phase and biotransformation.

Following biodegradation, the pollutants are extracted from the membrane of biofilters. The microorganism determines the biofilter's proficiency. They are accountable for the phase transition and degradation of pollutants in the contaminated air or water. Thus, to achieve effective removal of contaminants, the optimal growth of microorganisms must be maintained. Recent studies are developing and refining genetically modified bacteria to shield environmental sources from dangerous chemicals and metals (Stottmeister et al. 2003, Srivastava and Majumder 2008, Pachaiappan et al. 2022). Thus, the genetically modified microbial agents are immobilized in the biofilter bed. Microbial immobilization seeks to achieve a high production rate of desired microbial products by increasing microbe survival rates through increased metabolism and cell loading in the filter bed material (Stottmeister et al. 2003).

Mechanism of Biofiltration

The basic processes that enable biofilters to work are intricate and need to be closely watched to ensure effectiveness. The biofilter membrane has a porous medium with microorganisms and water all over its surface. The pollutant may combine with the organic substances in the water to form complexes, which may then be absorbed by the support medium. Eventually, the pollutant is transformed into carbon dioxide, biomass, water, or metabolic byproducts through biotransformation. In a biofilter, three primary biological processes can take place (Srivastava and Majumder 2008):

- The attachment of microbes:
- The growth and development of microbes.
- Microorganisms' detachment and decay.

Mathematical Modeling of Biofilters

The earliest work on submerged biofilm models served as the foundation for biofilter modeling, which began in the early 1980s. The models made the assumptions of a plug flow stream, simple reaction kinetics, and fundamental mass balance concepts. Quantitative structure-activity connections have been used more recently to estimate the effectiveness of biofilters using data characterizing the remediation of a few known contaminants. These models are fundamentally different but possibly useful. The intricacy of the fundamental mechanisms makes biofilter modeling challenging. Several chemical, physical, and microbiological processes are involved in biofiltration. In order to replicate the efficacy of biofilters under different operating settings, a model needs to incorporate these diverse phenomena. Moreover, the definition of equations for a biofilter model has several unknowns or challenges (Srivastava and Majumder 2008). The performance of a biofilter can only be predicted by a small number of models that have been documented in the literature. The majority of these models operate under the presumption of uniform and stationary flow. In most biofilter models, a process of diffusion is considered and is explained by Fick's law. The biofilms used in filtration are incredibly erratic and filamentous, which contradicts the diffusion theory. Additionally, research conducted on both aerobic and anaerobic biofilms reveals that response related to the removal of substrate only takes place in a relatively small portion of the biofilm's edge, with an estimated active thickness around 10 μm, but the actual biofilm's thickness can exceed 500 μm (Srivastava and Majumder 2008).

Treatment Methods

Remediation of Heavy Metals by Biofilters using Microbial Agents

As a trickling filter, biofiltration was first used in the UK in 1893 (Srivastava and Majumder 2008). Thus far, it has been extensively employed in the remediation of polluted water that contains biodegradable organic molecules. Its use in the removal of contaminants like non-biodegradable heavy metals from wastewater is recent research. An investigation by Erik et al. in 2002 was conducted on iron removal through filters (sand) of three different freshwater plants—one abiotic and two biotic—in a similar region of Denmark, and compared the efficacy of iron precipitation in both the biotic and abiotic settings (Søgaard et al. 2000). The observations state that biotic sludge is 7 to 9 times deeper and 60 times more efficient in iron precipitation than in the abiotic setting. *Gallionella ferruginea* served as the study's microorganism of choice. The physicochemical parameters controlling precipitation and precipitated iron sludge have also been studied. It is also mentioned that native bacteria may be utilized to form a biolayer in situations where organic carbon is not necessary for the microbes to use as an energy source (Srivastava and Majumder 2008). Recirculating, lab-grown strains of *Leptothrix discophora* SP-6 could potentially be used as the inoculum for a new manganese biofilter, given the comparatively unproblematic scale-up to the commercial biofilter in the removal of Mn. While the percentage removal of arsenic is only 80%, it is still high in many cases. This observed 80% removal can also be used successfully in a polishing stage of water treatment within a lower range of pollutants. Additionally, the efficiency of adsorbents can be improved chemically, and microbes can be engineered to have higher capabilities (Sekhar et al. 2003).

Improving the Functional Operation of Biofilters with the Aid of Genetically Modified Microorganisms

The process of bioremediation degrades or transforms contaminants into less dangerous or non-hazardous substances. Although fungi, plants, and algae have all been employed in bioremediation, bacteria are the predominant agent (Pachaiappan et al. 2022). Three categories exist for bioremediation:

- Biotransformation: The process by which harmful molecules are transformed into non-toxic or non-hazardous ones.
- Biodegradation: The biotransformation of organic substances into smaller non-toxic organic or inorganic molecules.
- Mineralization: The process by which organic molecules break down into smaller organic or inorganic molecules termed as biodegradation.

These three kinds of bioremediations can take place *in situ* or *ex situ*. There are benefits and drawbacks to both *in-situ* and *ex-situ* procedures. Previously, the pollutants were extracted and confined within a controlled environment, which expedites the cleanup process by simplifying the monitoring and maintenance of settings and development. The major downfalls of the procedure are that it is expensive, time-consuming, and perhaps hazardous to remove the contamination from the contaminated site (Srivastava and Majumder 2008). The *in-situ* method, on the other hand, does not call for the pollutant to be removed from the contaminated location. Alternatively, bioaugmentation or biostimulation are employed. The fact that heavy metals are not transformable but can be chemically converted via methylation, sorption, and valence state alterations is a crucial component to remediating heavy metals from wastewater. While adopting bioremediation is a promising idea, the active bacteria in the bioremediation process are frequently sensitive to the contaminants. It may be exceedingly challenging to maintain a high rate of bioremediation due to this issue. Genetically modified microorganisms with bioremediation capabilities and resistance to the harsh surroundings of the polluted site offer a potential solution to this issue (Srivastava and Majumder 2008, Pachaiappan et al. 2022).

Improving the Functional Operation of Biofilters with the Pretreatment of Surfactants

Apart from incorporating genetically modified organisms to boost the remediation process, the pretreatment of biomass with a variety of available surfactants can also efficiently improve the remediation process. For instance, it was discovered that the biosorption efficiency of *P. chrysogenum* fungal biomass was enhanced by pretreatment with conventional surfactants (such as dodecyl amine and hexadecyl-trimethyl ammonium bromide) and a cationic polyelectrolyte. Furthermore, in comparison to traditional methods, this biosorption technique saves principal costs by 20%, operating expenses by 36%, and overall treatment costs by 28% (Volesky 2001).

Factors Influencing the Biofiltration Process

The effectiveness of the biofiltration process is determined by several factors that fall under the physical, chemical, and biological categories.

Biological Agents

A biological entity that functions as a bio-catalyst to start the process is essential to biofiltration. To create biofilm, microorganisms such as fungi, bacteria, protozoa, and invertebrates are employed. The microorganisms in the bed of bioreactors are in charge of controlling odors and breaking down both organic and inorganic pollutants that enter the system. The type of microbe used is entirely based on the nature and chemistry of the pollutants. For instance, fungal-based biofiltration was found to be more successful compared to bacterial biofiltration in the removal of toluene (Pachaiappan et al. 2022). In a second study, the degrading species belonging to the genus *Enterobacter* were able to remove benzene from waste gas up to almost 90% at a rate of 21.46 g/m^3h (Pachaiappan et al. 2022). *Actinobacteria*, which are aerobic microorganisms, have demonstrated their effectiveness in eliminating odors and degrading butyric acid (Márquez et al. 2021). Therefore, the concentration and choice of bacteria involved are key factors in the breakdown of pollutants in the biofilter.

Biofilter Bedding

Apart from microbial agents (biological creatures), the biofilter bed or packing material is the core of the biofiltration setting. Microbial agents are grown and immobilized on the bed of biofilters, which serves as a growth medium for the development of biofilm (Márquez et al. 2021). The packing material for biofilter beds consists of both inorganic and organic materials. Packing media was created cheaply using readily available materials like soil, wood chips, cocopeat, ceramics, perlite, compost, and polyurethane foam. The following qualities are anticipated to be present in these packing materials (Márquez et al. 2021):

1. To ensure the uniform distribution of influent, the bedding should be porous and have a specific surface area (300–1000 m^2m^{-3}).
2. Packed with inherent nutrients that promote microbial growth and the formation of biofilm.
3. Intense availability of a diverse range of microorganisms.
4. Greater ability to retain water (40–60%) to improve microbial metabolism.
5. Structural stability to prevent the filter bed from packing down.

Supply of Nutrients

Nutrition is another important component that determines biofiltration efficiency. The contaminants are degraded by the microorganisms and fed into the biofilter. These contaminants function as the microbial agents' chief source of nitrogen and carbon. The supporting pack materials utilized in the biofilter bed supply the macro- and micronutrients required for microbial growth. The chief macronutrients are nitrogen and phosphorus, together with potassium and sulfur; the principal micronutrients used in biofilter beds are vitamins and metals (Pachaiappan et al. 2022). Additionally, mineral salts are frequently used with water and applied as a nutrient solution in biofilter beds. $CaCl_2$, KNO_3, $(NH_4)_2SO_4$, $FeSO_4$, $MnSO_4$, KH_2PO_4, $MgSO_4$, NH_4Cl, NH_4HCO_3, and Na_2MoO_4 are mineral salts that are generally utilized (Delhoménie and Heitz 2005). According to the findings, biofilters that received nutrients produced 97% removal efficiency, while biofilters that did not receive nutrients produced 86% (Srivastava and Majumder 2008).

Effect of pH

pH is another crucial component that is just as significant as biofiltration efficiency. At the ideal pH level, nutrients can play a part in the efficient breakdown of contaminants by microorganisms. Because the living environment in the biofilters has a neutral pH of 7, the heterotrophic microorganisms that live there are neutrophilic species. Numerous studies support the idea that pH affects how well pollutants are removed from the environment (Delhoménie and Heitz 2005).

Operating Temperature

The degradation of contaminants is determined by the quantity of active microorganisms in the biofilter, which is influenced by its operating temperature. It has been documented that temperature affects how well biofilters remove organic materials from wastewater. Biofilters running at lower temperatures have demonstrated superior results, such as decreased microbe activity (e.g. *Legionella*) and prevention of the harmful effects of water vapor. The building's internal air quality is harmed when the biofilter is kept at an increased temperature (Darlington et al. 2001).

Moisture Contents

The activity of microorganisms provides the foundation for the efficiency of a biofilter. Microorganisms would inevitably spread or develop uniformly as biofilm over the biofilter bed to achieve a sufficient biodegradation rate (Pachaiappan et al. 2022). The only thing that could make this happen is the presence of water. The water content aids in the microorganisms' ability to absorb nutrients. Therefore, volumetric analysis of the biofilters was performed to determine moisture content (Bohn and Bohn 1999). The ideal range for the moisture content of the biofilter is between –0.2 and –3 bars to get a decent degradation rate (Bohn and Bohn 1999).

Pressure Drops

An additional factor in the biodegradation of contaminants is the pressure drop in the biofilter bed. The proficiency and stability of the biofilter were predicted using the relationship between biomass content and pressure drops (Pachaiappan et al. 2022).

Biofiltration Technique in the Remediation of Heavy Metals

As mentioned, contaminated soil and water supplies contain heavy metals released from conventional and commercial processes. Thus, the removal of such metal pollutants is critical. The efficacy of conventional technologies in terms of heavy metal accumulation, release, and detoxification was constrained. Hence, they cannot attain the permitted ppm limits of heavy metals in water and land resources. However, green technology was taken into account because of its numerous advantages, most notably its eco-friendliness and ability to protect the environment by eliminating heavy metals (Srivastava and Majumder 2008, Pachaiappan et al. 2022). For the remediation of heavy metals, this area provides biomimetic membranes, microbes-based biofilters, and phytoremediation (botanical biofilters) (Srivastava and Majumder 2008).

Macrophyte Biofilters

Many plants, like macrophytes—aquatic plants that grow in ponds and streams—can absorb heavy metal pollutants from waterbodies and store them in different parts of the plants. Even though heavy metals are necessary for aquatic plants to develop, a significant amount of heavy metals in water resources has detrimental consequences on all living things. Macrophytes capture heavy metals from contaminated water sources while functioning as biofilters or natural filters. The affinity of the heavy metal ions determines the absorption capability of the aquatic plants (Pachaiappan et al. 2022).

The Accumulates procedure is easy to use and reasonably priced. Heavy metals from polluted water have been demonstrated to be passively absorbed by *Azolla* fern (Bennicelli et al. 2004). *Azolla* plants can hold 3–5% copper and cadmium, 5–11% chromium, and 5–12% lead. The heavy metals bound to *Azolla* biomass are heated at extreme temperatures (incineration process) to transform into ash and are acid-washed for reuse of the biofilter. *Azolla caroliniana* was also used to remove chromium and mercury ions from municipal water (Bennicelli et al. 2004). The highest entrapment of chromium (III) in the range of 200–48,000 mg/dm^3 and the highest absorption of mercury, up to 578 mg/dm^3, were found in the tissue of *Azolla caroliniana*. Plants store heavy metals in their tissues or roots, where they eventually break down safely. Therefore, an increased concentration of heavy metals can be stored using plants in phytoremediation processes (Bennicelli et al. 2004).

Microorganism-based Biofilters

Heavy metals are converted or degraded into less or non-toxic contaminants via microbial activities termed bioremediation by microorganisms. Additionally, the microorganisms may be used to regulate odor in the biofiltration process. For bioremediation, plants, algae, fungi, and bacteria are employed. Bacterial enzymes are mostly used to break down materials during the metabolization process (Kang et al. 2016). The bacteria employed in bioremediation include *Alcaligenes*, *Citrobacter*, *Bacillus*, *Klebsiella*, *Pseudomonas*, *Escherichia*, *Rhodococcus*, and *Staphylococcus*. Fungi like *Penicillium*, *Pleurotus*, *Aspergillus*, *Rhizopus*, and *Saccharomyces* are generally involved in bioremediation. Lead and zinc metals were eliminated from rainwater using vertically designed biofilters of laboratory-scale conditioned with sulfate-reducing bacteria (SRB) (Rafida and Sallis 2005). Therefore, given the right conditions, microbes can efficiently absorb heavy metals (Rafida and Sallis 2005).

Biomimetic Membranes

Scientists have worked on creating biological membranes that resemble cellular membranes to filter water. These biological membranes, referred to as biomimetic membranes, filter water in the same way that cells have for a billion years. The biomimetic membrane's composition and capabilities are similar to those of real cellular membranes (Fuwad et al. 2019). Water is transported via channels in the cell membrane created by aquaporin in living cells. For his in-depth research into aquaporin's water channel, Peter Agre was awarded the Nobel Prize in 2003 (Agre

2006). The development of improved aquaporin-associated biomimetic membranes with cutting-edge manufacturing technologies has been proven through research efforts. These membranes outperform traditional technology in the selection and transport of molecules in water purification systems.

Key Challenges and Future Perspective in the Application of Biofiltration for Heavy Metal Removal

Microbial-based filter technologies have demonstrated significant promise in the wastewater remediation process for eliminating heavy metals and other volatile organic compounds. Nevertheless, there are issues and constraints with the biofiltration process that need to be addressed (Pachaiappan et al. 2022).

1. Firstly, the functioning of biofilters greatly depends on the microbial community and its development. Research has been conducted on the role of microbes in the degradation of pollutants in water. It is necessary to consider expanding the degradation activity to a mass scale. Many works are executed at the laboratory scale with a pilot-level technique but are never upgraded into commercial applications (Srivastava and Majumder 2008).
2. The process underlying the metabolic pathways utilized by microorganisms to degrade contaminants is not clearly understood. Observing the metabolic pathways of microorganisms throughout the complete process of pollutant elimination is essential. This, in turn, depends on microbial genera and their community within the biofilter bed. Therefore, one must fully comprehend the mechanics underpinning microbial proliferation and degradation to apply biofiltration technology effectively (Wani et al. 1997).
3. The next issue is the formation of a thick biofilm layer, which occurs after the increased quantities of pollutants. The bacteria on the surface of the biofilm adsorb the contaminants. It also reaches the depths of the biofilm. After a certain period, the contaminant's diffusion to the biofilm's depth ceases. Consequently, microbial biofilm becomes dormant and with varied pore sizes. This is how multi-microbial cultures can be utilized to more efficiently and quickly degrade various pollutants (Swanson and Loehr 1997).
4. The contaminants originate from different sources, such as water treatment facilities, oil refineries, municipal waste and industrial effluent, and pharmaceutical waste. The nature and concentration of contaminants contained in wastewater must be analyzed to develop and optimize biofiltration settings and methods. Additionally, when it comes to recycling or the safe disposal of valuable products, the acquired metals and treated water should be considered (Srivastava and Majumder 2008).
5. By tackling the aforementioned issues, biofilters that are more reasonably priced and have superior technical designs at minimal cost will be produced in the future. This is all possible due to technological interventions like artificial intelligence. These days, water treatment is one of the many industries where artificial intelligence (AI) is becoming a contender. Therefore, the best bio-based water treatment method controlled by AI may be achieved in the days ahead (Srivastava and Majumder 2008, Pachaiappan et al. 2022).

Microbial Processes in Heavy Metal Remediation

Hazardous contaminants from distinct sources involving electroplating, poor industrial waste management, and mining pollute our environment with a large impact due to the ever-increasing human population and needs. Activities resulting in environmental degradation and resource depletion threaten human health and the equilibrium of the ecosystem. With the boom in industrialization, magnanimous manufacturing and utilization of metal contaminants, which can be accreted via the food web, can potentially deteriorate individual well-being. Even trace amounts of heavy metals administered via the gastrointestinal tract, inhalation, or topical exposure can significantly contribute to physiological and biochemical changes, including enzyme activity, membrane, and DNA impairment (Witkowska et al. 2021). For example, arsenic facilitates anomalous gene expression, and copper plays a role in BRAF gene signaling, which can cause mutations leading to cancer. Henceforth, heavy metals' passive consumption results in numerous potential diseases involving arrhythmia, Alzheimer's, renal and hepatic damage, etc.

There are numerous technologies available, viz. membrane filtration, chemical precipitation, reverse osmosis, soil washing, ion exchange, solidification, phytoremediation, and electrochemical treatments, etc., which are widely used for heavy metal remediation from contaminated water. However, these traditional techniques of heavy metal remediation result in the production of toxic chemical remnants, which further require vigilant management to lessen further environmental harm. Bioremediation is emerging as a promising and comparatively economical technology with less production of environmentally polluting toxicants, thus countering the negative effects of pollution by decreasing the persistence of toxic compounds.

Bioremediation leverages natural biological processes as an innovative and eco-friendly approach for the complete elimination of hazardous pollutants. This advanced technology utilizes microorganisms, fungi, green plants, or their enzymatic systems to restore contaminated environments to their original, unpolluted state. Bioremediation methods are broadly categorized into *in situ* and *ex situ* techniques. In *ex situ* bioremediation, contaminated soil is excavated and processed in a controlled, lined, above-ground treatment facility, where it undergoes natural aeration to enhance the microbial degradation of organic pollutants. Conversely, *in situ* bioremediation treats contaminated soil and groundwater directly at the site without the need for excavation or pumping, relying on biological processes to degrade or immobilize pollutants within the subsurface environment. This multidisciplinary technology integrates principles from microbiology, chemistry, geology, and engineering. Key techniques include landfarming, bioreactors, bioventing, bioaugmentation, composting, rhizofiltration, and biostimulation. The organisms responsible for these processes, commonly referred to as "bioremediators," play a pivotal role in breaking down or stabilizing contaminants within the affected ecosystems (Sanjana et al. 2024).

Microbial-mediated Mechanisms of Heavy Metal Detoxification

With long-term effects on the ecosystem, microbes help restore degraded surroundings to their native state. In addition to facilitating heavy metal extraction, microbial remediation stops heavy metals from leaching and mobilizing. Thus, in this particular context, recent technological advancements have made the use of bioremediation an essential method for cleaning up contaminated soils possible (Jørgensen 2007). To lessen the harmful impacts of heavy metals, microbes employ various processes, including biosorption, bioaccumulation, bioleaching, biotransformation, biovolatilization, and biomineralization. The most common are discussed further in this chapter.

Biosorption

Industrial effluents often contain microorganisms that have evolved diverse mechanisms to mitigate the toxic effects of heavy metals. These mechanisms may operate through metabolic pathways or independently of metabolism. One notable mechanism is biosorption, which involves the binding of metal ions to metal-binding proteins located in the cell walls. This phenomenon is exhibited by various organisms, including yeasts, fungi, algae, and bacteria. Biosorption refers to the ability of biological materials to sequester heavy metals from wastewater through either passive, spontaneous uptake or metabolically mediated pathways requiring ATP. It can also occur in non-living microbial biomass, which is capable of binding and concentrating heavy metals even from highly diluted aqueous solutions. Both living and non-living organisms exhibit biosorptive properties, with agricultural by-products such as husks, seeds, peels, and stalks of crops serving as effective biosorbents. Several factors influence the efficiency of biosorption, including temperature, pH, biosorbent type, surface area-to-volume ratio, biomass concentration, initial metal ion concentration, and the specific affinity of the metal to the biosorbent (Shamim 2018). Several models, such as the Freundlich and Langmuir models, can be employed to explain biosorption. Agents like thiosulfate, organic acids, and mineral acids can be used to recover metals that have been biosorbed. It is important to choose a desorption agent carefully to avoid changing a biosorbent's physical characteristics.

Bioaccumulation

The term "bioaccumulation" refers to the buildup of contaminants by organisms from the abiotic environment and food sources (trophic transfer). This means that this mechanism causes toxins to be absorbed by species more quickly than they can be eliminated. Sites contaminated with metals can be bioremediated by using plants' and microorganisms' bioaccumulation. Through importer complexes that create a lipid bilayer translocation pathway, bacteria can absorb metals into the intracellular space (i.e. import system). Furthermore, metals can be sequestered inside the intracellular space by proteins and peptide ligands. Microorganisms' capacity to gather, utilize, and get rid of contaminants depends on ambient factors—temperature, moisture content, and pH—that are suitable for their growth and metabolism. Phytoaccumulation is the process by which toxins from a contaminated environment build up and become concentrated in the roots of plants (Nnaji et al. 2023).

After that, the pollutants are moved and stored in the plant's aerial sections (leaves and shoots). Phytoabsorption, phytosequestration, or phytoaccumulation are terms used to describe the process of plant roots absorbing pollutants, followed by translocation and accumulation in shoots and leaves. The aerial parts of plants, including shoots and leaves, can be harvested and incinerated to generate energy, with the residual ash serving as a source for metal recovery. This approach is a key aspect of phytoremediation and has been extensively applied using hyperaccumulator plants such as *Helianthus annuus* (sunflower) and *Thlaspi caerulescens*. These plants effectively extract and concentrate metals like lead, zinc, copper, nickel, and cadmium from contaminated environments, offering a sustainable solution for metal removal and recycling (Milner and Kochian 2008).

Bioleaching

Bioleaching is a hydrometallurgical technique that employs microorganisms to extract metals and metalloids from solid materials. Unlike conventional hydrometallurgy, which typically relies on acids or bases to dissolve metals from solid matrices, bioleaching achieves this process without the use of such chemical agents. Leaching microorganisms' metabolism or the byproducts of their metabolism can dissolve nutrients and heavy metals from solid substrates, which is the basis of the long-term response process known as microbial leaching. The material's chemical and mineralogical makeup, as well as microorganisms, have the most effects on the bioleaching process. The microorganism's ideal growth conditions may result in the highest metal extraction yields (Dong et al. 2023).

Therefore, it is important to take into account the effects of several parameters on microorganisms, including water, heavy metal types, pH, temperature, mineral substrate, and concentrations of O_2 and CO_2 (Haferburg and Kothe 2007). A constant supply of oxygen is necessary for biological leaching since the microorganism's activity is correlated with the concentration of carbon dioxide and oxygen. The development and activity of leaching bacteria are significantly influenced by the regulation of pH levels. The ideal leaching rates for various metals require varying pH values; therefore, acidic conditions are frequently employed. Through microbial activity, temperature also has an impact on the leaching of heavy metals (Nguyen et al. 2021).

Bioremediation Techniques Employing Microbial Consortia

Microbes can exploit a wide variety of carbon sources when they are part of a microbial consortium. It gives microorganisms resilience against environmental stresses. When microbes work together, they can carry out intricate tasks that would be impossible for one organism to do alone. Thanks to technological advancements, consortium formation and comprehension of microbial interaction mechanisms are now achievable.

In an effort to create better strains, scientists have attempted to take advantage of the genetic and metabolic underpinnings of the diverse metal resistance mechanisms that microorganisms have developed. Genetic modification for strain enhancement is both straightforward and practical, as bacterial resistance to heavy metals is frequently governed by genes located on plasmids rather than exclusively on chromosomal DNA. These plasmid-encoded genes enable the production of proteins that confer resistance. A promising strategy for developing microorganisms with enhanced heavy metal resistance involves monitoring microbial metabolism and growth under conditions of progressively increasing toxicant concentrations. This gradual exposure facilitates the adaptation and selection of bacterial strains capable of tolerating elevated levels of heavy metals. Researchers have studied this property to see if tolerance to higher concentrations can help such toxicants biodegrade. Using organisms resistant to metals helps solve the issue of metal toxicity in developing cells. It is possible to leave the continuously self-replenishing system running for extended periods of time. To enhance the metal-accumulative strain, the living cells do, however, have the capacity for mutant isolation or general recombination (Adams et al. 2014).

Tolerance and absorption capacities are crucial features of live biomass utilized in a metal ion removal method. Yeast is one of the most common biomass types that can be used for the bioremediation of heavy metals at lower pH levels. A cheap and easily accessible source of biomass is yeast, such as *Saccharomyces cerevisiae*, *Schizosaccharomyces pombe*, *Kluyveromyces marxianus*, and *Candida* sp. (Ayele et al. 2021). Furthermore, under a variety of external settings, yeast cells continue to be able to accumulate a broad range of heavy metals to various degrees. It has been demonstrated that heavy metals like copper can accumulate in yeast cells (Wang and Chen 2006). There is a first, fast accumulation step that is believed to involve cation binding at the surface and is independent of temperature and metabolism. A second, metabolism-dependent mechanism that is considerably slower than the first and capable of accumulating greater amounts of cations comes after this stage. It is thought that cation internalization into the cell is involved in this second step. Furthermore, the absence of an energy source or the presence of competing anions and cations can affect the metabolism-dependent internalization of metal ions. Dönmez and Aksu serially subcultured *Candida* species strains in growth media supplemented with copper and nickel in an effort to adapt the strains to these metals. Higher metal concentrations allowed adapted cells to develop well, but non-adapted cells died. Additionally, at every concentration tested, the adapted cells' specific metal uptake capacity and metal elimination percentage (%) were greater than those of the non-adapted cells (Dönmez and Aksu 1999).

Metalloid tolerance in the genes of microbial consortia and their corresponding mechanisms are essential for bioremediation. The arsABC operon, for example, is a group of genes that confers tolerance to arsenic oxyanion in bacteria, both Gram-positive and Gram-negative (including *Staphylococcus*, *E. coli*, *Bacillus*, and *Rhodococcus*). This operon, which may be found on both plasmids and chromosomes, codes for the enzyme arsenate reductase arsC, which converts AsO_4^{3-} into AsO_3^{3-}. Research revealed that the SulP-type permease and sulfate-thiosulfate permease were frequently employed by microbes to transfer selenium compounds. Apart from these proteins, additional studies have shown that YedE, a membrane protein, may also be useful for selenium oxyanion transfer (Bhatia et al. 2018).

At least three different systems were involved in the transport of SeO_3^{2-} in *E. coli*: sulfate transport complex ABC, sulfate permease, and one or more unidentified transporters that continued to function even after the other two transporters were deactivated. The sulfate transport complex ABC is encoded by the *cysAWTP* operon, which is made up of *CysT* and *CysP*. SeO_3^{2-} is transferred into cells by a polyol transporter that is present in Rhodobacter sphaeroides (Lashani et al. 2023).

Challenges and Opportunities in Utilizing Microbial Processes for Heavy Metal Removal

The great diversity of bacteria in the living world makes them an important resource for the bioremediation of many contaminants. The effectiveness of heavy metal remediation can be assessed by exploring a wide range of potential materials, including both living and non-living biomass. In the context of bioremediation, microbial genetic engineering plays a pivotal role in enhancing the capacity of microorganisms to tolerate and absorb heavy metals. This approach involves the modification of genetic pathways to optimize metal uptake and resistance, thereby improving the efficiency and applicability of bioremediation strategies (Fatima et al. 2024). In addition, the immobilization of bacterial biomass on an appropriate carrier might be studied in order to enhance the chemical, physical, and porosity stability. Particularly, extremophiles have drawn attention because they include a genetic and metabolic reservoir of potential uses that might be utilized as microfactories to remove different kinds of contaminants and heavy metals. However, further research into the physiological characteristics of these extremophilic microfactories can support other heavy metal bioremediation devices that have been conceptualized but not yet recognized or integrated in the literature. Furthermore, the limited proliferation and stressful community interactions of extremophiles make *in situ* application difficult. Extremophiles may experience more stress in order to stably improve heavy metal bioremediation due to the complexity of their cellular systems and interspecies linkages under complicated, co-occurring harsh environments. Current technologies still require a great deal more work in conjunction with conventional methods to overcome the limitations related to the metabolism of bacteria itself. The complex interplay between microorganisms and biotic environmental factors is influenced by several key mechanisms. These include the production of organic acids by specific bacterial strains to modulate environmental pH, the presence of organic solutes in water to counteract external osmotic pressure, and the secretion of polysaccharides that contribute to biofilm formation in polluted sites. Moreover, the concentration of toxic pollutants plays a significant role in shaping microbial interactions and adaptive responses, highlighting the intricate balance between microbial activity and environmental stressors. Using readily available natural resources, global changes can be sustained while still meeting the treatment process's cost-efficiency needs in a bio-green manner. Studies on proteomics and metabolomics may be ideal tools for illuminating the complex mechanisms involved in bioremediation (Ojuederie and Babalola 2017). Additionally, algorithms can be developed to forecast the right combination of nutrients, electron shuttles, enzyme cofactors, microbe types, and physiochemical elements needed for effective *ex situ* bioremediation of contaminated locations.

Case Studies and Applications of Nature-based Solutions

The environmental remediation industry holds immense potential for leveraging natural solutions, particularly given the global scale of contamination. For instance, China faces the challenge of managing tens of millions of hectares of heavy metal-contaminated land, Europe has identified approximately 342,000 contaminated sites, and the United States has registered 450,000–600,000 brownfield sites, with the actual figure likely exceeding one million. These sites often harbor organic pollutants and hazardous substances, posing significant risks to human health and ecological stability. Conventional remediation technologies, which rely heavily on chemical reagents, thermal processes, and electrical inputs, entail substantial social, financial, and environmental costs. High remediation expenses often result in brownfield sites remaining abandoned and neglected for extended periods. However, various governments have introduced regulatory measures to promote brownfield redevelopment. For example, the UK government initiated a pilot program for planning grants in 2016 to facilitate site restoration. Furthermore, under the Paris Agreement, countries have committed to ambitious targets for reducing greenhouse gas (GHG) emissions, underscoring the need for sustainable and cost-effective remediation technologies that align with global climate goals. The redevelopment of brownfield sites and vacant urban land has enormous potential to mitigate climate change (Song et al. 2019).

There are many benefits to using nature-based remediation methods, like bioremediation and phytoremediation, including enhanced resilience to environmental changes, lower energy use, and higher material efficiency. The potential of technology has been limited by the fact that its broader socioeconomic and environmental benefits have not always been fully integrated into evaluation processes. Often, an inclusive approach that optimizes social, economic, and environmental sustainability has been absent, with the primary emphasis instead placed on the effectiveness of pollution control. This narrow focus has constrained the ability to realize the full potential of technological solutions in achieving comprehensive sustainability goals.

Real-world Applications of NBS for Heavy Metal Remediation

Today's hope for heavy metal remediation currently rests on biotechnologies and NBS, which are the preferred strategies, as certain ways of removing heavy metals from the land and aquatic environment are sophisticated and expensive. NBS can be defined as the planning, organizing, and carrying out of initiatives related to the preservation, sustainable management, and restoration of natural or altered ecosystems that will be able to adjust to the changing climate and social demands of the modern world while also benefiting biodiversity and human well-being. Nature-based solutions are much more affordable, natural, aesthetically pleasing, and ecologically friendly. Furthermore, because the solutions are naturally occurring, bio- and phytoremediation techniques are low-tech and require minimal automation. They are also reasonably priced. Plant species known as hyperaccumulators—plants of the genera Pteris, Pityrogramma, Noccaea, Sedum, Arabidopsis, Biscutella, and

Iberis, for example—are used in remediation when heavy metal contamination occurs in water or sewage. These species have the capacity to store hazardous metals in their tissues (Piwowarska et al. 2024). Transporters or carriers in the plasma membrane of plant root cells take up heavy metal ions from a specific environmental matrix; these combine with chelators to create a complex that permits the harmless transfer of the heavy metal ions to the shoot and other plant organs. NBS provides a viable method for heavy metal remediation in contaminated areas. By utilizing ecosystems and natural processes to reduce pollution, these solutions provide viable and affordable substitutes for traditional techniques. Wetlands are an efficient way to remove heavy metals from industrial effluent because of their intricate interactions between plants and microbes. Another NBS technique that makes use of fungi is called mycoremediation, and it takes advantage of the innate capacity of fungal networks to absorb and sequester heavy metals from contaminated substrates. These natural approaches support habitat restoration, biodiversity development, and the provision of ecosystem services in addition to aiding in the detoxification of contaminated places. Green infrastructures, such as bioretention systems and green roofs, can collect and immobilize heavy metals from stormwater runoff in urban contexts, lowering the amount of pollution that reaches water bodies (Lin et al. 2023). All things considered, NBS for heavy metals remediation is a prime example of how incorporating ecological concepts into environmental management can have multiple advantages, reducing pollution and fostering ecological resilience and health.

Success Stories and Lessons Learned from Implemented Projects

The investigation of novel and sustainable cleanup techniques has become imperative due to the increasing apprehension regarding heavy metal pollution in diverse ecosystems. NBS has become a viable strategy for addressing environmental issues by utilizing ecosystems and natural processes. The effectiveness, sustainability, and potential for wider use of NBS for heavy metals remediation are highlighted in this essay as it explores a number of success stories and the lessons discovered from projects that have been put into practice. Phytoremediation of contaminated environments is a noteworthy success story in the field of NBS for heavy metal cleanup. Utilizing plants to absorb, gather, and detoxify contaminants from soil and water is known as phytoremediation. One noteworthy instance is the use of hyperaccumulator plants in China to remediate soils poisoned with cadmium (Wang et al. 2023b). Utilizing species like *Pteris vittata* and *Sedum plumbizincicola* has proven remarkably effective in lowering cadmium levels, regaining the health of the soil, and allowing for the restart of agricultural operations (Wei and Zhou 2006). This study demonstrates the potential of phytoremediation as an economical and sustainable option. The lessons learned highlight how crucial it is to choose suitable plant species that are either native to the area or well-adapted there in order to guarantee the effectiveness of such interventions.

In line with this, mycoremediation has demonstrated encouraging outcomes in degrading or sequestering heavy metals using fungi. White rot fungi (*Pleurotus ostreatus*) were used in a Polish study to effectively remediate soil that had been contaminated with heavy metals from industrial processes. Because the fungi could create extracellular enzymes, the contaminants were easier to break down and became immobile, which greatly decreased their bioavailability (Baldrian 2003). This method improved microbial diversity and soil fertility in addition to addressing pollution. Important takeaways from this study include the necessity of comprehending the particular relationships that exist between fungi and pollutants and the significance of maximizing environmental conditions in order to improve remediation efficiency (Harms et al. 2011). Another effective use of NBS for heavy metal removal is in constructed wetlands. In Spain, the installation of a built wetland system in a mining-affected area successfully lowered the wastewater's concentrations of heavy metals like lead, zinc, and copper. The vegetation of the marsh, especially *Phragmites australis*, was essential in filtering and stabilizing the pollutants. The experiment showed that artificial wetlands might be used to remediate contaminated water in a sustainable and low-maintenance manner (Vymazal 2010). It also made clear how crucial it is to incorporate these systems into larger watershed management plans and how ongoing maintenance and monitoring are required to guarantee long-term efficacy (Kadlec and Wallace 2008). The application of biochar, a carbon-rich substance made from organic matter, in heavy metal cleanup has also demonstrated a great deal of promise. Applying biochar made from agricultural waste to contaminated soils successfully immobilized heavy metals and decreased their mobility and bioavailability, according to a Ghanaian project (Beesley et al. 2011). This enhanced soil fertility and structure while also reducing pollution and increasing agricultural output. The accomplishment of this project highlights the adaptability and dual purpose of biochar as an NBS. The most effective outcomes can only be obtained by customizing the manufacture and application of biochar to the unique soil and pollution circumstances (Gupta and Saleh 2013). Restoring mangrove ecosystems in Southeast Asia is another noteworthy application of NBS in heavy metal cleanup. Mangroves are naturally able to filter out heavy metals by capturing and stabilizing them in their root systems and the sediments that surround them. In an experiment conducted in Thailand, the amounts of heavy metals like arsenic and mercury in the soil and water were considerably lowered by reforesting mangroves in regions impacted by industrial pollution. Through improved fishing and tourism prospects, this project not only restored ecological balance but also benefited local populations socioeconomically. The initiative made clear how crucial multi-stakeholder cooperation and community involvement are to the effective implementation of NBS.

The successful implementation of NBS in urban settings has also had favorable results. The implementation of green infrastructure, such as bioretention systems and green roofs, has successfully reduced the amount of heavy metal pollution caused by stormwater runoff in Melbourne, Australia. By capturing and treating contaminants with soil and vegetation, these systems keep them out of water bodies. The accomplishment of this project shows how NBS can improve water quality and urban resilience. Important takeaways include the necessity of integrating NBS into urban planning and design, in addition to the significance of upkeep and observation of these systems to guarantee their continuous efficacy. NBS implementation for heavy metal remediation is not without difficulties, notwithstanding these achievements. Understanding the degree and kind of contamination requires comprehensive site evaluations and baseline research, which is one important lesson learned. This data is essential for choosing the right NBS and creating successful repair plans. Long-term observation and flexible management are also necessary to guarantee the effectiveness and durability of NBS actions. This includes monitoring the health and productivity of biological elements like plants and fungi, as well as the quality of the soil and water on a regular basis. Furthermore, a key factor in the accomplishment of NBS initiatives is the socioeconomic environment. Involving stakeholders and local communities in the design and execution phase guarantees that the solutions are both commercially and culturally feasible while also fostering a sense of ownership. In order to provide communities with the information and abilities required to sustain and profit from NBS, capacity building and education are also essential. These features were especially noticeable in initiatives like the restoration of Thailand's mangroves, where community participation was crucial to the project's success (MacFarlane and Burchett 2002).

In many different contexts and ecosystems, the use of nature-based methods for heavy metal remediation has shown to be highly successful. Phytoremediation, mycoremediation, biochar, artificial wetlands, and mangrove restoration projects have demonstrated how NBS may successfully reduce heavy metal pollution while also yielding extra ecological and socioeconomic advantages (Hossain et al. 2022). The insights gained from these projects highlight the significance of stakeholder engagement, site-specific strategies, ongoing monitoring, and adaptive management. NBS offers a viable and comprehensive strategy for heavy metal remediation, opening the door for resilient communities and healthier ecosystems as the world community searches for long-term solutions to environmental problems.

Comparative Analysis of NBS with Conventional Treatment Methods

Remediation of heavy metals from contaminated ecosystems is a serious challenge that requires sustainable and efficient methods to protect human and ecological health. Heavy metal contamination has traditionally been addressed by conventional treatment procedures such as chemical precipitation, ion exchange, electrochemical treatment, and physical removal methods. NBS are becoming more and more popular, nevertheless, because of their capacity to offer long-term and multipurpose advantages. This analysis offers a comparison and contrast between NBS and traditional treatment approaches, emphasizing the benefits, drawbacks, and optimal situations for each. For example, chemical precipitation involves adding chemicals to soluble heavy metals to change them into insoluble forms that can be extracted from the aqueous phase. High metal concentrations can be effectively treated with this approach, which also yields results rather quickly. However, the process produces a lot of sludge, which needs to be treated and disposed of further, presenting logistical and environmental problems. Furthermore, especially for large-scale applications, the cost of chemicals and the requirement for continuous chemical inputs can be significant. A further common technique is ion exchange, which uses resin materials to swap ions with heavy metals in solution to effectively remove them from the aqueous phase. This method can obtain low residual metal concentrations and is very effective. However, ion exchange resins need chemical renewal and are prone to clogging, which raises further environmental issues. Additionally, the upfront expenses of resin ingredients and the infrastructure needed for ion exchange systems may be unaffordable. Electrical currents are used in electrochemical treatment techniques like electroflotation and electrocoagulation to remove heavy metals from water. These methods have the benefit of requiring fewer chemicals and can be quite successful for some metals. On the other hand, they can be difficult to maintain and operate and demand large energy inputs. Their applicability is further limited by the requirement for specialized equipment and knowledge, especially in environments with limited resources.

To eliminate heavy metals from water, chemical treatments are frequently combined with physical removal methods like filtration and sedimentation. These techniques can be useful for eliminating metals that are particulate-bound, but they are less successful for dissolved forms and may be constrained by the physical filters' capacity. Filtration media replacement and maintenance can be expensive and labor-intensive (Vidu et al. 2020). On the other hand, nature-based solutions leverage ecosystems and natural processes to address heavy metal contamination, offering a sustainable and often cost-effective alternative to conventional methods. An example of this approach is phytoremediation, where plants are utilized to absorb, accumulate, and detoxify heavy metals from soil and water (Bhat et al. 2022). This approach is environmentally benign and can enhance biodiversity and soil health. However, phytoremediation is limited by the bioavailability of metals and plant tolerance to elevated pollution levels, resulting in a prolonged remediation timeframe. Mycoremediation, which employs fungi to degrade or immobilize heavy metals, is another exciting NBS. Fungi create organic acids and extracellular enzymes that can convert heavy metals into less toxic forms. This technique is particularly helpful for remediating soil since it can enhance the soil's fertility and structure (Akhtar and Mannan 2020). To guarantee these systems' long-term efficacy, meticulous planning is needed for both their design and upkeep. The capacity of biochar, a carbon-rich substance made from organic materials, to adsorb heavy metals from soil and water has drawn attention. Applying biochar can enhance soil characteristics and encourage plant development, making it a multipurpose NBS.

When NBS is contrasted with traditional treatment techniques, each strategy has unique benefits and drawbacks. Because they may be used quickly and reliably to achieve regulatory compliance, conventional procedures are frequently chosen, especially in industrial and municipal contexts where high pollution levels necessitate quick action. Conversely, NBS provides comprehensive advantages that go beyond simple correction. They increase ecosystem services, boost biodiversity, and, once established, frequently have cheaper operating costs. However, longer remediation timelines, the requirement for site-specific modifications, and possible fluctuations in efficacy due to environmental conditions can all pose challenges to the application of NBS.

In the cleanup of heavy metals, both of these treatment modalities—conventional and nature-based solutions—have important roles to play, depending on the particular environment and project goals. Conventional procedures have higher costs and environmental trade-offs but yield fast and dependable results. Though they can need more time and meticulous site-specific adjustments, NBS provides long-term, multifunctional advantages and are especially useful for ecological restoration. The most efficient and long-lasting solutions for heavy metal cleanup may come from an integrated strategy that incorporates the advantages of conventional and naturally occurring techniques. This would address both short-term contamination issues and long-term environmental health. The combined use of these methods has great potential for building robust and healthy ecosystems as the world community looks for new and sustainable ways to combat environmental pollution.

Emerging Trends and Future Directions

Recent Advancements in NBS Technologies for Heavy Metal Mitigation

The arsenal of instruments for dealing with environmental pollution has been greatly increased by recent developments in NBS technologies for heavy metal mitigation. These creative methods use the innate powers of microbial, fungal, and plant systems to clean up contaminated areas in an environmentally friendly and sustainable manner. This paper examines the most recent advancements in NBS for heavy metal mitigation, emphasizing the technological developments that are propelling their advancement as well as their potential and effectiveness. The creation of improved phytoremediation methods is one of the most noteworthy developments in NBS for heavy metal mitigation. Conventional phytoremediation depends on some plants' innate capacity to draw heavy metals out of the soil and water. The goal of recent research has been to genetically modify hyperaccumulator plants to increase their capacity for absorbing metals and their resistance to elevated levels of pollution. For instance, metal-binding proteins and transporters have been overexpressed in transgenic *Arabidopsis thaliana* and *Brassica juncea*, greatly enhancing their capacity to remove heavy metals like cadmium, lead, and arsenic from contaminated areas. These plants, which have undergone genetic modification, show a greater accumulation of heavy metals together with improved growth and biomass production, which makes them more suitable for large-scale cleanup efforts. Apart from genetic engineering, developments in the interactions between plants and microbes have created new opportunities for phytoremediation. It has been demonstrated that using plant growth-promoting rhizobacteria (PGPR) increases the phytoremediation effectiveness of plants (Gangola et al. 2023). PGPR can reduce stress caused by metals, enhance metal bioavailability in the soil, and help plants absorb metals. Certain strains of PGPR have been found in recent research to be able to solubilize heavy metals and generate phytohormones, which in turn promote plant growth and metal accumulation. A viable tactic for maximizing heavy metal mitigation is the combination of PGPR with phytoremediation plants. Additionally, there have been notable developments in mycoremediation, the use of fungi to clean up the environment. Finding and using fungal species with strong metal-binding and degrading abilities has been the focus of recent studies. For example, the capacity of the white rot fungus *Phanerochaete chrysosporium* to bind heavy metals and break down complex organic contaminants has been thoroughly investigated. Fungal biosorbents are materials based on fungal biomass that are used in mycoremediation processes to remove heavy metals from aqueous solutions. Several functional groups have been added to these biosorbents to improve their selectivity and ability to bind metals. Furthermore, the stability and reusability of mycoremediation systems have been enhanced by the development of fungal biofilms, which are composed of immobilized fungal cells on solid supports, making them more useful for practical uses in the real world (Wang et al. 2021).

Recent technological advancements have also helped in the development of constructed wetlands, an NBS that uses natural wetland processes to clean wastewater. Advanced modeling and simulation technologies have improved the design and optimization of artificial wetlands. With the use of these techniques, it is possible to accurately forecast how wetlands would behave in various environmental settings, facilitating the construction of wetlands that are suited to certain pollution scenarios. Moreover, it has been demonstrated that combining constructed wetlands with other treatment methods, including electrocoagulation systems and biochar filters, increases the total effectiveness of heavy metal removal. By combining the benefits of engineering and natural processes, these hybrid systems produce remediation solutions that are stronger and more efficient. Significant advancements have been made in the use of biochar for heavy metal mitigation, especially in the creation of functionalized biochars. Because of its high surface area and porous structure, biochar—a carbon-rich substance obtained by the pyrolysis of organic matter—is a powerful heavy metal adsorbent. The goal of recent developments has been to improve biochar's ability to bind metals by using different modification methods. For instance, to improve biochar's ability to adsorb particular heavy metals, nanoparticles like iron and manganese oxide have been functionalized into the material. Furthermore, the creation of biochar composites—a mixture of activated carbon and clay minerals—has led to the development of adsorbents with improved stability and performance. These developments in the field of biochar technology present viable ways to remediate heavy metal-contaminated soils and waterways sustainably. As a viable and effective NBS for heavy metal mitigation, the use of algae in bioremediation has also drawn interest. Algae, especially microalgae, can develop quickly under a range of conditions and have a high capacity for metal uptake. Utilizing nutrient-rich wastewater as a growth medium and other strategies to optimize the environment for algae growth and metal buildup has been the subject of recent studies. This method tackles problems with wastewater treatment and resource recovery in addition to reducing heavy metal contamination. The ability of algae to accumulate metals has been significantly improved by developments in genetic engineering. For instance, genetically modified strains of Chlorella and Scenedesmus have been developed to express metal-binding proteins, resulting in higher metal uptake efficiencies (Nyika and Dinka 2023). The integration of algal bioremediation with biofuel production also offers a sustainable pathway for energy recovery, adding economic value to the remediation process.

Another promising NBS technology for heavy metal mitigation is the use of biopolymers and biosurfactants (Mpentshu 2018). Biopolymers, such as chitosan and alginate, are natural materials that can be derived from renewable resources. These biopolymers have shown excellent metal-binding properties and can be used to develop adsorbent materials for heavy metal removal. Recent advancements include the modification of biopolymers with functional groups to enhance their selectivity and adsorption capacity. Microorganisms produce surface-active chemicals called biosurfactants, which can help remove heavy metals from contaminated locations by making them more bioavailable. The potential applications of biopolymers and biosurfactants in heavy metal cleanup have increased due to the development of scalable and reasonably priced production technologies.

Using fungal biosorbents and biofilms, developing genetically modified plants, and optimizing plant-microbe interactions have all contributed to the increased effectiveness of mycoremediation and phytoremediation. These technological advancements show how NBS may offer long-term, multipurpose solutions to environmental pollution, opening the door to better ecosystems and communities. The development of hybrid systems and the combination of NBS with traditional treatment techniques will probably significantly improve the efficacy and sustainability of heavy metal remediation initiatives as research into these areas progresses.

Integration of NBS with Other Sustainable Wastewater Treatment Approaches

The incorporation of NBS alongside other sustainable wastewater treatment methodologies offers a paradigm shift in the way the increasing issues of resource scarcity, water pollution, and ecosystem degradation are addressed. The potential of NBS to improve the resilience and sustainability of wastewater treatment systems has come to light. NBS uses ecosystems and natural processes to reduce environmental impacts. The synergies between NBS and other sustainable approaches are examined in this essay, with an emphasis on how their integration might result in comprehensive and multipurpose wastewater management systems. The use of modern filtration and bioreactor technologies in conjunction with built wetlands is one of the most promising ways to integrate NBS with sustainable wastewater treatment. By employing plant, soil, and microbial populations to treat wastewater through physical, chemical, and biological processes, constructed wetlands are designed environments that resemble natural wetlands (Yin et al. 2019). Constructed wetlands can provide improved removal efficiency for a wide range of contaminants, including heavy metals, viruses, and nutrients when combined with modern filtering systems like sand filters or membrane bioreactors (MBRs).

Together with offering extra ecosystem services including habitat construction, carbon sequestration, and aesthetic value, the integration of these systems improves treatment capacity and effluent quality. Using phytoremediation techniques in conjunction with biochar-based adsorption techniques is another cutting-edge method for treating wastewater sustainably. Because of its high porosity and surface area, biochar, which is produced by pyrolyzing organic biomass, is a powerful adsorbent for eliminating impurities from water. Using biochar in conjunction with phytoremediation can improve the treatment system's overall effectiveness (Qiu et al.

2022). In order to enhance plant growth and pollutant uptake, biochar can be added to the soil. Alternatively, it can be employed as a stand-alone filter medium to remove pollutants before they enter the phytoremediation zone. By improving soil health and agricultural production in addition to enhancing the removal of contaminants, this integrated strategy fosters a circular economy where waste products are recycled for the benefit of the environment (Wei et al. 2020). Combining microbial fuel cells (MFCs) with algal-based treatment systems is another notable integration. The capacity of algae to absorb nutrients and pollutants from wastewater and produce biomass that may be utilized for bioenergy or bioproducts has been the subject of much research. On the other hand, microbial fuel cells are bioelectrochemical devices that use bacteria's metabolic processes to transform organic matter in wastewater into electrical energy. Wastewater treatment systems can accomplish simultaneous nutrient removal, energy recovery, and biomass production by combining algae cultivation with MFCs. MFC performance can be improved by the oxygen that algae create during photosynthesis, and algal cultivation systems can be powered by the electricity produced by MFCs, resulting in a self-sufficient and cooperative treatment process. Enhancing sustainability and resilience in urban and peri-urban regions is a big potential benefit of integrating NBS with decentralized wastewater treatment systems. The overall liveability and sustainability of urban environments are enhanced by these integrated systems, which also bring co-benefits like enhanced public spaces, enhanced biodiversity, and urban cooling. Additionally, NBS-based hybrid systems that incorporate conventional chemical and physical treatment techniques can improve wastewater treatment efficiency and sustainability. For example, combining coagulation-flocculation procedures with constructed wetlands can enhance the removal of heavy metals and suspended particles while lowering the need for chemical coagulants. Likewise, pharmaceuticals and persistent organic pollutants in wastewater can be more effectively degraded by combining advanced oxidation processes (AOPs) with phytoremediation (Badawi et al. 2023). By combining the best features of artificial and natural processes, these hybrid systems offer dependable, adaptable solutions that can be customized to meet unique treatment requirements and environmental constraints. Furthermore, wastewater treatment plants can become resource recovery facilities by integrating NBS with resource recovery technology. Wastewater treatment can play a significant role in the circular economy by combining NBS with technologies for the recovery of nutrients, energy, and water. For instance, biogas recovery for energy production can be facilitated by combining anaerobic digestion with constructed wetlands, and treated wastewater can be recycled for industrial or agricultural uses. These systems generate biosolids that can be processed into compost or biochar to improve soil fertility and sequester carbon. This all-encompassing strategy encourages sustainability and resource efficiency in addition to addressing water contamination. NBS integration opens up new possibilities for wastewater treatment system optimization by combining smart water management technologies, including automation, data analytics, and sensors. Smart technologies can improve NBS monitoring and control by allowing for real-time performance optimization and ambient condition adaptation. In constructed wetlands, for instance, sensors can be used to track nutrient levels and water quality indicators, enabling precise control over microbial activity and plant development.

By analyzing this data, data analytics and machine learning algorithms can forecast treatment outcomes and spot possible problems, allowing for proactive and flexible management of NBS. This integration can provide useful data for research and policy creation, as well as increase the efficacy and dependability of wastewater treatment systems. These integrated systems may address a variety of treatment requirements and offer co-benefits to communities and ecosystems by utilizing the advantages of both engineered and natural processes.

Research Gaps and Potential Areas for Further Exploration

Even though NBS for heavy metal cleanup has come a long way, there are still many unanswered questions and possible study topics. To maximize NBS's effectiveness, scalability, and sustainability in a variety of environmental contexts, these shortcomings must be met. With an emphasis on improving the comprehension, utilization, and integration of NBS in heavy metal remediation, this essay highlights important research gaps and makes recommendations for prospective areas for further study A significant area of unmet research need in NBS for heavy metal remediation is the intricate interplay of microbes, plants, and soil chemistry. For example, phytoremediation is highly dependent on these relationships; nevertheless, the precise processes by which plants and the microbiomes they are connected with absorb, move, and remove heavy metals remain incompletely understood. The molecular and genetic underpinnings of these processes, as well as the identification of important genes, proteins, and signalling pathways involved in metal uptake and tolerance, should be the main areas of future research. Proteomics, metabolomics, transcriptomics, and genomics are examples of advanced omics technologies that may offer deep insights into these interactions and aid in the development of genetically modified plants with improved remediation capacities (Yin et al. 2019).

The heterogeneity of NBS effectiveness across various environmental conditions and contamination situations represents an important research need. The efficacy of NBS can be greatly impacted by variables such as co-contaminant presence, pH, temperature, moisture content, and type of soil (Zheng et al. 2020). Systematic investigations that assess NBS's performance in a variety of environmental settings and pollution levels are required. This entails field testing and extended observation to evaluate the robustness and longevity of NBS in practical environments. Optimizing the design and use of NBS for particular sites and pollution profiles can also be achieved by creating prediction models that take these environmental factors into account. Important research needs also exist in the areas of NBS economic viability and scalability. Although NBS have shown promise in small-scale studies and pilot programs, larger-scale implementation is still difficult. The goal of research should be to create scalable, reasonably priced NBS systems that can be used in sizable, polluted regions. This involves investigating the application of accessible and nearby plant species in addition to creative culture methods that improve plant development and metal uptake. To provide a better understanding of NBS's financial viability and potential for wider adoption, economic assessments comparing the costs and benefits of NBS with traditional remediation techniques are also required (Verma and Kuila 2019).

Another interesting subject to investigate further is the integration of NBS with other sustainable remediation methods. Further study is required to optimize these integrated techniques, even though hybrid systems that combine NBS with physical, chemical, and biological therapies have demonstrated potential. This involves looking into the synergistic benefits of mixing various technologies, including integrating constructed wetlands with cutting-edge filtration systems or using biochar or nanomaterials to improve phytoremediation (Sounthararajah et al. 2016). Studies should assess these hybrid systems' sustainability and effects on the environment, taking into account things like resource use, energy use, and secondary waste production. The creation of efficient monitoring and assessment systems for assessing NBS performance represents a significant research gap. Conventional monitoring techniques frequently depend on expensive and time-consuming laboratory analysis and periodic sampling. Innovative monitoring devices that offer *in situ*, real-time assessments of heavy metal concentrations and other critical parameters are desperately needed. This includes tracking the development and efficacy of NBS in real-time through the use of sensor networks, remote sensing technologies, and advanced data analytics. Creating standardized procedures and measurements to evaluate NBS performance is also essential for comparing outcomes between studies and guaranteeing the validity and repeatability of study findings. The successful implementation of NBS depends on public acceptance and policy backing; nevertheless, little is known about the social and institutional aspects of these technologies. Creating successful communication and engagement strategies requires an understanding of how the public views NBS and how they behave (Dasgupta et al. 2022).

Policy and regulatory frameworks that either facilitate or impede the implementation of NBS should also be investigated in research. Finding best practices and legislative tools that can encourage the incorporation of NBS into currently running environmental management and remediation projects is part of this. The integration of government, industry, academia, and local communities in collaborative research can aid in bridging the knowledge gap between science and policy, as well as expediting the adoption of NBS. Thus, it can be said that NBS in heavy metal remediation has new potential and problems as a result of climate change. The resilience and effectiveness of NBS may be impacted by changing meteorological conditions, such as elevated temperatures, modified precipitation patterns, and extreme weather occurrences.

Conclusion and Future Perspectives

Conclusion

Heavy metal pollution in wastewater is a pressing environmental and public health issue, originating from various industrial, agricultural, and urban sources. Traditional wastewater treatment methods, while effective to some extent, often fall short in terms of efficiency, cost, and sustainability. Nature-based solutions (NBS) emerge as promising alternatives, leveraging natural processes and ecosystems to mitigate heavy metal contamination. NBS, including phytoremediation, bioremediation, constructed wetlands, and green infrastructure, offers several advantages over conventional methods. They are cost-effective, sustainable, enhance biodiversity, and can engage local communities in environmental stewardship. These solutions harness the inherent capabilities of plants, microbes, and natural materials to remove, stabilize, or transform heavy metals, thereby reducing their bioavailability and toxicity.

The integration of NBS into wastewater management strategies not only addresses the issue of heavy metal pollution but also contributes to the restoration and enhancement of natural ecosystems. By promoting the use of renewable and naturally occurring processes, NBS align with the principles of sustainability and environmental conservation.

Future Perspectives

While the potential of nature-based solutions for heavy metal remediation is well-recognized, several areas require further research and development to enhance their effectiveness and applicability. Firstly, optimization of NBS is necessary through further research to refine the selection and combination of plants, microbes, and materials. This involves identifying species with high metal uptake capacities, enhancing microbial communities' performance, and improving the design of constructed wetlands and green infrastructure.

Scalability and implementation are also critical. Developing scalable models and guidelines for implementing NBS in various contexts, from small-scale urban applications to large industrial sites, is crucial. This includes addressing practical challenges such as land availability, maintenance requirements, and integration with existing infrastructure. Additionally, long-term performance and monitoring are essential to assess the durability and sustained performance of NBS. Continuous monitoring and evaluation will help understand the long-term fate of heavy metals in treated ecosystems and ensure the effectiveness of NBS over extended periods.

Supportive policies and regulatory frameworks are needed to encourage the adoption of NBS. Governments and regulatory bodies should promote NBS through incentives, funding, and inclusion in environmental management plans. Community engagement and education are vital for the acceptance and success of NBS projects. Raising awareness and involving local communities can enhance their acceptance and success. Educational programs and participatory approaches can empower communities to take an active role in environmental protection and sustainable wastewater management.

Finally, interdisciplinary collaboration is essential to address the complex issue of heavy metal pollution through NBS. Scientists, engineers, government and private bodies, and policymakers must work together. Collaborative efforts can drive innovation, share knowledge, and develop integrated solutions that are effective and sustainable.

By addressing these future perspectives, nature-based solutions can play a pivotal role in mitigating heavy metal pollutants in wastewater, contributing to cleaner water, healthier ecosystems, and improved public health. As we continue to advance our understanding and implementation of NBS, they hold the promise of transforming wastewater management and fostering a more sustainable and resilient environment.

Reference

Adams, G., Tawari-Fufeyin, P., Igelenyah, E. and Odukoya, E. 2014. Assessment of heavy metals bioremediation potential of microbial consortia from poultry litter and spent oil contaminated site. Int J Environ Bioremed Biodegrad, 2(2): 84-92.

Agre, P. 2006. The aquaporin water channels. Proceedings of the American Thoracic Society, 3(1): 5-13.

Akhtar, N. and Mannan, M.A.-u. 2020. Mycoremediation: Expunging environmental pollutants. Biotechnology Reports, 26: e00452.

Akpor, O.B., Ohiobor, G.O. and Olaolu, D. 2014. Heavy metal pollutants in wastewater effluents: Sources, effects and remediation. Advances in Bioscience and Bioengineering, 2(4): 37-43.

Ali, H. and Khan, E. 2018. Trophic transfer, bioaccumulation, and biomagnification of non-essential hazardous heavy metals and metalloids in food chains/webs—Concepts and implications for wildlife and human health. Human and Ecological Risk Assessment: An International Journal, 25(6): 1353-1376.

Ali, H., Khan, E. and Ilahi, I. 2019. Environmental chemistry and ecotoxicology of hazardous heavy metals: environmental persistence, toxicity, and bioaccumulation. Journal of Chemistry, 2019(1): 6730305.

Ang, S.Y., Goh, H.W., Mohd Fazli, B., Haris, H., Azizan, N.A., Zakaria, N.A. and Johar, Z. 2023. Heavy metals removal from domestic sewage in batch mesocosm constructed wetlands using Tropical Wetland plants. Water, 15(4): 797.

Ayele, A., Haile, S., Alemu, D. and Kamaraj, M. 2021. Comparative utilization of dead and live fungal biomass for the removal of heavy metal: A concise review. The Scientific World Journal, 2021(1): 5588111.

Badawi, A.K., Salama, R.S. and Mostafa, M.M.M. 2023. Natural-based coagulants/flocculants as sustainable market-valued products for industrial wastewater treatment: A review of recent developments. RSC Advances, 13(28): 19335-19355.

Balali-Mood, M., Naseri, K., Tahergorabi, Z., Khazdair, M.R. and Sadeghi, M. 2021. Toxic mechanisms of five heavy metals: Mercury, lead, chromium, cadmium, and arsenic. Frontiers in Pharmacology, 12: 643972.

Baldrian, P. 2003. Interactions of heavy metals with white-rot fungi. Enzyme and Microbial Technology, 32(1): 78-91.

Beesley, L., Moreno-Jiménez, E., Gomez-Eyles, J.L., Harris, E., Robinson, B. and Sizmur, T. 2011. A review of biochars' potential role in the remediation, revegetation and restoration of contaminated soils. Environmental Pollution, 159(12): 3269-3282.

Bennicelli, R., Stępniewska, Z., Banach, A., Szajnocha, K. and Ostrowski, J. 2004. The ability of Azolla caroliniana to remove heavy metals (Hg (II), Cr (III), Cr (VI)) from municipal waste water. Chemosphere, 55(1): 141-146.

Bhat, S.A., Bashir, O., Haq, S.A.U., Amin, T., Rafiq, A., Ali, M. et al. 2022. Phytoremediation of heavy metals in soil and water: An eco-friendly, sustainable and multidisciplinary approach. Chemosphere, 303: 134788.

Bhatia, S.K., Bhatia, R.K., Choi, Y.-K., Kan, E., Kim, Y.-G. and Yang, Y.-H. 2018. Biotechnological potential of microbial consortia and future perspectives. Critical Reviews in Biotechnology, 38(8): 1209-1229.

Bohn, H. 1992. Consider biofiltration for decontaminating gases. Chemical Engineering Progress, 88(4): 34-40.

Bohn, H.L. and Bohn, K.H. 1999. Moisture in biofilters. Environmental Progress, 18(3): 156-161.

Camacho, J.V., Romero, L.R., Marchante, C.F., Morales, F.F. and Rodrigo, M.R. 2017. The salinity effects on the performance of a constructed wetland-microbial fuel cell. Ecological Engineering, 107: 1-7.

Darlington, A.B., Dat, J.F. and Dixon, M.A. 2001. The biofiltration of indoor air: Air flux and temperature influences the removal of toluene, ethylbenzene, and xylene. Environmental Science & Technology, 35(1): 240-246.

Dasgupta, S., Badola, R., Ali, S.Z., Jiju, J. and Tariyal, P. 2022. Adaptive capacity and vulnerability of the socio-ecological system of Indian Himalayan villages under present and predicted future scenarios. Journal of Environmental Management, 302: 113946.

Davis, L. 1995. A handbook of constructed wetlands: A guide to creating wetlands for agricultural wastewater, domestic wastewater, coal mine drainage, stormwater. In the Mid-Atlantic Region. Volume 1: General Considerations. USDA-Natural Resources Conservation Service.

Delhoménie, M.-C. and Heitz, M. 2005. Biofiltration of air: A review. Critical Reviews in Biotechnology, 25(1-2): 53-72.

Dimitroula, H., Syranidou, E., Manousaki, E., Nikolaidis, N.P., Karatzas, G.P. and Kalogerakis, N. 2015. Mitigation measures for chromium-VI contaminated groundwater – The role of endophytic bacteria in rhizofiltration. Journal of Hazardous Materials, 281: 114-120.

Dong, Y., Zan, J. and Lin, H. 2023. Bioleaching of heavy metals from metal tailings utilizing bacteria and fungi: Mechanisms, strengthen measures, and development prospect. Journal of Environmental Management, 344: 118511.

Dönmez, G. and Aksu, Z. 1999. The effect of copper (II) ions on the growth and bioaccumulation properties of some yeasts. Process Biochemistry, 35(1-2): 135-142.

Fatima, Z., Azam, A., Iqbal, M.Z., Badar, R. and Muhammad, G. 2024. A comprehensive review on effective removal of toxic heavy metals from water using genetically modified microorganisms. Desalination and Water Treatment, 100553.

Fuwad, A., Ryu, H., Malmstadt, N., Kim, S.M. and Jeon, T.-J. 2019. Biomimetic membranes as potential tools for water purification: Preceding and future avenues. Desalination, 458: 97-115.

Gangola, S., Joshi, S., Bhandari, G., Bhatt, P., Kumar, S., Bhandari, N.S. and Mittal, A. 2023. Remediation of heavy metals by rhizospheric bacteria and their mechanism of detoxification. *In*: Advanced Microbial Technology for Sustainable Agriculture and Environment. Elsevier, pp. 31-46.

Ghosh, S., Selvakumar, G., Ajilda, A.A.K. and Webster, T.J. 2021. Microbial biosorbents for heavy metal removal. *In*: New Trends in Removal of Heavy Metals from Industrial Wastewater. Elsevier, pp. 213-262.

Gill, S.S., Anjum, N.A., Ahmad, I., Thangavel, P., Sridevi, G., Pacheco, M. et al. 2012. Metal hyperaccumulation and tolerance in Alyssum, Arabidopsis and Thlaspi: An overview. The Plant Family Brassicaceae: Contribution Towards Phytoremediation, 99-137.

Girdhar, M., Tabassum, Z., Singh, K. and Mohan, A. 2022. A review on the resistance and accumulation of heavy metals by different microbial strains. Biodegradation Technology of Organic and Inorganic Pollutants, 219.

Gupta, V.K. and Saleh, T.A. 2013. Sorption of pollutants by porous carbon, carbon nanotubes and fullerene – An overview. Environmental Science and Pollution Research, 20: 2828-2843.

Haferburg, G. and Kothe, E. 2007. Microbes and metals: Interactions in the environment. Journal of Basic Microbiology, 47(6): 453-467.

Hammond, C.M., Root, R.A., Maier, R.M. and Chorover, J. 2018. Mechanisms of arsenic sequestration by Prosopis juliflora during the phytostabilization of metalliferous mine tailings. Environmental Science & Technology, 52(3): 1156-1164.

Harms, H., Schlosser, D. and Wick, L.Y. 2011. Untapped potential: Exploiting fungi in bioremediation of hazardous chemicals. Nature Reviews Microbiology, 9(3): 177-192.

Hossain, M.B., Masum, Z., Rahman, M.S., Yu, J., Noman, M.A., Jolly, Y.N. et al. 2022. Heavy metal accumulation and phytoremediation potentiality of some selected mangrove species from the world's largest mangrove forest. Biology, 11(8): 1144.

Hua, T. and Haynes, R.J. 2016. Constructed wetlands: Fundamental processes and mechanisms for heavy metal removal from wastewater streams. International Journal of Environmental Engineering, 8(2-3): 148-178.

Huang, X., Glick, B. and Greenberg, B. 2000. Combining remediation techniques increases kinetics for removal of persistent organic contaminants from soil. Environmental Toxicology and Risk Assessment: Science, Policy, and Standardization—Implications for Environmental Decisions: Tenth Volume. ASTM International.

Ince, M. and Ince, O.K. 2017. An overview of adsorption technique for heavy metal removal from water/wastewater: A critical review. International Journal of Pure and Applied Sciences, 3(2): 10-19.

Jørgensen, K.S. 2007. In situ bioremediation. Advances in Applied Microbiology, 61: 285-305.

Kadlec, R. 2009. Comparison of free water and horizontal subsurface treatment wetlands. Ecological Engineering, 35(2): 159-174.

Kadlec, R.H. and Wallace, S. 2008. *Treatment Wetlands*. CRC Press.

Kang, C.-H., Kwon, Y.-J. and So, J.-S. 2016. Bioremediation of heavy metals by using bacterial mixtures. Ecological Engineering, 89: 64-69.

Kaur, I., Behl, T., Aleya, L., Rahman, M.H., Kumar, A., Arora, S. and Akter, R. 2021. Role of metallic pollutants in neurodegeneration: Effects of aluminum, lead, mercury, and arsenic in mediating brain impairment events and autism spectrum disorder. Environmental Science and Pollution Research, 28: 8989-9001.

Khayatzadeh, J. and Abbasi, E. 2010. The effects of heavy metals on aquatic animals. The 1st International Applied Geological Congress, Department of Geology, Islamic Azad University – Mashad Branch, Iran, 1: 26-28.

Kosolapov, D., Kuschk, P., Vainshtein, M., Vatsourina, A., Wiessner, A., Kästner, M. and Müller, R. 2004. Microbial processes of heavy metal removal from carbon-deficient effluents in constructed wetlands. Engineering in Life Sciences, 4(5): 403-411.

Kumar, N., Kardam, A., Rajawat, D.S. and Jain, V. 2019. Carboxymethyl nanocellulose stabilized nano zero-valent iron: An effective method for reduction of hexavalent chromium in wastewater. Materials Research Express, 6(11): 1150-1153.

Kumar, N., Kardam, A., Jain, V. and Nagpal, S. 2020. A rapid, reusable polyaniline-impregnated nanocellulose composite-based system for enhanced removal of chromium and cleaning of waste water. Separation Science and Technology, 55(8): 1436-1448.

Kumar, P., Asthana, S., Purty, R.S. and Chatterjee, S. 2020. Process design for removal of heavy metals by a bio-sorbent trickle bed system: A proof of concept. Journal of Physics: Conference Series: IOP Publishing, 1531: 012119.

Lashani, E., Amoozegar, M.A., Turner, R.J. and Moghimi, H. 2023. Use of microbial consortia in bioremediation of metalloid polluted environments. Microorganisms, 11(4): 891.

Li, H., Sheng, G., Sheng, W. and Xu, O. 2002. Uptake of trifluralin and lindane from water by ryegrass. Chemosphere, 48(3): 335-341.

Lin, H., Zhou, M., Li, B. and Dong, Y. 2023. Mechanisms, application advances and future perspectives of microbial-induced heavy metal precipitation: A review. International Biodeterioration & Biodegradation, 178: 105544.

MacFarlane, G. and Burchett, M. 2002. Toxicity, growth and accumulation relationships of copper, lead and zinc in the grey mangrove Avicennia marina (Forsk.) Vierh. Marine Environmental Research, 54(1): 65-84.

Madhu, P.M. and Sadagopan, R.S. 2020. Effect of heavy metals on growth and development of cultivated plants with reference to cadmium, chromium and lead – A review. Journal of Stress Physiology & Biochemistry, 16(3): 84-102.

Márquez, P., Herruzo-Ruiz, A., Siles, J., Alhama, J., Michán, C. and Martín, M. 2021. Influence of packing material on the biofiltration of butyric acid: A comparative study from a physico-chemical, olfactometric and microbiological perspective. Journal of Environmental Management, 294: 113044.

Martin, C.D. and Johnson, K.D. 1995. The use of extended aeration and in-series surface-flow wetlands for landfill leachate treatment. Water Science and Technology, 32(3): 119-128.

Milner, M.J. and Kochian, L.V. 2008. Investigating heavy-metal hyperaccumulation using Thlaspi caerulescens as a model system. Annals of Botany, 102(1): 3-13.

Mitsch, W.J. 1992. Combining ecosystem and landscape approaches to Great Lakes wetlands. Journal of Great Lakes Research, 18(4): 552-570.

Mpentshu, Y.P. 2018. Biosurfactant Producing Biofilms for the Enhancement of Nitrification and Subsequent Aerobic Denitrification. Cape Peninsula University of Technology.

Nguyen, T.H., Won, S., Ha, M.-G., Nguyen, D.D. and Kang, H.Y. 2021. Bioleaching for environmental remediation of toxic metals and metalloids: A review on soils, sediments, and mine tailings. Chemosphere, 282: 131108.

Nnaji, N.D., Onyeaka, H., Miri, T. and Ugwa, C. 2023. Bioaccumulation for heavy metal removal: A review. SN Applied Sciences, 5(5): 125.

Nyika, J. and Dinka, M.O. 2023. Factors and mechanisms regulating heavy metal phycoremediation in polluted water. Discover Water, 3(1): 14.

Odumbe, E., Murunga, S. and Ndiiri, J. 2023. Heavy metals in wastewater effluent: Causes, effects, and removal technologies. Trace Metals in the Environment. IntechOpen. Available at: http://dx.doi.org/10.5772/intechopen.1001452.

Ojuederie, O.B. and Babalola, O.O. 2017. Microbial and plant-assisted bioremediation of heavy metal polluted environments: A review. International Journal of Environmental Research and Public Health, 14(12): 1504.

Pachaiappan, R., Cornejo-Ponce, L., Rajendran, R., Manavalan, K., Femilaa Rajan, V. and Awad, F. 2022. A review on biofiltration techniques: Recent advancements in the removal of volatile organic compounds and heavy metals in the treatment of polluted water. Bioengineered, 13(4): 8432-8477.

Piwowarska, D., Kiedrzyńska, E. and Jaszczyszyn, K. 2024. A global perspective on the nature and fate of heavy metals polluting water ecosystems, and their impact and remediation. Critical Reviews in Environmental Science and Technology, 1-23.

Qasaimeh, A., Al Sharie, H. and Masoud, T. 2015. A review on constructed wetlands components and heavy metal removal from wastewater. Journal of Environmental Protection. 6(07): 710.

Qin, R. and Chen, H. 2016. The procession of constructed wetland removal mechanism of pollutants. 2016 4th International Conference on Mechanical Materials and Manufacturing Engineering: Atlantis Press, 487-489.

Qiu, B., Shao, Q., Shi, J., Yang, C. and Chu, H. 2022. Application of biochar for the adsorption of organic pollutants from wastewater: Modification strategies, mechanisms and challenges. Separation and Purification Technology, 300: 121925.

Rafida, A. and Sallis, P. 2005. Removal of heavy metals from rainwater in vertical flow biofilters conditioned with sulphate reducing bacteria (SRB). In XII International Rainwater Catchment Systems Conference: Citeseer.

Raza, H., Bibi, H., Bibi, T. and Bilal, H. 2023. Potential of constructed wetlands for the treatment of heavy metals from wastewater. Journal of Bioresource Management, 10(1): 15.

Reddy, K. and DeLaune, R.D. 2022. Biogeochemistry of wetlands science and applications. (2nd ed.). CRC Press. https://doi.org/10.1201/9780429155833.

Sahu, O. 2014. Reduction of heavy metals from waste water by wetland. International Letters of Natural Sciences, 7.

Salido, A.L., Hasty, K.L., Lim, J.-M. and Butcher, D.J. 2003. Phytoremediation of arsenic and lead in contaminated soil using Chinese brake ferns (Pteris vittata) and Indian mustard (Brassica juncea). International Journal of Phytoremediation, 5(2): 89-103.

Sanjana, M., Prajna, R., Katti, U.S. and Kavitha, R. 2024. Bioremediation – The recent drift towards a sustainable environment. Environmental Science: Advances, 3(8): 1097-1110.

Sekhar, K.C., Kamala, C., Chary, N. and Anjaneyulu, Y. 2003. Removal of heavy metals using a plant biomass with reference to environmental control. International Journal of Mineral Processing, 68(1-4): 37-45.

Shakir, E., Zahraw, Z. and Al-Obaidy, A.H.M. 2017. Environmental and health risks associated with reuse of wastewater for irrigation. Egyptian Journal of Petroleum, 26(1): 95-102.

Shamim, S. 2018. Biosorption of heavy metals. Biosorption, 2: 21-49.

Sharma, M., Kant, R., Sharma, A. and Sharma, A. 2024. Exploring the impact of heavy metals toxicity in the aquatic ecosystem. International Journal of Energy and Water Resources, 1-14.

Sheoran, A. and Sheoran, V. 2006. Heavy metal removal mechanism of acid mine drainage in wetlands: A critical review. Minerals Engineering, 19(2): 105-116.

Si, W.-T., Zhang, W.-Y., Lv, Y., Yang, F., Liu, J.-M. and Zhang, Y.-M. 2014. Heavy metal removal in a constructed wetland and benefits for the development of the Toad Bufo raddei. Polish Journal of Environmental Studies, 23(6).

Sobolewski, A. 1999. A review of processes responsible for metal removal in wetlands treating contaminated mine drainage. International Journal of Phytoremediation, 1(1): 19-51.

Sochacki, A., Yadav, A.K., Srivastava, P., Kumar, N., Fitch, M.W. and Mohanty, A. 2018. Constructed wetlands for metals: Removal mechanism and analytical challenges. Constructed Wetlands for Industrial Wastewater Treatment, 223-247.

Søgaard, E.G., Medenwaldt, R. and Abraham-Peskir, J.V. 2000. Conditions and rates of biotic and abiotic iron precipitation in selected Danish freshwater plants and microscopic analysis of precipitate morphology. Water Research, 34(10): 2675-2682.

Song, Y. 2003. Mechanisms of Lead and Zinc Removal from Lead Mine Drainage in Constructed Wetland. University of Missouri-Rolla.

Song, Y., Kirkwood, N., Maksimović, Č., Zheng, X., O'Connor, D., Jin, Y. and Hou, D. 2019. Nature based solutions for contaminated land remediation and brownfield redevelopment in cities: A review. Science of the Total Environment, 663: 568-579.

Sounthararajah, D.P. 2016. Mitigation of pollutants for beneficial use of stormwater. University of Technology Sydney (Australia) ProQuest Dissertations & Theses, 2016.31100358.

Sounthararajah, D.P., Loganathan, P., Kandasamy, J. and Vigneswaran, S. 2016. Column studies on the removal of dissolved organic carbon, turbidity and heavy metals from stormwater using granular activated carbon. Desalination and Water Treatment, 57(11): 5045-5055.

Srivastava, N.K. and Majumder, C. 2008. Novel biofiltration methods for the treatment of heavy metals from industrial wastewater. Journal of Hazardous Materials, 151(1): 1-8.

Stottmeister, U., Wießner, A., Kuschk, P., Kappelmeyer, U., Kaestner, M., Bederski, O. et al. 2003. Effects of plants and microorganisms in constructed wetlands for wastewater treatment. Biotechnology Advances, 22(1-2): 93-117.

Sukumaran, D. 2013. Phytoremediation of heavy metals from industrial effluent using constructed wetland technology. Applied Ecology and Environmental Sciences, 1(5): 92-97.

Sunding, D. and Zilberman, D. 2002. The economics of environmental regulation by licensing: An assessment of recent changes to the wetland permitting process. Natural Resources Journal. 59-90.

Swanson, W.J. and Loehr, R.C. 1997. Biofiltration: Fundamentals, design and operations principles, and applications. Journal of Environmental Engineering, 123(6): 538-546.

Thullen, J.S., Sartoris, J.J. and Nelson, S.M. 2005. Managing vegetation in surface-flow wastewater-treatment wetlands for optimal treatment performance. Ecological Engineering, 25(5): 583-593.

Varma, G., Singh, R. and Sahu, V. 2013. A comparative study on the removal of heavy metals by adsorption using fly ash and sludge: A review. International Journal of Application or Innovation in Engineering and Management, 2(7): 45-56.

Verma, S. and Kuila, A. 2019. Bioremediation of heavy metals by microbial process. Environmental Technology & Innovation, 14: 100369.

Vidu, R., Matei, E., Predescu, A.M., Alhalaili, B., Pantilimon, C., Tarcea, C. and Predescu, C. 2020. Removal of heavy metals from wastewaters: A challenge from current treatment methods to nanotechnology applications. Toxics, 8(4): 101.

Vijayaraghavan, K., Reddy, D.H.K. and Yun, Y.-S. 2019. Improving the quality of runoff from green roofs through synergistic biosorption and phytoremediation techniques: A review. Sustainable Cities and Society, 46: 101381.

Volesky, B. 2001. Detoxification of metal-bearing effluents: Biosorption for the next century. Hydrometallurgy, 59(2-3): 203-216.

Vymazal, J. 2010. Constructed wetlands for wastewater treatment. Water, 2(3): 530-549.

Wang, J. and Chen, C. 2006. Biosorption of heavy metals by Saccharomyces cerevisiae: A review. Biotechnology Advances, 24(5): 427-451.

Wang, S. and Liu, J. 2014. The effectiveness and risk comparison of EDTA with EGTA in enhancing Cd phytoextraction by Mirabilis jalapa L. Environmental Monitoring and Assessment, 186: 751-759.

Wang, S., Liu, T., Xiao, X. and Luo, S. 2021. Advances in microbial remediation for heavy metal treatment: A mini review. Journal of Leather Science and Engineering, 3: 1-10.

Wang, Y., Zhang, Z. and Ma, J. 2023. Nature-based solutions to restoration of contaminated soils and enhance biodiversity and human health. E3S Web of Conferences: EDP Sciences, 393: 02020.

Wang, Y., Zhou, T., Wang, W., Zhao, J., Li, Z., Ge, Y. et al. 2023. Phytoextraction of highly cadmium-polluted agricultural soil by Sedum plumbizincicola: An eight-hectare field study. Science of the Total Environment, 905: 167216.

Wani, A.H., Branion, R.M. and Lau, A.K. 1997. Biofiltration: A promising and cost-effective control technology for Odors, VOCs and air toxics. Journal of Environmental Science & Health Part A, 32(7): 2027-2055.

Wei, S. and Zhou, Q.-X. 2006. Phytoremediation of cadmium-contaminated soils by Rorippa globosa using two-phase planting (5 pp). Environmental Science and Pollution Research, 13: 151-155.

Wei, S., Luo, H., Zou, J., Chen, J., Pan, X., Rousseau, D.P. and Li, J. 2020. Characteristics and removal of microplastics in rural domestic wastewater treatment facilities of China. Science of the Total Environment, 739: 139935.

Witkowska, D., Słowik, J. and Chilicka, K. 2021. Heavy metals and human health: Possible exposure pathways and the competition for protein binding sites. Molecules, 26(19): 6060.

Yan, D.Y. and Lo, I.M. 2012. Pyrophosphate coupling with chelant-enhanced soil flushing of field contaminated soils for heavy metal extraction. Journal of Hazardous Materials, 199: 51-57.

Yin, K., Wang, Q., Lv, M. and Chen, L. 2019. Microorganism remediation strategies towards heavy metals. Chemical Engineering Journal, 360: 1553-1563.

Yu, G., Li, P., Wang, G., Wang, J., Zhang, Y., Wang, S. et al. 2021. A review on the removal of heavy metals and metalloids by constructed wetlands: Bibliometric, removal pathways, and key factors. World Journal of Microbiology and Biotechnology, 37(9): 157.

Yu, G., Wang, G., Chi, T., Du, C., Wang, J., Li, P. et al. 2022. Enhanced removal of heavy metals and metalloids by constructed wetlands: A review of approaches and mechanisms. Science of The Total Environment, 821: 153516.

Yu, H., Yan, X., Zheng, X., Xu, K., Zhong, Q., Yang, T. et al. 2020. Differential distribution of and similar biochemical responses to different species of arsenic and antimony in Vetiveria zizanioides. Environmental Geochemistry and Health, 42: 3995-4010.

Zaghloul, M. 2020. Phytoremediation of heavy metals principles, mechanisms, enhancements with several efficiency enhancer methods and perspectives: A review. Middle East J. 9(1): 186-214.

Zheng, X., Liu, T., Guo, M., Li, D., Gou, N., Cao, X. et al. 2020. Impact of heavy metals on the formation and properties of solvable microbiological products released from activated sludge in biological wastewater treatment. Water Research, 179: 115895.

Zhou, W. and Qiu, B. 2005. Effects of cadmium hyperaccumulation on physiological characteristics of Sedum alfredii Hance (Crassulaceae). Plant Science, 169(4): 737-745.

Zhu, Y. and Costa, M. 2020. Metals and molecular carcinogenesis. Carcinogenesis, 41(9): 1161-1172.

CHAPTER

3

Nature-based Solutions for the Management of Urban Wastewater

Namratha B.[1] **and Santosh L. Gaonkar**[2*]

[1] Department of Chemistry, The Yenepoya Institute of Arts, Science, Commerce and Management, Yenepoya (Deemed to be University), Mangaluru - 575002, Karnataka, India

[2] Department of Chemistry, Manipal Institute of Technology, Manipal Academy of Higher Education, Manipal - 576014, Karnataka, India

Introduction

Urbanization is identified as the primary consequence of human-induced activities that have significantly impacted the hydrological cycle. An increase in impervious surfaces leads to a decrease in infiltration, resulting in heightened runoff and flooding, even during minor rainfall events. The flow of stormwater across impermeable surfaces ultimately reaches surface water reservoirs. Pollution in urban environments stems from sources such as vehicles, industries, atmospheric deposition, and soil erosion. This has led to the introduction of various contaminants, such as metals, organic substances, nutrients, oil, grease, detergents, and surfactants, into the environment. These pollutants have the potential to enter surface waters through stormwater runoff. Consequently, the implementation of stormwater treatment is crucial in urban areas, with cost-effective and eco-friendly methods being more sustainable (Akanksha et al. 2020).

Water circularity embodies a more sustainable approach in sponge cities, where reclaimed water is utilized, and excess water is only drained as a last resort. This approach still offers opportunities to enhance urban greenery and liveability. Surfaces that are sealed off and hydraulically smooth, coupled with minimal vegetation, lead to reduced rates of infiltration and evaporation. The repercussions of this disturbed water balance are anticipated to be exacerbated by climate change. Nature-based solutions emerge as a versatile approach, offering advantages such as creating habitats, establishing urban wildlife refuges, and providing recreational and landscaping spaces, rather than solely focusing on wastewater treatment (Susana and Davide 2022).

*Corresponding author: sl.gaonkar@manipal.edu

DOI: 10.1201/9781003527350-3

Urban water challenges are complex and cannot be adequately addressed through conventional water management systems. The reuse of treated effluent for irrigating commercial crops, the local utilization of rainwater and graywater for various purposes such as toilet flushing, car washing, and garden irrigation, and the separation of urine and feces from graywater to optimize nutrient retrieval are gradually gaining prominence in discussions regarding circular shifts in the water sector (Lise and Patience 2019). Rainwater, in particular, serves as a valuable resource for local activities, including gardening, car washing, and laundry. Intercepting and harvesting precipitation before it turns into runoff proves to be an effective strategy for reinstating the pre-developed water cycle within urban catchments. This concept can be seamlessly integrated with stormwater management practices and additional water supply systems. The persistent trends of climate change and urbanization, combined with the inability of city soil to adequately filter rainwater, have led to urban runoff entering drainage networks and discharging into water bodies in an untreated state, further deteriorating the environment when mixed with sewage. Extreme environmental events such as floods and droughts result in human casualties and infrastructure damage (Tomáš et al. 2012).

Constructed wetlands, green roofs, permeable pavements, bioretention systems, rain gardens, and riparian buffers are identified as optimal management practices for addressing stormwater issues in urban areas. These systems have demonstrated their ability to mitigate flooding and pollution by mimicking the natural environment's mechanisms, consequently positively impacting public health and diverse habitats. Such solutions have the potential to enhance water quality and yield socioeconomic benefits. They fall under the overarching concept of nature-based solutions (NBSs), encapsulating ecosystem-centered approaches. It is crucial to recognize that transitioning to circular cities faces challenges encompassing technical, design, and environmental aspects, as well as policy, governance, politics, and regulations (Qi et al. 2020).

The interconnectedness of cities, water, nature, and circularity remains an underexplored area in academic discourse. Urban renaturing has emerged as a promising strategy for the radical transformation of landscapes, involving spatial alterations that expand natural elements and restore ecological functions within human settings (Lehmann 2021).

Management Practices

Constructed Wetlands

The establishment of constructed wetlands for the treatment of urban wastewater is predicated on the identification of an appropriate site, its proximity to the wastewater source, its soil characteristics, and other relevant topographical attributes. Engineers and environmental experts are tasked with a comprehensive evaluation of factors such as the volume of urban wastewater and local weather conditions prior to commencing the construction process (Agaton and Guila 2024). Mimicking natural wetlands, these wetlands typically feature shallow foundations and impermeable liners to minimize seepage. The introduction of wetland vegetation aids in soil stabilization and the purification of contaminants. The liquid fraction of wastewater is conveyed through a narrow sewer network to a primary reservoir for the retention and separation of solids (Vymazal 2010). Wastewater is then introduced into the wetland via either gravitational flow or mechanical pumping. Subsequently, the wastewater may be pumped into the header tanks, from which it traverses through the constructed wetlands. Adhering to this fundamental design, the creation of constructed wetlands can be tailored based on treatment efficacy and spatial constraints at the designated site (Figure 1). This endeavor also necessitates the supervision of plant growth, constant monitoring of water levels, and regular evaluation of operational efficiency (Hassan et al. 2021).

Projects on constructed wetlands necessitate collaboration among diverse stakeholders, such as governmental agencies, local communities, and environmental organizations. The social, economic, and environmental aspects must be considered to guarantee the enduring sustainability of such projects (Vymazal et al. 2006). The financial components of the construction ought to be grounded in the lifecycle of the facility. According to the literature, individuals residing near wetlands have raised concerns about the unpleasant odor experienced during operations, particularly at night. This issue is attributed to the presence of ammonia in sewage water, a problem that can be alleviated by vegetation within constructed wetlands (Oscar and Caren 2021). Challenges hindering the flawless execution of constructed wetland projects include insufficient widespread public support, coupled with obstacles related to public acceptance, technical complexities, and economic viability. Changes in hydrology and increasing temperatures have the potential to alter the biogeochemistry

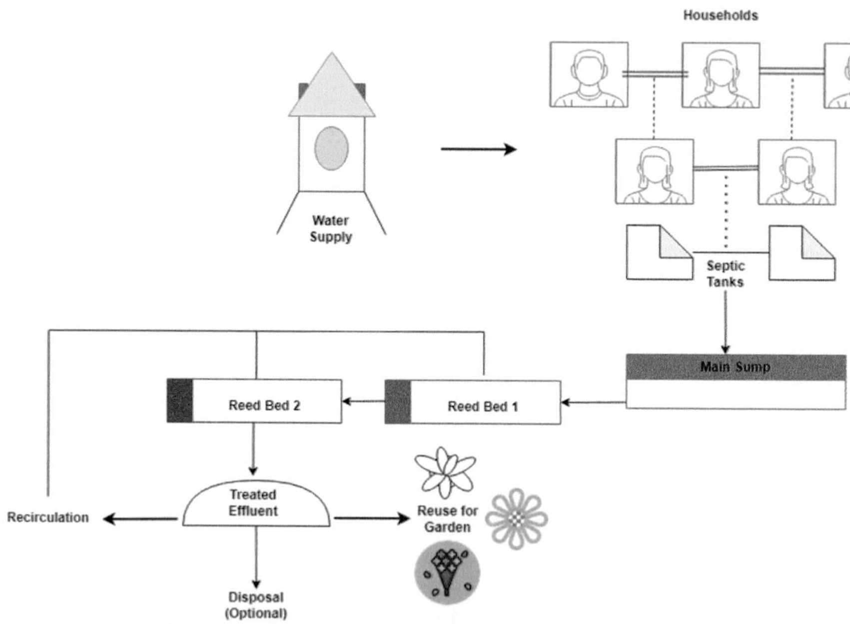

Figure 1: A representative flowchart of a constructed wetland.

and functionality of a wetland, thereby posing a threat to constructed wetlands. Furthermore, consistent monitoring of technical parameters and external influences is imperative to ensure the sustainability of such projects (Justino et al. 2023).

In contrast to traditional methods of treating wastewater, constructed wetlands require minimal upkeep and demonstrate reduced susceptibility to fluctuations in inflow. Furthermore, they offer numerous advantages, such as water provision, the creation of habitats, nutrient cycling, flood prevention, erosion control, and opportunities for recreational activities and scientific investigation (Agaton and Guila 2023).

Green Roofs

A green roof is a stratified structure comprising a waterproofing membrane, a growing medium containing a soil mixture, and a layer of vegetation. Within green roofs, one can find a root barrier layer, a drainage layer, and an irrigation system, the latter being contingent on the prevailing climate conditions (Figure 2). Termed as vegetated roofs, ecoroofs, or living roofs, these structures play a crucial role in

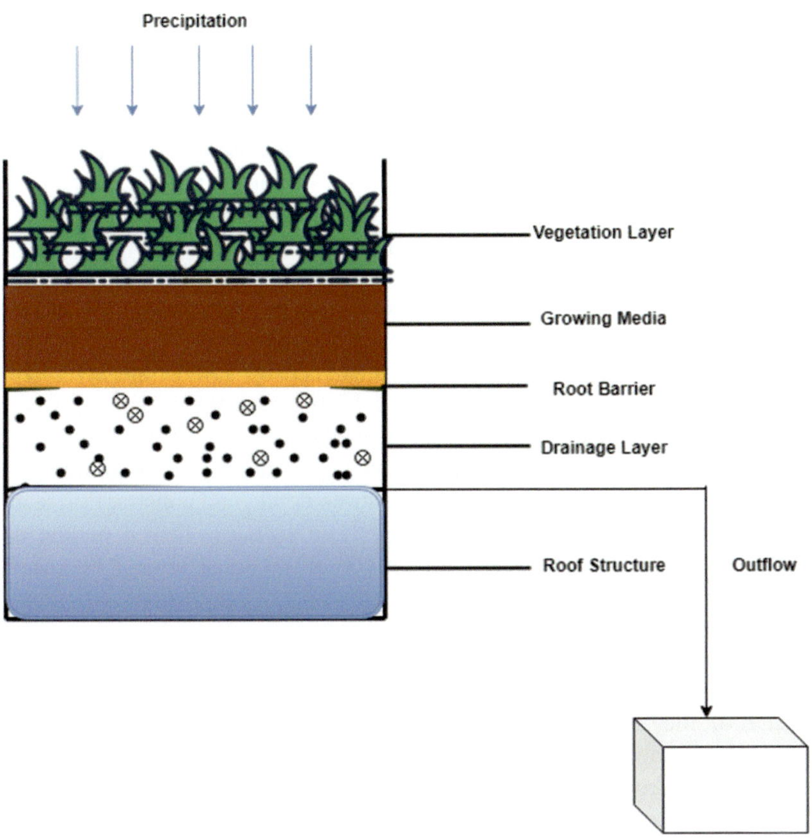

Figure 2: Green roof design.

urban settings where roof surfaces make up approximately 40–50% of impermeable urban areas (Ahmet and Erdem 2018). Therefore, such locations are ideally suited for the implementation of green roofs, which offer significant environmental advantages and enhance energy efficiency. To prevent roof overburdening, the soil (optional) and vegetation layers are relatively thin compared to those of alternative approaches (Rezaei et al. 2021). These systems facilitate the movement of energy, water, nutrients, and specific organisms. The installation of green roofs on existing buildings can be viewed as a strategy for lowering carbon emissions, mitigating the environmental impact of urban areas, and addressing the effects of climate change. Hence, it is advisable to incorporate green roofs into the initial design phase of new construction to ensure that their structural stability is duly considered (Umberto et al. 2014).

Green roofs have the potential to incorporate water storage for nonpotable purposes and irrigation during droughts (Rizzo et al. 2023). This adaptation is commonly referred to as a green-blue roof. The introduction of an extra storage layer in this type of roof serves to mitigate peak flow and decrease heat retention on the roof surface. Furthermore, a biosolar roof presents another beneficial alteration that not only prevents water runoff but also generates renewable energy using photovoltaic panels (Vijayaraghavan 2016).

There are three categories—intensive, semi-intensive, and extensive—contingent upon their structural composition and maintenance requirements. Intensive green roofs encompass a diverse array of plant species, ranging from grasses and small shrubs to trees, with substrate thicknesses typically exceeding 25 cm. The maintenance demands parallel those of a garden in terms of fertilization, irrigation, and plant accommodation. These spaces can serve as areas for recreation and communal interactions. Extensive green roofs are distinguished by plant species such as herbs, grasses, aromatic plants, mosses, and succulents, with a substrate layer measuring 8–15 cm in thickness. These areas are not open to the public and necessitate minimal maintenance. Finally, semi-intensive green roofs are typified by small plants such as shrubs and grass, sustained by a substrate layer typically ranging from 15 to 25 cm. While they require limited maintenance, there is potential for public access (Cristina and Alexandros 2021). The weight-bearing capacity of the roof structure, in conjunction with the structural integrity of the building and its age, can dictate the choice of vegetation, substrate depth, and construction technique utilized. The topography varies in each of these scenarios based on the design. To attain optimal outcomes, local climatic conditions must always be considered (Monteiro et al. 2023).

The uppermost stratum comprises the vegetation cover placed on a technical substrate. This stratum necessitates support from ecosystem services, climatic factors, and irrigation needs (Stovin 2010). Given that plants constitute the visible aspect of the green roof, they serve as crucial indicators of the system's health and operational conditions. Vegetation offers protection against substrate erosion and reflects a portion of solar radiation. Proper irrigation management becomes vital in regions experiencing prolonged periods of drought. Additionally, more water is typically required during the initial phase of plant root establishment, with a subsequent significant decrease in water demand (Andenaes et al. 2018). The substrate is an integral component of the roof, housing essential microfauna and macrofauna. The technical characteristics of the substrate layer include water retention, drainage capacity, thermal insulation, and optimal conditions for plant growth. Chemically, these substrates consist of specialized blends of lightweight materials, primarily mineral components with some organic content (Shahmohammad et al. 2022). The introduction of soil at this stage is usually omitted unless an intensive roof is constructed based on the conventional model, albeit with stringent restrictions and adherence to standard protocols. A geotextile filter sheet is specifically engineered to prevent substrate particles from infiltrating the drainage layer and causing blockage. To prevent water accumulation, a drainage layer is positioned beneath the filter membrane (Shafique et al. 2018). This layer helps regulate moisture levels, sustain a healthy microfauna population, and prevent oversaturation of the substrate. Common materials used for this layer include high-density polyethylene, polystyrene, or lightweight expanded clay. Ongoing research aims to introduce more environmentally friendly alternatives to the prevalent polyolefin reference product. The drainage layer is a protective stratum designed to shield against mechanical harm, serving as a waterproofing and insulating barrier for the green roof, typically crafted from polyurethane, rubber, or polystyrene. Subsequently, a roofing membrane is installed to create a physical partition between the upper layer and the underlying structural support. If a particular stratum has not been previously mentioned in the upper sequences, it may function as an impermeable membrane with anti-root properties (Pearlmutter et al. 2020).

In a general sense, green roofs provide a safeguard to the structure against physical harm and solar exposure while also serving to moderate temperature fluctuations and extend the durability of the roof membrane (Walters and Midden 2018). Consequently, the implementation of green roofs can lead to a reduction in future upkeep requirements and associated expenses throughout the lifespan of buildings, thereby increasing the longevity of rooftops (Cristiano et al. 2021). Their application is particularly well-suited for managing stormwater to diminish peak flow and runoff. Moreover, there is a growing interest in utilizing green roofs for urban agriculture, a trend that has gained momentum following the recent global pandemic caused by the novel coronavirus 2019 (COVID-19). This crisis prompted cities to reconsider the demand for food among the population, the reliance on food imports, and the repercussions of unexpected disruptions in food supply chains (Khan et al. 2020). To ensure the effectiveness of this approach, a collective effort involving various stakeholders, such as designers, engineers, constructors, builders, contractors, building operators, owners, end-users, and government representatives, is essential (Cristina et al. 2022).

Permeable Pavement

Permeable pavement is considered a prevalent nature-based solution utilized in engineering infrastructure for pedestrian pathways, driveways, and parking areas. It is particularly well-suited for managing stormwater in urban settings, effectively addressing issues arising from traditional road construction. As an eco-technology, it enhances on-site stormwater infiltration and enables runoff storage within an aggregate reservoir. Moreover, these pavements support groundwater replenishment, promote ecological well-being, and offer possibilities for capturing and reusing stormwater (Fini et al. 2017).

A typical permeable pavement system comprises various layers, including a wear layer, bedding layer, base layer, subbase layer, and underdrain, along with geotextiles and small aggregate filters (Figure 3). The selection of layer depth and materials is significantly influenced by traffic loads, thereby enhancing the overall structural robustness of the pavement system. Well-known porous pavements typically encompass three layers, namely, a top layer of 0.2 meters consisting of paver blocks, a second layer of 0.1 meters of sand, and a final layer of 0.3 meters comprising 20–40 mm gravels (Alam et al. 2019).

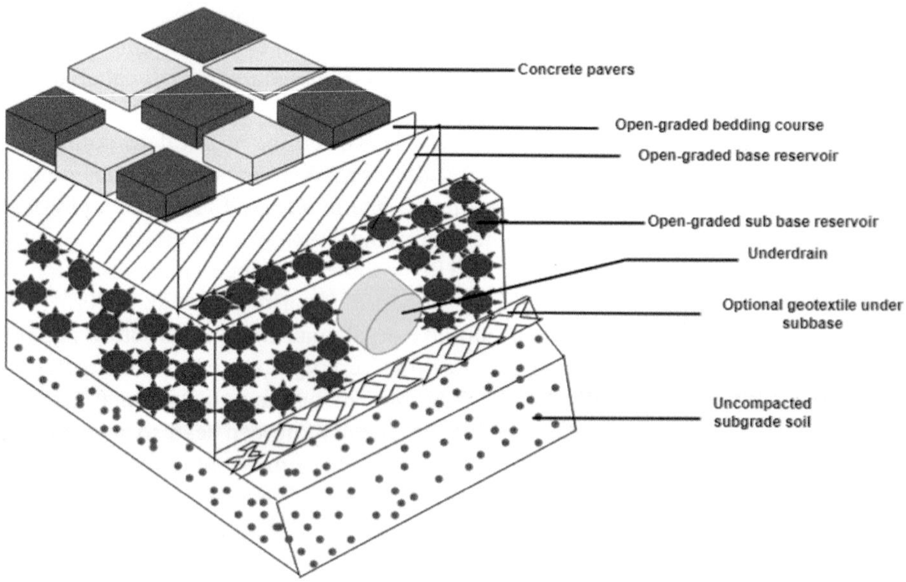

Figure 3: Cross-sectional view of a typical permeable pavement.

Permeable pavements can be categorized based on the types of surface materials employed, such as porous concrete (PC), porous asphalt (PA), permeable interlocking concrete pavers (PICP), concrete grid pavers (CGP), and plastic reinforcing grid pavers. The effectiveness of these pavements is primarily determined by the particle size distribution of the bedding material and water retention within the surface blocks. They serve as superior alternatives to impermeable surfaces, offering a replacement for urban roads, sidewalks, and parking lots, thereby increasing drainage area coverage during runoff events. Notably, these pavements aid in reducing surface runoff, decreasing peak flow, and extending the time taken to reach peak flow (Imran et al. 2013).

Porous concrete (PC) is recognized as a valuable surface paving material due to its composition comprising a Portland cement binder, coarse aggregates, water, and admixtures. The incorporation of air voids in the matrix is typically achieved by excluding fine aggregates, thus facilitating infiltration within the pavement structure. Stormwater has the capacity to permeate the void spaces present in the concrete-aggregate matrix of porous concrete. The literature indicates that it represents a more cost-effective life cycle option than other alternatives on the market and is believed to outperform porous asphalt (PA) in various applications (Yuxin et al. 2021).

Conversely, permeable interlocking concrete pavers are constructed from precast concrete block units that are interlocked during installation. This material is characterized by its robust nature and visual appeal, serving as a durable permeable pavement system that is easily maintainable. These pavers are available in a variety of shapes, sizes, and configurations. The arrangement of two interlocking blocks results in void spaces at the center and edges of the paver, which are commonly filled with crushed stone, pea gravel, or topsoil for landscaping purposes (Meysam et al. 2017).

The effectiveness of a permeable pavement system is predominantly governed by factors such as the type of pavement, thickness of the pavement layer, particle size distributions, and porosity of the bedding materials. Additionally, variables such as block dimensions, spacing, cross-sectional design, aggregate properties, prevailing climatic conditions, and soil characteristics play pivotal roles in influencing the performance of permeable pavements (Miklas and Piotr 2017)

Apart from hydrological and pavement-related aspects, variables such as rainfall intensity, initial moisture levels, pavement aging, and upkeep practices can significantly impact the functionality of permeable pavement systems. Research indicates that the performance of such systems will be influenced by climate change and alterations in precipitation patterns in the future (Scholz 2014).

Furthermore, it is imperative to investigate the effects of temperature variations, maintenance routines, and initial moisture conditions on pavement behavior. Subsequent studies should focus on conducting laboratory experiments or computer simulations to analyze diverse designs of permeable pavements and elucidate the mechanisms underlying pollutant removal at different layers. The use of field-scale modeling tools can provide insights into the hydrological and environmental performance of permeable pavements across varying drainage areas and installation dimensions (Sun et al. 2024).

Bioretention Systems

Bioretention systems are designed to remove dissolved pollutants and solid particles from runoff water, thus reducing the water discharge volume and velocity. This progression of treatment involves a variety of chemical, biological, and physical mechanisms, such as sedimentation, filtration, absorption, reduction, uptake by vegetation, and assimilation by microbial biomass. The efficiency of these systems in cleansing stormwater relies on factors such as substrate conditions, plant types, local climate conditions, and water infiltration and discharge volume and rate (Liu et al. 2014).

Bioretention systems possess the capacity to manage stormwater and improve water quality by capturing and treating pollutants in urban settings. However, these systems have limited ability to retain contaminants due to their finite capacity. Hence, planting suitable vegetation alongside bioretention systems represents an effective phytoremediation strategy to address the challenges encountered by these remedial systems (Vijayaraghavan et al. 2021).

A bioretention system is a constructed depression in terrain that gathers runoff from impermeable surfaces situated uphill. It consists of multiple layers of filtration materials, vegetation, an overflow structure, and, optionally, an underdrain (Figure 4). Typically, with small dimensions, bioretention cells usually cater to catchment areas of less than 2 hectares. These systems mimic the natural water cycle by capturing runoff, thereby reducing flow volume and speed. Additional benefits include improving the aesthetic appeal of the area, creating wildlife habitats, diminishing soil erosion, and recharging groundwater, consequently augmenting water flow to nearby streams (Zhang and Guo 2014).

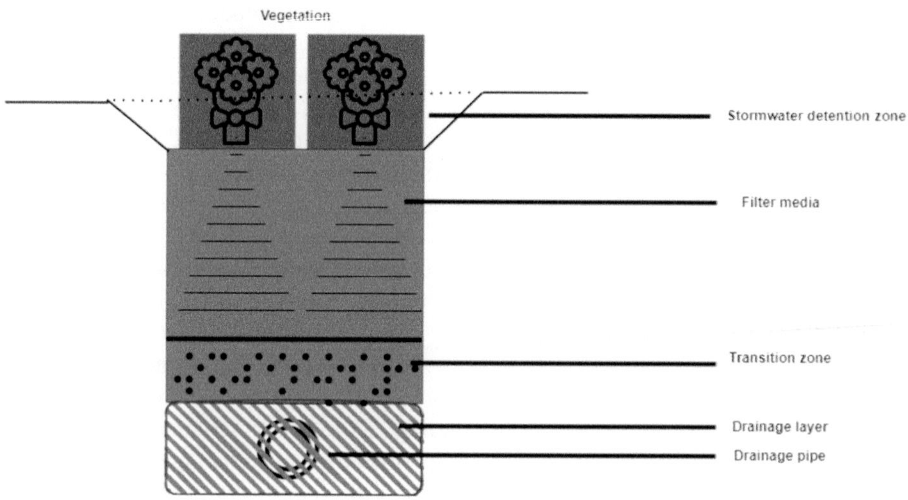

Figure 4: Cross-sectional view of a classical bioretention system.

The essential elements of a bioretention system include a layer of sand, soil, and organic material approximately 0.7–1 m thick, which is recognized for its exceptional infiltration capabilities. This is complemented by a surface mulch layer designed to prevent erosion and maintain suitable soil moisture, while vegetation helps in regulating flow, filtering out large particles, purifying pollutants, and supporting microbial activity. Inlet and outlet pipes are incorporated to control water entry and exit, with overflow pipes installed to divert excess water during high-flow periods. Typically, these systems allow for temporary ponding of 100–300 mm of water on the surface. The filter media can consist of natural or modified soils, or completely engineered materials such as zeolite and synthetic geotextiles, all of which are carefully selected to be free of debris, harmful substances, toxins, and weeds. For natural or amended soils, loamy sand is preferred due to its rapid infiltration rate. The inclusion of small amounts of silt and clay can further assist in reducing pollutants during infiltration, owing to their extensive surface area and high surface charge potential (Chunbo et al. 2019).

In terms of media composition, an optimal bioretention system typically comprises approximately 50%–60% sand and a combination of 40%–50% loam/sandy loam/loamy sand per unit volume. Maintaining a low clay content, ideally between 5% and 8%, is crucial to ensure adequate hydrology within the system. The presence of excessive clay can hinder media infiltration. It is essential to carefully manage soil phosphorus levels to satisfy plant nutritional requirements while minimizing nutrient leaching. Studies have proposed that increased media depths could improve the hydrological efficiency of bioretention systems. Employing two media layers could be beneficial, with the upper layer intended for vegetation support and the lower layer optimized for filtration (Li and Davis 2009).

The selection of vegetation in bioretention systems varies depending on the region. Recommendations often suggest the use of indigenous plants due to their ability to adapt to local environmental conditions, yet the soils in these systems are frequently brought in and modified. Plant choice may also take into account visual and landscaping aspects; for example, tall and densely populated species such as reeds and rushes can act as barriers to foot traffic. Additionally, plants can hide unsightly areas, such as thick layers of gray sediments and contaminants from melted snow runoff. Nevertheless, the main focus remains on selecting noninvasive plants that can endure different levels of flooding and exposure while also having extensive root systems to enhance biological processes and optimize water quality treatment. Vegetation within the bioretention cell absorbs water and nutrients from the media. If the media becomes saturated, overflow may occur, leading to water accumulation in a small storage area until it reaches a specific elevation, at which point discharge begins (Stefanakis et al. 2024).

Initially, the capacity of a bioretention system to absorb water decreases as the filter material settles and becomes denser. However, as the plant community becomes established and the root depth expands, the system's capacity gradually improves. The vegetated surface layer reduces the speed of runoff and captures sediment. Within the bioretention cell, treatment involves various unit processes that leverage the chemical, biological, and physical properties of plants, microbes, and soils to remove pollutants from urban runoff. Bioretention reduces peak flows, runoff volume, and pollutant loads while increasing evapotranspiration through vegetation uptake and extending lag time (Dagenais et al. 2018).

An important issue impacting the efficiency of bioretention systems is surface clogging resulting from the accumulation of fine silts and sediments from urban runoff. If these systems experience significant sediment buildup, their hydrological effectiveness can substantially decrease, leading to reduced water storage capacity and infiltration rates. A crucial part of hydrological restoration involves water exfiltration into adjacent soils, eventually percolating into the groundwater table (Goh et al. 2019).

Nitrogen treatment encompasses various processes, including ammonification, volatilization, nitrification, denitrification, and vegetative uptake. Ammonification decomposes organic nitrogen compounds into ammonium. Volatilization primarily causes ammonia loss in bioretention systems. Ammonium ions can transform into ammonia gas under alkaline conditions, despite bioretention media typically having lower pH values. Nitrification occurs in water-saturated soils, particularly in the aerobic zone surrounding plant roots and other aerobic regions. Denitrification entails conversion of nitrate (NO_3) into gaseous nitrogen by microorganisms under anaerobic conditions. Plants absorb ammonium and nitrate through their roots for physiological purposes. The rate of nitrogen uptake is influenced by the plant growth rate and the concentrations of inorganic nitrogen forms (Fan et al. 2019, Zhou et al. 2023).

Phosphorus plays a crucial role in regulating the productivity of most freshwater ecosystems, and excessive inputs can lead to eutrophication. In bioretention systems, essential processes for phosphorus elimination include precipitation, adsorption, filtration, and vegetation uptake. Precipitation can efficiently eliminate phosphorus from stormwater with a high metal ion content. Many soils readily adsorb phosphorus through ion exchange or ligand exchange mechanisms. The factors affecting phosphorus uptake by plants include the proportion of exposed plant roots, plant and root age, and environmental conditions such as temperature and soil pH (Roy-Poirier et al. 2010).

Metals pose significant concerns due to their potential accumulation and ecotoxic effects. Observations indicate that the surface layer of bioretention systems plays a vital role in metal retention. Field studies suggest that bioretention is effective at eliminating heavy metals such as copper (Cu), zinc (Zn), and lead (Pb) from runoff (Hsieh and Davis 2005).

Total suspended solids can be effectively removed by bioretention layers, mainly through sedimentation in the basin and media filtration. However, it is crucial not to use bioretention solely as a sediment trap to prevent possible clogging problems, despite its efficient sediment removal capabilities. Pathogenic bacteria, which present a considerable threat to water quality due to their ability to cause infections, can be effectively managed through the utilization of bioretention systems. These findings indicate that within urban environments, bioretention systems have the potential to decrease the concentrations of various targeted pollutants, including markers for pathogenic bacteria (Carpenter and Hallam 2010).

The functionality of bioretention extends to providing thermal regulation by reducing and managing runoff. A different research investigation delved into the dimensions of bioretention and its thermal regulation capability, suggesting that smaller bioretention cells could offer more efficient mitigation of thermal impacts (Brown and Hunt 2012).

The expenses linked to bioretention exhibit significant variation and are strongly impacted by the objectives of the design and the characteristics of the site. The utilization of numerical modeling for bioretention systems is a field of active exploration. Computational models can aid in comprehending the diverse physical, chemical, and biological processes occurring within a bioretention cell. By integrating these processes with site hydrology, it becomes feasible to anticipate treatment efficacy for a specific design (Premarathna et al. 2023).

While bioretention systems and constructed wetlands share similar purposes, the distinguishing factor is the absence of stagnant water between rainfall events in bioretention systems. Moreover, they require less space for construction and utilize a smaller portion of the surface catchment to achieve comparable water quality (Li et al. 2009).

An analysis based on citations revealed that the majority of primary studies on bioretention originate from a limited number of countries and specific research clusters. As a result, many bioretention systems worldwide adhere to guidelines from regions with different climates, often resulting in operational challenges (Lloyd and Wong 2008).

Bioretention systems possess a compact yet intricately designed structure. The physical and biological processes they incorporate imitate natural ecological processes. These systems represent one of the most sophisticated methods for hydrologic and ecological restoration in urban settings. When implemented in a cost-effective manner and operated reliably for runoff treatment, they have the potential to serve as a truly sustainable treatment approach. Improved estimations of performance will contribute to meeting water quality goals downstream. Ongoing research is crucial for enhancing bioretention design, boosting performance, and providing sustainable solutions to drainage challenges in urban areas (Stefanakis et al. 2024).

Rain Gardens

Although frequently utilized interchangeably, rain gardens and bioretention systems exhibit subtle yet significant variations. Bioretention facilities, characterized by a combination of soil and vegetation for stormwater management, adhere to more rigorous design criteria. These meticulously engineered systems are specifically designed to meet project goals related to hydraulic conductivity, pollutant removal, and strong plant vitality. They often materialize as larger-scale installations with customized hydraulic controls and optimized media for vegetation health and stormwater control (Yuan et al. 2022).

Rain gardens and bioretention systems represent crucial stormwater management techniques that leverage natural hydrological cycle processes such as infiltration and evapotranspiration (Figure 5). Originating as a residential innovation in Maryland in 1990, rain gardens emerged as an environmentally friendly substitute for traditional urban infrastructure, gaining popularity globally in nations such as Japan, China, Australia, and the USA to promote sustainable development. With the rapid growth of urbanization leading to increased impervious land cover, ecosystems are at risk due to excessive surface runoff and pollutants. This underscores the importance of rain gardens in reducing flood peaks, recharging groundwater, and enhancing biodiversity, as they act as natural stormwater filters. As a sustainable development approach, rain gardens demonstrate long-lasting effectiveness and visual appeal, blending ecological resilience with aesthetic charm. This article explores the numerous advantages and historical deployment of bioretention systems in urban settings, with the goal of educating the scientific community on efficient stormwater management strategies (Karabay et al. 2024).

Rain gardens, which resemble shallow ditches gracefully nestled on earth, serve as hosts for carefully selected vegetation in nurturing soil and are topped with a protective mulch layer. As protectors of urban waterways, they cradle runoff gently, guiding its descent through a soil-based sanctuary. Within this lush refuge, aquifers are replenished, and ecosystems thrive, as rain gardens diligently cleanse

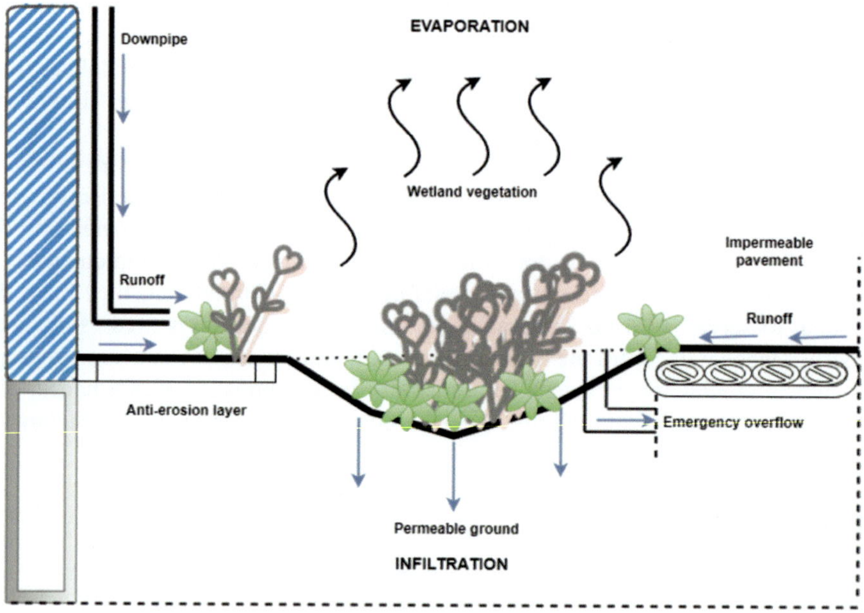

Figure 5: Scheme of a rain garden.

stormwater containing pathogens, nutrients, metals, and organic impurities. Here, the symbiotic interaction of vegetation and soil orchestrates the harmonious removal of contaminants, ensuring the vitality of our natural environments (Magda et al. 2022).

The notion of a rain garden is broad, yet its core is often distilled into several defining features: a shallow depression with permeable soil, which is meticulously planted to intercept, absorb, and purify stormwater runoff, particularly in urban settings. Embracing layers of soil and mulch promotes a thriving microbial community while advocating for native flora, which is finely attuned to local climate and soil conditions, and avoids artificial fertilizers. Beyond mere practicality, a rain garden is a refuge—a sanctuary for precipitation—which is meticulously crafted to filter pollutants, recharge groundwater, and withstand periods of flooding. Adaptable in its incorporation, it seamlessly integrates with surrounding surfaces, providing a smooth overflow mechanism to redirect excess water to nearby drains or other rain gardens. Although size is important, with larger rain gardens providing improved runoff control and biodiversity, each of these areas holds potential for wildlife habitat and human enjoyment. Whether functioning as an infiltration solution or a retention basin, a rain garden's effectiveness is highlighted by its capacity to minimize outflow through the elegant process of evapotranspiration (Yamada and Shibata 2017).

Through careful plant selection and layered arrangements, a rain garden has emerged as a frontline defender, skillfully treating stormwater runoff from impermeable pavements. Here, water finds its solace, captured within the landscape, undergoing purification through the graceful processes of evapotranspiration and infiltration. While predominantly found in urban areas, these green sanctuaries are equally essential in rugged terrains and thrive especially in the lush environments of tropical regions (Zhang et al. 2020b).

Rain gardens serve as green guardians, enhancing environmental beauty, cleansing the air, combating soil erosion, and strengthening thermal resilience, while reducing both runoff volume and pollutant loads. Within their gentle embrace, water is pure, free from chemical contamination, and ready for various uses (Hanbae et al. 2013).

In the realm of stormwater, phosphorus (P) primarily originates from agricultural runoff, fertilizers, atmospheric deposition, soil erosion, vehicular emissions, animal waste, and detergent residues. Nevertheless, rain gardens present a promising solution to this issue. By incorporating phosphorus-sorbing materials such as red mud, along with aluminum or iron oxide amendments, rain gardens can transform into effective purifiers, achieving enhanced phosphorus removal. Additionally, those equipped with an anoxic layer demonstrate exceptional capability for extracting total phosphorus from water (Zhang et al. 2020a).

In regard to nitrogen, its source in urban areas can be traced back to fertilizers and vehicular emissions, with its excess contributing to eutrophication and harmful algal blooms. Various nitrogen compounds are found within the chemical composition of stormwater, each undergoing microbial transformation within the boundaries of rain gardens. Although the basic design of rain gardens may hinder nitrate removal, the combination of carbon-rich soil, optimal pH levels, and deep, moist substrates enhances nitrate removal effectiveness. Through controlled wetting patterns, rain gardens reveal their full potential, with nitrogen removal rates exceeding 80%, highlighting their significant impact on urban water management (Chao et al. 2019).

In addition to the lush surroundings of rain gardens, a complex array of mechanisms and processes orchestrates the elegant removal of contaminants, creating a network of exfiltration, cation exchange, adsorption, reduction, microbial degradation, mineralization, nitrification, and denitrification (Good et al. 2012).

However, among their virtues are challenges; stagnant waters can become breeding grounds for pests, attracting rodents, raccoons, and insects, while the danger of accidental drowning lurks at pooled depths. Groundwater also risks contamination if rain gardens are not adequately isolated. Nevertheless, these drawbacks, though present, are overshadowed by numerous benefits, showcasing the enduring appeal and effectiveness of rain gardens in contemporary landscapes (Chen et al. 2023).

Nevertheless, a lingering concern arises regarding the hydraulic performance of these sanctuaries as they strive to purify stormwater. The continuous influx of fine particles and stormwater poses a risk of clogging, reducing infiltration rates and compromising their effectiveness. Maintenance emerges as a critical aspect of this intricate interplay between nature and infrastructure, serving as a vital duty to safeguard the purity of rain gardens. Through the removal of deposited sediments, the rejuvenation of unhealthy vegetation, and the replenishment of soil media, their legacy as a symbol of cleanliness is ensured. Municipalities play a key role in ensuring the harmonious coexistence of rain gardens and urban environments via diligent monitoring and performance evaluations, preserving their ability to purify stormwater and maintain a delicate ecological balance (Chuan et al. 2023).

At the core of rain garden efficiency lies the essential process of water infiltration through soil, a crucial factor in managing runoff volume and protecting the integrity of landscapes. An elaborate assessment protocol is employed to evaluate their performance, including visual inspections, capacity tests, synthetic runoff assessments, and continuous monitoring (Asleson et al. 2009).

While traditional rain gardens effectively handle stormwater runoff, concerns persist regarding groundwater pollution and nutrient leaching. To address these issues, a shift towards biphasic rain garden design is proposed. By creating a balance between saturated and unsaturated environmental conditions, biphasic gardens transition from water-saturated anaerobic areas to aerated recharge zones. Enclosed by impermeable liners and reverse drainage systems, these gardens effectively filter contaminants through sedimentation and biological processes. As purified water flows into recharge zones, replenishes aquifers, and cleans the environment, biphasic rain gardens emerge as symbols of progress, reducing peak flows, minimizing runoff, and revolutionizing stormwater management. With increased organic content, extended pollutant retention times, and superior hydraulic capabilities, these engineering materials have the potential to transform urban landscapes, providing a refuge amidst the challenges of modernity.

Riparian Buffers

The riparian landscape, where aquatic and terrestrial realms intertwine, possesses a distinctive charm. It serves as a refuge where the lush embrace of riparian buffers protects aquatic environments from the impacts of surrounding land use, thus maintaining the cleanliness of water bodies (Venkatachalam et al. 2005). These buffers are recognized as pillars of environmental stewardship, utilizing advanced technology to manage water quality in agricultural areas, and are often acknowledged for their effectiveness in denitrification and nitrogen removal processes (Heneghan et al. 2021).

In urban settings, riparian buffers emerge as unnoticed champions, diligently purifying polluted runoff, halting erosion along water edges, and cleansing the atmosphere, particularly near chemically treated farmlands. The canopy provides cooling shade, controlling the excessive growth of aquatic plants, while also providing nearby fields with a rejuvenating microclimate. They establish a network of interconnectedness, facilitating the transfer of energy and nutrients, promoting habitat variety, and supporting diverse biological communities, thereby enhancing ecosystem functioning (Graziano et al. 2022).

Within these dynamic corridors, biodiversity thrives, providing a haven for various specialized species. Riparian ecosystems emerge as bastions of resilience, offering safe spaces for populations to recover and repopulate disturbed habitats, thus perpetuating the circle of life (Jyoti et al. 2024).

The natural richness provided by riparian vegetation, such as leaf litter and woody debris, acts as a defense against the erosive power of water currents, while enhancing habitat diversity through the accumulation of alluvial resources (Haase 2017).

Furthermore, these buffers possess a significant ability to filter out sediment, heavy metals, farm residues, and pathogens, preventing their movement towards groundwater reservoirs. Beneath the surface, the complex root systems of riparian plants strengthen the structure of the soil, ensuring soil stability (Masiero et al. 2024).

Over millennia, riparian habitats demonstrate resilience against the impacts of climate change, yet their attractiveness sometimes leads to exploitation. Despite their ecological significance, the allure of fertile land often results in their transformation into agricultural fields, encouraging excessive livestock grazing and subsequent degradation of riparian plants and soil quality (Ülo et al. 2005).

Conclusion

The importance of addressing water management in urban areas is crucial, particularly with the increase in urbanization and climate change. Moving towards circularity and adopting nature-based solutions present a promising path for reducing the negative impacts of urban growth on water resources and ecosystems. However, successful integration necessitates not only technological advancements but also strong policy frameworks and community involvement.

An essential element of this shift is education and awareness. Including nature-based solutions in university curricula and offering training for teachers can empower individuals with the necessary knowledge to support these efforts. Concrete illustrations and scientific proof are key in gaining public support and nurturing a shared willingness to invest in nature-based methodologies.

Moreover, the success of nature-based solutions relies on a comprehensive grasp of local circumstances, encompassing environmental, social, and economic elements. Customizing strategies to specific urban environments guarantees that these solutions are not only environmentally advantageous but also socially and economically feasible.

Fundamentally, through the adoption of nature-based treatments and the promotion of widespread education and awareness, cities can lead the way towards more sustainable and resilient urban prospects. This collaborative endeavor is crucial for safeguarding water resources, improving urban quality of life, and conserving the well-being of both people and ecosystems in our increasingly urbanized world.

References

Agaton, C.B. and Guila, P.M.C. 2023. Ecosystem services valuation of constructed wetland as a nature-based solution to wastewater treatment. Earth, 4: 78-92. https://doi.org/10.3390/earth4010006.

Agaton, C.B. and Guila, P.M.C. 2024. Success factors and challenges: Implications of real options valuation of constructed wetlands as nature-based solutions for wastewater treatment. Resources, 13: 1-14. https://doi.org/10.3390/resources13010011.

Ahmet, B.B. and Erdem, C. 2018. Green roofs and facades: A comprehensive review. Renew. Sustain Energy Rev, 82: 915-939. https://doi.org/10.1016/j.rser.2017.09.106.

Akanksha, B., Bramha, D.V., Suneel, P. and Chander K.S. 2020. Predicting impact of urbanization on water resources in megacity Delhi. Remote Sens. Appl.: Soc. Environ, 20: 100361. https://doi.org/10.1016/j.rsase.2020.100361.

Alam, T., Mahmoud, A., Jones, K.D., Bezares-Cruz, J.C. and Guerrero, J. 2019. A comparison of three types of permeable pavements for urban runoff mitigation in the semi-arid south Texas, U.S.A. Water, 11: 1992. https://doi.org/10.3390/w11101992.

Andenaes, E., Kvande, T., Muthanna, T.M. and Lohne, J. 2018. Performance of blue-green roofs in cold climates: A scoping review. Buildings, 8: 55. https://doi.org/10.3390/buildings8040055.

Asleson, B.C., Nestingen, R.S., Gulliver, J.S., Hozalski, R.M. and Nieber, J.L. 2009. Performance assessment of rain gardens. J. Am. Water Resour. Assoc, 45: 1019-1031. https://doi.org/10.1111/j.1752-1688.2009.00344.x.

Brown, R.A. and Hunt, W.F. 2012. Improving bioretention/biofiltration performance with restorative maintenance. Water Sci. Techno, 65: 361-367. https://doi.org/10.2166/wst.2012.860.

Carpenter, D. and Hallam, L. 2010. Influence of planting soil mix characteristics on bioretention cell design and performance. J. Hydrol. Eng, 15: 404-416. https://doi.org/10.1061/(ASCE)HE.1943-5584.0000131.

Chao, G., Jiake, L., Huaien, L. and Yajiao, L. 2019. Influences of stormwater concentration infiltration on soil nitrogen, phosphorus, TOC and their relations with enzyme activity in rain garden. Chemosphere, 233: 207-215. https://doi.org/10.1016/j.chemosphere.2019.05.236.

Chen, C., Yanqi, L., Wencai, L., Chengyun, Y., Zhenzhong, L., Wei, L. and Ru, Z. 2023. Field performance of rain garden in red soil area in Southern China. Water, 15: 267. https://doi.org/10.3390/w15020267.

Chuan, L., Lingling, C., Huan, L., Zimei, M., Rui, Y., Lei, C. and Lingfeng, M. 2023. Effects of different types of plants on runoff reduction and suspended solids removal in rain gardens. J. Environ. Eng. Landsc. Manag, 31: 113-120. https://doi.org/10.3846/jeelm.2023.19016.

Chunbo, J., Jiake, L., Huaien, L. and Yajiao, L. 2019. An improved approach to design bioretention system media. Ecol. Eng, 136: 125-133, https://doi.org/10.1016/j.ecoleng.2019.06.014.

Cristiano, E., Deidda, R. and Viola, F. 2021. The role of green roofs in urban Water-Energy-Food-Ecosystem nexus: A review. Sci. Total Environ. 756: 143876. https://doi.org/10.1016/j.scitotenv.2020.143876.

Cristina, S.C.C. and Alexandros, I.S. 2021. Green roofs towards circular and resilient cities. Circ. Econ. Sustaina, 1: 395-411. https://doi.org/10.1007/s43615-021-00033-0.

Cristina, S.C.C., Beatriz, C. and Paulo, P. 2022. Nature-based solutions for socially and environmentally responsible new cities: The contribution of green roofs. pp. 235-255. *In*: Alexandros S. and Ioannis N. [eds.]. Circular Economy and Sustainability. Elsevier, Amsterdam, Netherlands. https://doi.org/10.1016/B978-0-12-821664-4.00015-7.

Dagenais, D., Brisson, J. and Fletcher, T.D. 2018. The role of plants in bioretention systems: Does the science underpin current guidance? Ecol. Engineering, 120: 532-545. https://doi.org/10.1016/j.ecoleng.2018.07.007.

Fan, G., Li, Z., Wang, S., Huang, K. and Luo, J. 2019. Migration and transformation of nitrogen in bioretention system during rainfall runoff. Chemosphere, 232: 54-62. https://doi.org/10.1016/j.chemosphere.2019.05.177.

Fini, A., Frangi, P., Mori, J., Donzelli, D. and Ferrini F. 2017. Nature based solutions to mitigate soil sealing in urban areas: Results from a 4-year study comparing permeable, porous, and impermeable pavements. Environ. Res, 156: 443-454. https://doi.org/10.1016/j.envres.2017.03.032.

Goh, H.W., Lem, K.S., Azizan, N.A., Chang, C.K., Talei, A., Leow, C.S. and Zakaria, N.A. 2019. A review of bioretention components and nutrient removal under different climates-future directions for tropics. Environ. Sci. Poll. Res, 26: 14904-14919. https://doi.org/10.1007/s11356-019-05041-0.

Good, J.F., O'Sullivan, A.D., Wicke, D. and Cochrane, T.A. 2012. Contaminant removal and hydraulic conductivity of laboratory rain garden systems for stormwater treatment. Water Sci. Technol, 65: 2154-2161. https://doi.org/10.2166/wst.2012.135.

Graziano, M.P., Deguire, A.K. and Surasinghe, T.D. 2022. Riparian buffers as a critical landscape feature: Insights for riverscape conservation and policy renovations. Diversity, 14: 172. https://doi.org/10.3390/d14030172.

Haase, D. 2017. Urban wetlands and riparian forests as a nature-based solution for climate change adaptation in cities and their surroundings. pp. 111-121. *In*: Kabisch, N., Korn, H., Stadler, J. and Bonn, A. [eds.]. Nature-based Solutions to Climate Change Adaptation in Urban Areas. Theory and Practice of Urban Sustainability Transitions. Springer, Cham. https://doi.org/10.1007/978-3-319-56091-5_7.

Hanbae, Y., Warren, A.D., Edward, L.M., Larry, P. and Parwinder, S.G. 2013. Field evaluation of a new biphasic rain garden for stormwater flow management and pollutant removal. Ecol. Eng, 54: 22-31. https://doi.org/10.1016/j.ecoleng.2013.01.005.

Hassan, I., Chowdhury, S.R., Prihartato, P.K. and Razzak, S.A. 2021. Wastewater treatment using constructed wetland: Current trends and future potential. Processes, 9(11): 1-27. https://doi.org/10.3390/pr9111917.

Heneghan, E., Marcus, C. and Mary, K. 2021. An evaluation of the potential applications of nature-based solutions for water quality protection: Ireland as a case study. Biol. Environ.: Proc. R. Ir. Acad, 121B: 147-162. https://doi.org/10.3318/bioe.2021.14.

Hsieh, C.H. and Davis, A.P. 2005. Evaluation and optimization of bioretention media for treatment of urban storm water runoff. J. Environ. Eng, 131: 1521-1531. https://doi.org/10.1061/(ASCE)0733-9372(2005)131:11(1521).

Imran, H.M., Akib, S. and Karim, M.R. 2013. Permeable pavement and stormwater management systems: A review. Environ. Technol, 34: 2649-2656. https://doi.org/10.1080/09593330.2013.782573.

Justino, S., Calheiros, C., Castro, P. and Goncalves, D. 2023. Constructed wetlands as nature-based solutions for wastewater treatment in the hospitality industry: A review. Hydrology, 10: 1-16. https://doi.org/10.3390/hydrology10070153.

Jyoti, V.N., Jyoti, V. and Victor, R.S. 2024. Chapter 9: Enriching and maintaining the riparian buffer zone. pp. 161-180. *In*: Victor R.S., Rajiv R.M., Uday B. and Hitesh V. [eds.]. Managing Urban Rivers. Elsevier, Amsterdam, Netherlands. https://doi.org/10.1016/B978-0-323-85703-1.00018-3.

Karabay, K., Öztürk, H., Ceylan, E. and Ayral, Ç.D. 2024. Assessment of urban rain gardens within climate change adaptation and circularity challenge. Nature-based Solutions for Circular Management of Urban Water Circular Economy and Sustainability, 51-72. https://doi.org/10.1007/978-3-031-50725-0_4.

Khan, M.M., Akram, M.T., Janke, R., Qadri, R.W.K., Al-Sadi, A.M. and Farooque, A.A. 2020. Urban horticulture for food secure cities through and beyond COVID-19. Renew. Sustain. Energy Rev, 12: 9592. https://doi.org/10.3390/su12229592.

Lehmann, S. 2021. Growing biodiverse urban futures: Renaturalization and rewilding as strategies to strengthen urban resilience. Sustainability, 13: 2932. https://doi.org/10.3390/su13052932.

Li, H. and Davis, A.P. 2009. Water quality improvement through reductions of pollutant loads using bioretention. J. Environ. Eng, 135: 567-576. https://doi.org/10.1061/(ASCE)EE.1943-7870.000002.

Li, H., Sharkey, L.J., Hunt, W.F. and Davis, A.P. 2009. Mitigation of impervious surface hydrology using bioretention in North Carolina and Maryland. J. Hydrol. Eng, 14: 407-415. https://doi.org/10.1061/(ASCE)1084-0699(2009)14:4(407).

Lise, H. and Patience, M. 2019. Examining urban water management practices – Challenges and possibilities for transitions to sustainable urban water management in Sub-Saharan cities. Sustain. Cities Soc, 48: 101573. 101573.https://doi.org/10.1016/j.scs.2019.101573.

Liu, J., Sample, D.J., Bell, C. and Guan, Y. 2014. Review and research needs of bioretention used for the treatment of urban stormwater. Water, 6: 1069-1099. https://doi.org/10.3390/w6041069.

Lloyd, S. and Wong, T. 2008. Paired catchment storm event monitoring: Assessing the performance of a bioretention system (rain garden). Australas. J. Water Resour, 12: 133-141. https://doi.org/10.1080/13241583.2008.11465341.

Magda, K., Wojciech, S., Eliza, P., Floris, C.B., Katarzyna, B. and Magdalena, G. 2022. Technical solutions and benefits of introducing rain gardens – Gdańsk case study. Sci. Total Environ, 835: 155487. https://doi.org/10.1016/j.scitotenv.2022.155487.

Masiero, M., Bottaro, G., Righetti, C., Nikolaidis, N.P., Lilli, M.A. and Pettenella, D. 2024. Riparian forests as nature-based solutions within the Mediterranean context: A biophysical and economic assessment for the Koiliaris River Watershed (Crete, Greece). Forests, 15: 760. https://doi.org/10.3390/f15050760.

Meysam, K., Madjid, D. and Massoud, T. 2017. Evaluation of permeable pavement responses to urban surface runoff. J. Environ. Manag, 187: 43-53. https://doi.org/10.1016/j.jenvman.2016.11.027.

Miklas, S. and Piotr, G. 2017. Review of permeable pavement systems. Build. Environ, 42: 3830-3836. https://doi.org/10.1016/j.buildenv.2006.11.016.

Monteiro, C.M., Mendes, A.M. and Santos, C. 2023. Green roofs as an urban NbS strategy for rainwater retention: Influencing factors—A review. Water, 15: 2787. https://doi.org/10.3390/w15152787.

Oscar, O.D. and Caren, N.A. 2021. Constructed wetlands in wastewater treatment and challenges of emerging resistant genes filtration and reloading. Inland Waters – Dynamics and Ecology. IntechOpen. http://dx.doi.org/10.5772/intechopen.93293.

Pearlmutter, D., Theochari, D., Nehls, T., Pinho, P., Piro, P., Korolova, A. et al. 2020. Enhancing the circular economy with nature-based solutions in the built urban environment: Green building materials, systems and sites. Blue-Green Syst, 2: 46-72. https://doi.org/10.2166/bgs.2019.928.

Premarathna, K.S.D., Jayanta, K.B., Manish, K., Sunita, V., Bede, M., Pau, L.S. et al. 2023. Biofilters and bioretention systems: The role of biochar in the blue-green city concept for stormwater management. Environ. Sci.: Water Res. Technol, 9: 3103-3119. https://doi.org/10.1039/D3EW00054K.

Qi, Y., Chan, F.K.S., Thorne, C., O'Donnell, E., Quagliolo, C., Comino, E. et al. 2020. Addressing challenges of urban water management in Chinese sponge cities via nature-based solutions. Water, 12: 2788. https://doi.org/10.3390/w12102788.

Rezaei, M., Emmanuel, N., Kim, J. and Kim, D.H. 2021. Analyzing the impact of green roof functions on the citizens' mental health in metropolitan cities. Iran. J. Public Health, 50: 900-907. https://doi.org/10.18502/ijph.v50i5.6107.

Rizzo, Gianfranco, Laura C., Maria L.G., Giorgia, P. and Gianluca, S. 2023. Green roofs' end of life: A literature review. Energies, 16: 596. https://doi.org/10.3390/en16020596.

Roy-Poirier, A., Champagne, P. and Filion, Y. 2010. Bioretention processes for phosphorus pollution control. Environ. Rev, 18: 159-173. https://doi.org/10.1139/A10-006.

Scholz, M. 2014. Permeable pavements and storm water management. pp. 247-260. *In*: Gopalakrishnan, K., Steyn, W. and Harvey, J. [eds.]. Climate Change, Energy, Sustainability and Pavements. Green Energy and Technology. Springer, Berlin, Heidelberg. https://doi.org/10.1007/978-3-662-44719-2_7.

Shafique, M., Kim, R. and Rafiq, M. 2018. Green roof benefits, opportunities and challenges – A review. Renew. Sustain. Energy Rev, 90: 757-773. https://doi.org/10.1016/j.rser.2018.04.006.

Shahmohammad, M., Hosseinzadeh, M., Dvorak, B., Bordbar, F., Shahmohammadmirab, H. and Aghamohammadi, N. 2022. Sustainable green roofs: A comprehensive review of influential factors. Environ. Sci. Poll. Res, 29: 78228-78254. https://doi.org/10.1007/s11356-022-23405-x.

Stefanakis, A., Oral, H.V., Calheiros, C. and Carvalho, P. [eds.] 2024. Nature-based Solutions for Circular Management of Urban Water. Springer, Cham.

Stovin, V. 2010. The potential of green roofs to manage urban stormwater. Water Environ. J, 24: 192-199. https://doi.org/10.1111/j.1747-6593.2009.00174.x.

Sun, C., Rao, Q., Chen, B., Liu, X., Adnan, I.R.M., Li, J. et al. 2024. Mechanisms and applications of nature-based solutions for stormwater control in the context of climate change: A review. Atmosphere, 15: 403. https://doi.org/10.3390/atmos15040403.

Susana, O.M. and Davide, G. 2022. What variables matter when designing nature-based solutions for stormwater management? A review of impacts on ecosystem services. Environ. Impact Assess. Rev, 95: 106802. https://doi.org/10.1016/j.eiar.2022.106802.

Tomáš, H., Oldřich N., Eduard H. and Mark, R. 2012. Possible impacts of floods and droughts on water quality. J. Hydro-environ. Res, 6: 145-150. https://doi.org/10.1016/j.jher.2012.01.008.

Ülo, M., Yoshihiko, H. and Valdo, K. 2005. Purification processes, ecological functions, planning and design of riparian buffer zones in agricultural watersheds. Ecol. Eng, 24: 421-432. https://doi.org/10.1016/j.ecoleng.2005.01.015.

Umberto, B., Amir Hosein, G. and Ali, G. 2014. State-of-the-art analysis of the environmental benefits of green roofs. Appl. Energy, 115: 411-428. https://doi.org/10.1016/j.apenergy.2013.10.047.

Venkatachalam, A., Jay, R. and Eiji, Y. 2005. Impact of riparian buffer zones on water quality and associated management considerations. Ecol. Eng, 24: 517-523. https://doi.org/10.1016/j.ecoleng.2004.01.007.

Vijayaraghavan, K. 2016. Green roofs: A critical review on the role of components, benefits, limitations and trends. Renew. Sustain. Energy Rev, 57: 740-752. https://doi.org/10.1016/j.rser.2015.12.119.

Vijayaraghavan, K., Biswal, B.K., Adam, M.G., Soh, S.H., Tsen-Tieng, D.L., Davis, A.P. et al. 2021. Bioretention systems for stormwater management: Recent advances and future prospects. J. Environ. Manag, 292: 112766. https://doi. org/10.1016/j.jenvman.2021.112766.

Vymazal, J., Greenway M., Tonderski K., Brix H. and Mander Ü. 2006. Constructed wetlands for wastewater treatment. pp. 69-96. *In*: Verhoeven, J.T.A., Beltman, B., Bobbink, R., Whigham, D.F. [eds.] Wetlands and Natural Resource Management. Ecological Studies. Springer, Berlin, Heidelberg . https://doi.org/10.1007/978-3-540-33187-2_5.

Vymazal, J. 2010. Constructed Wetlands for Wastewater Treatment. Water, 2: 530-549. https://doi.org/10.3390/w2030530.

Walters, S.A. and Midden K.S. 2018. Sustainability of urban agriculture: Vegetable production on green roofs. Agric. Rev, 8: 168. https://doi.org/10.3390/agriculture8110168.

Yamada, S. and Shibata S. 2017. Quantitative evaluation about rainwater-runoff characteristics of rain garden. J. Jpn. Soc. Reveget. Technl, 43: 251-254. https://doi.org/10.7211/jjsrt.43.251.

Yuan, M.H., Lo, F.C., Yu, C.P., Tung, H.H., Chang, Y.S., Chiueh, P.T. et al. 2022. Nature-based solutions for securing contributions of water, food, and energy in an urban environment. Environ. Sci. Pollut. Res, 29: 58222-58230. https://doi.org/10.1007/s11356-022-19570-8.

Yuxin, Z., Hui, L., Bing, Y., Xue, Z., Saifullah, M., Xiaochun, Z. et al. 2021. Permeable pavement design framework for urban stormwater management considering multiple criteria and uncertainty. J. Clean. Prod, 293: 126114. https://doi.org/10.1016/j.jclepro.2021.126114.

Zhang, L., Oyake, Y., Morimoto, Y., Niwa, H. and Shibata, S. 2020a. Flood mitigation function of rain gardens for management of urban storm runoff in Japan. Landsca. Ecolog Eng, 16: 223-232. https://doi.org/10.1007/s11355-020-00409-8.

Zhang, L., Zehao, Y. and Shozo, S. 2020b. Assessment of rain garden effects for the management of urban storm runoff in Japan. Sustainability, 12: 9982. https://doi.org/10.3390/su12239982.

Zhang, S. and Guo, Y. 2014. Stormwater capture efficiency of bioretention systems. Water Resour. Manag, 28: 149-168. https://doi.org/10.1007/s11269-013-0477-y.

Zhou, J ., Xiong, Ji., Ni, J., Xie, X. and Liu, Y. 2023. Nitrogen transfer and transformation in bioretention cells under low temperature conditions. Sci. Total Environ, 871: 162087. https://doi.org/10.1016/j.scitotenv.2023.162087.

CHAPTER

4

Role of Biochar in Wastewater Treatment and Management

Arunima Nayak[1*], Brij Bhushan[1], Garima Bhatnagar[1], Anshika Chauhan[1] and Navdha Bhatt[2]

[1] Department of Chemistry, Graphic Era Deemed to be University - 248002, Dehradun, India

[2] Department of Chemistry, Graphic Era Hill University, Dehradun, India

Introduction

In recent years, sustainable and environmentally friendly solutions to the mass-scale problem of wastewater treatment have also been the focus of research. Therefore, scientists have explored several methods to resolve the urgent need to remove multiple contaminants, which include ion exchange, adsorption, biological remediation, catalytic and photocatalytic degradation, chemical precipitation, and membrane filtration (Barakat 2011). All techniques except adsorption are highly complex and expensive to operate. Therefore, these techniques are not economically favourable for environmental remediation (Crini et al. 2009). Among all these techniques, the adsorption technique is highly preferred because it has the requirement of low investment, low infrastructural set-up, low operational requirement, etc., and has resulted in high separation efficiency, causing no extra pollution (Rashed et al. 2013). The success of the adsorption technique has been determined by the framework of the adsorbent. Various types of materials as adsorbents have been used to remove pollutants from polluted water, such as nanomaterials, hydrogels, zeolites, silica gel, and activated alumina, and that have demonstrated good efficiency for this technique (Danish et al. 2021, Darban et al. 2022, Sodha et al. 2022). Biological adsorbents and agricultural-based industrial wastes are inexpensive adsorbents that are helpful in the treatment of wastewater (Torres-pérez et al. 2015, Escudero et al. 2019, Mo et al. 2018).

In this context, the term biochar is worth mentioning to remove contaminants from wastewater. Biochar is considered a sustainable and suitable adsorbent because of its porosity, economy, significant amount of carbon, renewability, hydrophobicity,

*Corresponding author: arunima_nayak@yahoo.com

DOI: 10.1201/9781003527350-4

and biodegradability. For the first time, the biochar term was introduced by Peter Read in 2015 (Read et al. 2004). On the basis of the chemical composition of agricultural waste, the term biochar can be classified into two types – lignocellulosic group (are present in agricultural waste, forestry waste, etc.) and non-lignocellulosic (are present in algae, manure, sewage sludge) (Rashidi et al. 2020). Biochar is derived from green sources, viz. farming residues (such as sugarcane bagasse, manure, rice husk, weeds), industrial waste (such as peels, fibres, textile fibres, paper waste), aqua waste (such as seaweeds, periwinkle shells, fish scale, algae), and forestry residue (such as wood chips, sawdust, tree bark, logging residues) via pyrolysis or thermal treatment. Pyrolysis is the preferred method for the fabrication of biochar in which bio-waste materials are heated at different temperatures under oxygen and nitrogen atmospheric pressure (Prabhahar et al. 2019, Yan et al. 2020). After heating, the emission of volatile matters such as H_2, CO, H_2O, CH_4, etc., is known to occur, and the remaining carbon-rich residue is indicated as biochar (Prabhahar et al. 2020). Pyrolysis has been shown to create high porosity, an essential prerequisite to removing various pollutants from polluted water.

Furthermore, adsorption properties for biochar are known to be modified via different modification technologies such as physical modification, including ball milling and gas/steam activation; chemical modification, including acidic treatment/ basic treatment; and biological modification, including biofilms. Higher surface area and appreciable high porosity are known to be demonstrated in modified biochar as compared to pristine biochar.

In this book chapter, we have compiled various aspects of biochar, such as its fabrication and modification methods (pyrolysis, ball milling, gas/steam activation, carbonisation, acidic/basic treatment, etc.). As a result of modification/activation, engineered biochar has been known to demonstrate better adsorption properties than pristine, and it is also used in the treatment of contaminated aqueous liquid. Figure 1

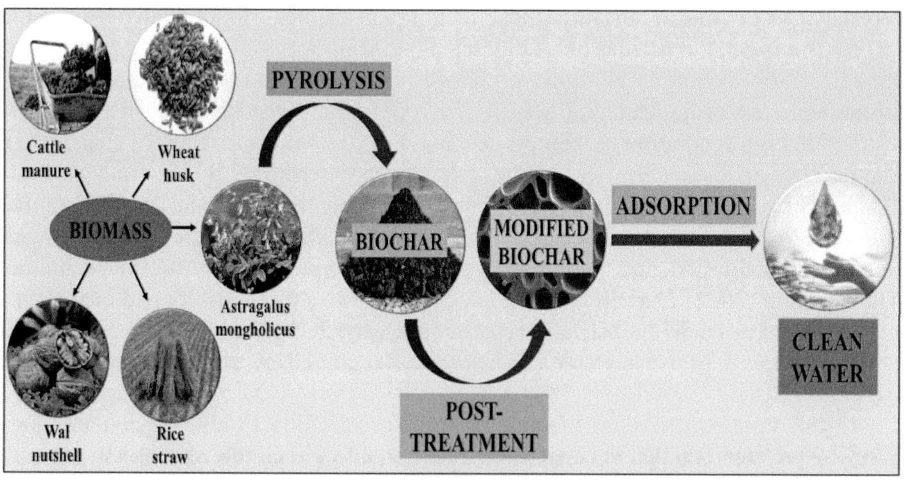

Figure 1: Biomass-derived pristine/modified biochar: An illustration of fabrication to application.

diagrammatically represents the fabrication of biomass-derived pristine/modified biochar also used in the treatment of contaminated aqueous liquid.

Fabrication Methodologies for Biochar

Methods Applied for Synthesis of Biochar

Physical Activation

The physically activated method has entailed the vapour, inert gas, and CO_2 to attain activation at relatively high temperatures. This procedure has assisted in giving the porous structure of biochar (Cha et al. 2016, Rajapaksha et al. 2015).

In this respect, literature has reported the activation of biochar derived from cattle manure via pyrolysis under an N_2 atmosphere at 200°C for about 30 mins in a tubular furnace. The synthesised biochar showed a high removal rate to remove MB from contaminated water (Zhu et al. 2018).

Hoslett et al. fabricated biochar derived from municipal waste (MMDM) that was pyrolysed at 300°C for 12 hours. The adsorptive capability of the biochar showed as 7.2 mg/g for MMDM biochar to remove MB dye (Hoslett et al. 2020).

In 2016, Shang et al. reported an *Astragalusmongholicus*-derived carbonised biochar via thermal activation at 800°C for 5 hours to remove ciprofloxacin from wastewater. With increased carbonisation temperature, the surface area and carbon content increased rapidly for the biochar and vice versa for hydrogen and oxygen content (Shang et al. 2016).

In another study, at temperatures over 700°C, the pyrolysed biochar was gasified via steam, CO_2, or both. The surface area, porosity, and pore size distribution of the biochar increased as a result of this treatment (Cha et al. 2016, Pawar et al. 2020).

Chemical Activation

The chemically activated method involves impregnating biochar with chemicals, followed by draining and filtering. The types and doses of the activating agent, as well as the temperature of activation, affect the characteristics of biochar (Cha et al. 2016).

In a study, dried biomass was carbonised at 673 K for 60 minutes under an inert atmosphere, followed by NaOH impregnation for the activation. This activation enhanced the surface area of biochar. FTIR studies indicated that the peak intensities corresponding to the oxygen-containing acidic groups in the biochar after activation decreased (Choudhary et al. 2020).

Angin et al. synthesised safflower seed press cake-derived biochar via pyrolysis at 600–900°C, followed by impregnation of $ZnCl_2$ as an activating reagent. The fabricated biochar was used to eliminate methylene blue from wastewater. The study also demonstrated that increasing the activation temperature and chemical impregnation ratio for the biochar increased porosity and surface area (Angin et al. 2013).

A study was also reported for the fabrication of biochar via thermal activation at different temperatures for 5 hours under oxygen-limited conditions. After activation, the biochar was impregnated with HCl to remove ciprofloxacin from wastewater. BET demonstrated a 176.33 m^2/g higher surface area for the activated biochar at an activation temperature of 800°C as compared to that of the precursor viz. natural herbal residue as 0.42 m^2/g (Shang et al. 2016).

Various studies reported the fabrication of biochar, which was derived from yak manure with H_2O_2 after pyrolysis at 350°C for 120 minutes under an inert atmosphere with a 900 ml/min flow rate, and the same was used to remove lead ions from polluted water. Fabricated biochar showed a 6.36 m^2/g surface area, higher than unmodified biochar at 1.03 m^2/g. FTIR studies revealed that after modification, the oxygen-containing surface functional group of the carboxyl group increased (Wang and Liu 2018). Ferric chloride hexahydrate ($FeCl_3 \cdot 6H_2O$) was used as an impregnated chemically modified wheat and rice husk biochar fabricated via pyrolysis at 600°C for 1 hour under a nitrogen atmosphere. BET studies demonstrated a 300 m^2/g surface area and 399 m^2/g surface area for iron oxide rice husk hybrid biochar (RHIOB) and iron oxide wheat husk hybrid biochar (WHIOB), respectively (Singh et al. 2020).

Physicochemical Activation

Physicochemical activation of biochar has involved the modification of its structure through physical (thermal treatment, steam activation) and chemical (impregnation with chemicals, acid/base treatment) methods simultaneously to enhance its properties for various applications across environmental, agricultural, and industrial areas.

In this respect, the fabrication of pine nut shells biochar modified via $FeCl_3$ and cetyl trimethyl ammonium bromide solution was used to remove acid chrome blue K (AK) from wastewater. Firstly, the pyrolysis was carried out at 700°C for precursor in a furnace with 10°C/min flow rate, followed by its impregnation with $FeCl_3$ solution for 120 mins. After impregnation, pyrolysis was again done at a temperature of 700°C for 180 mins with a 10°C/min flow rate, followed by adding cetyl trimethyl ammonium bromide to obtain synthesised biochar. SEM image revealed that the surface of CTAB-magnetic modified biochar was smoother than that of modified magnetic biochar (Wang et al. 2020).

Physico-chemical Characteristics of Biochar

The summary of physico-chemical characteristics of biochar derived from different sources of biochar is presented in Table 1.

Pyrolysis and the source of biochar play major roles in explaining the physico-chemical characteristics of biochar. In a study, Jindo et al. (2014) fabricated woody biochar such as Apple tree branch (AB) and Oak tree (OB), and non-woody biochar such as Rice husk (RH) and Rice straw (RS) via pyrolysis at various temperatures (400°C, 500°C, 600°C, and 700°C) (Jindo et al. 2014). BET studies demonstrated that when the temperature increased from 400°C to 800°C, the surface area of biochars was found to increase for AB and OB compared to RH and RS. At 800°C, higher surface area values of 545 m^2/g, 398 m^2/g, 295 m^2/g, and 256 m^2/g were reported for AB, OB, RH, and RS, respectively, as shown in Table 1 (Jindo et al. 2014). Studies have also observed that when the temperature increased from 400°C to 800°C, the pH value of biochars also increased (Novak et al. 2009).

Table 1: Detailed overview of physiochemical features of biochar derived from different sources of biochar (Jindo et al. 2014)

S. No.	Source of Biochar	Pyrolysis Temperature (°C)	Yield of Biochar (%)	Volatile Content (%)	pH	BET Surface Area (m^2/g)	C (%)	H (%)	O (%)
1.	Apple tree branch (AB)	400	28.3	32.6	7	11.9	70	4.13	20.5
		500	16.7	18.2	9.6	58.6	79	2.6	11.9
		600	16.6	11	10	208.6	81.4	1.9	13.63
		700	15.8	7.7	10	418.6	82	1.2	16.3
		800	15.5	6.8	10	545.4	84.8	0.6	5.8
2.	Oak tree (OB)	400	35.8	32	6.4	5.6	70.5	3.7	21.47
		500	28.6	19	8	103.1	77.5	2.5	17.7
		600	22	12	8.8	228.5	81	1.9	15.96
		700	20	8	9.5	335.6	83	1.1	14.97
		800	19.1	7.8	9.6	398.1	82.8	0.6	17.29
3.	Rice husk (RH)	400	48.6	22	6.8	193.7	44.5	2.5	16.32
		500	42.4	10.5	8.9	262	45	1.2	7.12
		600	37.3	6	9.4	243	40	0.8	9.23
		700	32.8	3.8	9.5	256	38.8	0.4	12.69
		800	32	3	9.6	295.5	40.4	0.2	2.69
4.	Rice straw (RS)	400	39.3	22.4	8.6	46.6	49.9	2.8	12.02
		500	32.6	12.8	9.8	59.9	37.4	0.9	8.64
		600	23.4	8.3	10.1	129	33.7	0.6	13.7
		700	18.4	5.3	10.3	149	36.2	0.5	17.38
		800	18.3	4.4	10.4	256.9	29.1	0.25	3.71

The type of feedstock was another feature that influenced the yield of biochar and volatile matter composition. The yield of biochar decreased for AB and OB compared to RH and RS. In contrast, AB and OB reported higher volatile matter composition than RH and RS due to pyrolysis temperatures from 400°C to 800°C. Woody biochar consisted of high volatile matter composition at minimum temperature, which was mainly observed due to the presence of lignin in woody feedstock, which can stop the pyrolysis at 400°C but not at high temperatures such as 950°C. Also, at 950°C, ash content was observed. In contrast, rice residue biochar showed high ash content at all temperature ranges because of high silica amount in rice residues (Enders et al. 2012, Mukome et al. 2013).

The literature revealed that when pyrolysis temperature increased, the H and O content in biochar decreased. For example, the % of H was reported as 4.1 to 0.6, 3.7 to 0.6, 2.5 to 0.2, and 2.8 to 0.25 for AB, OB, RH, and RS, respectively, with an increase in pyrolysis temperature from 400ºC to 800ºC. Also, the % of O was reported as 20.5 to 5.8, 21.4 to 17.2, 16.3 to 2.69, and 12 to 3.71 for AB, OB, RH, and RS, respectively. This showed that the % of H and % of O decreased for different feedstock with an increase in pyrolysis temperature (from 400°C to 800°C), and the % of C (for AB from 70 to 84.8 and for OB from 70.5 to 82.8) was found to increase at higher pyrolysis temperatures, but the % of C (for RH from 44.5 to 40.4 and for RS from 49.9 to 29.1) decreased, as shown in Table 1 (Jindo et al. 2014).

On the basis of FTIR and zeta potential studies, various researchers reported physico-chemical characteristics of the biochar. In this respect, Lee et al. synthesised biochar via pyrolysis at two different temperatures, 450°C and 700°C, respectively. FTIR study revealed that C-H stretching (2950–2850 cm^{-1}) of the aliphatic carbon group decreased as the temperature increased. Various literature sources showed that on increasing the temperature, the aliphatic C group decreased, but the aromatic C group increased (Lee et al. 2010). Zhu et al. fabricated cattle manure-derived biochar at 200°C via pyrolysis. The zeta potential studies showed that at low pyrolysis, the surface of biochar carried more negative charge. It concluded that a more negative charge on the surface of biochar showed excellent adsorption capability for the fabricated adsorbent. FTIR study revealed that 2924 cm^{-1} and 2854 cm^{-1} intense bands indicated the presence of CH_2 and CH_3 groups at low temperatures (200°C), but such peaks showed diminished intensity at high temperatures (800°C) (Bulut and Aydin 2006, Tsang et al. 2016, Zhu et al. 2018).

Engineered Biochar and their Advanced Characteristics

Engineered biochar is known as modified biochar which has shown enhanced adsorptive properties and functionalities as compared to pristine biochar. Biochar can be modified via physical, chemical and biological processes.

Physical Method

In physical method, ball milling, gas/steam activation, and magnetization techniques are used for the modification of adsorptive properties of biochar.

The ball milling method involves the modification of the surface area and particle size of biochar. In this respect, Qin et al. (2020) reported bamboo biochar (BB) and ball-milled bamboo biochar (BMBB) via pyrolysis at 450ºC to remove ammonium from polluted water. The adsorptive capability was 22.9 mg/g for BMBB, higher than that of BB (7.0 mg/g). The particle size of BMBB was observed to be finer after using the ball milling method. BET studies demonstrated a 298.6 m^2/g higher surface area for BMBB than BB (18.2 m^2/g). This study reported zeta potential values of −13 to −46.2 mV for BB and −29.9 to −49.2 mV for BMBB at pH 3–12. The more negative charge in BMBB showed the presence of more acidic oxygen-containing functional groups, which enhanced the adsorption properties of ammonium (Qin et al. 2020).

The gas steam activation method improves the surface area and porous nature of biochar. In this respect, Cho et al. (2017) reported that magnetic biochar was obtained via pyrolysis of $FeCl_3$-impregnated spent coffee grounds (SCG) in an inert atmosphere to remove AS (V) from polluted water. BET studies demonstrated a 512 m^2/g higher surface area and 0.249 cm^3/g pore volume for Fe-biochar in CO_2 atmosphere compared to 8.3 m^2/g surface area and 0.018 cm^3/g pore volume for Fe-biochar in N_2 atmosphere (Cho et al. 2017).

Chemical Method

In the chemical method, acid modification, basic modification, and chemical impregnation are used to improve the adsorptive properties of biochar.

In this respect, Jin et al. (2018) reported wheat straw (WH) acid-oxidised biochar (WHAO) and cow manure (CW) acid-oxidised biochar (CWAO) via HNO_3 acidic treatment at 90ºC for 4 hours to remove U(VI) from polluted water. The adsorptive capability was 355.6 mg/g for WHAO, compared to 8.7 mg/g for untreated WH. The adsorptive capability was 73.3 mg/g for CWAO compared to 64 mg/g for CW under a pH range of 4.5 and temperature of 298 K. XPS study revealed that the presence of oxygenated functional groups (C=O and COO) on the oxidised biochar enhanced the surface polarity and resulted in stronger adsorption removal of U(VI) from water than the untreated counterpart (Jin et al. 2018).

Choudhary et al. (2020) reported cladodes of *Opuntia Ficus-indica* (OFI) cactus-derived activated biochar via pyrolysis at 400°C for 1 hour, followed by chemical activation by NaOH solution. The fabricated engineered biochar was used to remove malachite green from polluted water. BET studies demonstrated 32.07 m^2/g surface area, 0.093 cm^3/g pore volume, and 11.51 nm pore diameter of activated biochar, compared to the untreated counterpart (1.15 m^2/g surface area, 0.018 cm^3/g pore volume, and 27.46 nm pore diameter) (Choudhary et al. 2020).

Biological Method

In the biological method, microorganisms are used to form a biofilm on the inner and outer surfaces of biochar. Biofilms are involved in modifying the porosity of biochar. Dalahmeh et al. (2018) reported using a biochar-active film to remove pharmaceutically active compounds (PhACs) from sewage (Dalahmeh et al. 2018).

Frankel et al. synthesised a biochar to remove naphthenic acids from oil sand. The enhancement in removal efficiency was found to increase from ~30% to ~87%

due to the presence of biofilms (Frankel et al. 2016). A clear vision of different methods used for the fabrication of engineered biochar and their effect on the adsorption properties of biochar is demonstrated in Figure 2.

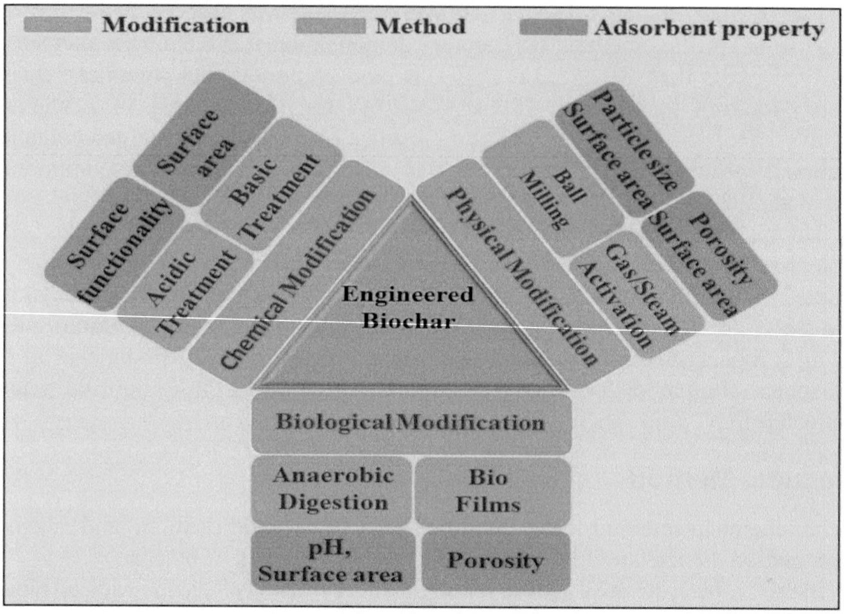

Figure 2: A representation of different methods used for fabrication of engineered biochar and their effect on the adsorption properties of biochar.

Application as Adsorbent

Removal of Metal Ions

Table 2 shows the performance assessment of different biochar for metal ions from wastewater.

Wang et al. used H_2O_2 for the modification of yak manure and synthesised yak manure biochar (AC YMB) to remove Pb (II) from polluted water. Adsorption capacity was shown as 76.41 mg/g with a duration time of 48 hours. BET studies demonstrated a 6.36 m^2/g surface area for AC YMB. SEM images revealed the smoother surface of AC YMB (Wang et al. 2018).

Chen et al. reported hardwood biochar via pyrolysis at 450°C to remove Cu (II) and Zn (II) from polluted water. The adsorptive capability was reported as 6.79 mg/g and 4.54 mg/g for Cu (II) and Zn (II), respectively, at pH 6.67 with a duration time of 12 days (Chen et al. 2011).

Achieng et al. synthesised fish scale-derived biochar (FSB) via pyrolysis at 600°C to remove Cu (II) from polluted water. The adsorptive capability of 39.39 mg/g was shown for a dose of adsorbent as 2 g/L with a duration time of 24 hours. XRD revealed the crystalline nature of FSB. FTIR studies showed bands at

3600–3300 cm^{-1} and 1700–1600 cm^{-1} due to the presence of –OH and C=O stretching vibrations (Achieng et al. 2020).

Dairy manure-derived biochar (MBC) was reported by Chen et al. in 2019 via pyrolysis at 300°C to remove Pb(II) and Cd(II) from polluted water. Adsorptive capability was shown as 175.53 mg/g and 68.08 mg/g for Pb(II) and Cd(II), respectively. Under optimum conditions at pH 9.9, the dose of adsorbent was 1.5 g/L for 24 hours. BET studies demonstrated a 25.94 m^2/g surface area for MBC (Chen et al. 2019).

Nworie et al. fabricated plantain peel-based biochar (PPB) to remove Pb (II) from polluted water. The adsorptive capability was shown as 4 mg/g under optimum conditions at pH 9.9, the dose of adsorbent as 10 g/L with a duration time of 150 mins. BET studies demonstrated a 16.69 m^2/g surface area and 8.79 cc/g pore size for PPB and plantain peel. XRD revealed the crystalline nature of PPB (Nworie et al. 2020).

Georgieva et al. synthesised walnut shell biochar (WSBA) to remove Ni (II) from polluted water. The adsorptive capability was reported as 13.92 mg/g at pH 6.5, with a dose of adsorbent as 3 g/L and adsorbate dosage as 150 mg/L, with a duration time of 120 mins (Georgieva et al. 2020).

Singh et al. synthesised iron oxide rice husk hybrid biochar (RHIOB) and iron oxide wheat husk hybrid biochar (WHIOB) via pyrolysis of $FeCl_3$-impregnated precursors at 600°C for 1 hour. Under optimum conditions at pH 7.5, the dose of adsorbent was 2 g/L for 24 hours. BET revealed 300 m^2/g and 399 m^2/g surface areas for RHIOB and WHIOB, respectively. Removal rates of 85% and 90% were observed for RHIOB and WHIOB, respectively, even after four consecutive cycles (Singh et al. 2020).

Removal of Dyes

Table 3 exhibited the performance assessment of different biochars for dyes from wastewater. Zhu et al. synthesised low-temperature cattle manure-derived biochar (CMB) via pyrolysis at 200°C for 30 mins to remove methylene blue (MB) from polluted water. The maximum adsorptive capability shown via batch studies for CMB was 241.9 mg/g. Under optimum conditions at pH 3–4 and 5–25°C temperature, 200 mg/L as the concentration of adsorbate, a dose of adsorbent as 1.25 g/L with a duration time of 24 hours. Furthermore, BET studies demonstrated a 0.3276 m^2/g surface area and 10.0 cm^3/g pore volume for the synthesised biochar. FTIR studies exhibited two peaks at 3280 cm^{-1} and 3647 cm^{-1}, which correspond to the hydroxyl stretch vibration bands of free phenols (Zhu et al. 2018).

Municipal waste-derived biochar (MMDM biochar; the precursor consisted of paper, cardboard, food scraps, and garden waste) was reported by Hoslett et al. fabricated via pyrolysis at 300ºC for 12 hours. The adsorptive capability of the biochar was 7.2 mg/g for MMDM biochar to remove MB dye at pH of 5, equilibrium time of 360 mins, dose of adsorbent as 5 g/L, and 100 mg/L as concentration of adsorbate. SEM images revealed that biochar made from paper/cardboard appeared to have a lattice-like structure that showed a significantly different structure from biochar made from paper. The surface morphology of the biochar made from food

Table 2: Performance evaluation of different biochar sources to remove metal ions from wastewater

S. No.	Source of Biochar	Adsorbent	Chemical Used/ Pyrolysis	Metal ions	Operating Conditions	Adsorbent Properties	Batch Results	Reference
						Adsorbent Surface Area (m^2/g)	Maximum Capacity (q_{max}) (mg/g)/ % removed	
1.	Yak manure	Yak manure biochar (AC-YMB)	H_2O_2 modified 350°C	Pb (II)	Equilibrium time – 48 hours	6.36 m^2/g	76.41 mg/g	Wang et al. 2018
2.	Hard wood	HW450	Pyrolysis – 450°C	Cu (II) Zn (II)	Equilibrium time – 12 days pH – 5.57	NA	6.79 mg/g 4.54 mg/g	Chen et al. 2011
3.	Fish scale	Fish scale biochar (FSB)	Pyrolysis – 600°C	Cu (II)	Dose of adsorbent – 2 g/L Equilibrium time – 24 hours	NA	39.39 mg/g	Achieng et al. 2020
4.	Dairy manure	Daily manure biochar (MBC)	Pyrolysis – 300°C	Pb (II) Cd (II)	Dose of adsorbent – 1.5 g/L pH – 9.9 Equilibrium time – 24 hours	25.94 m^2/g	175.53 mg/g 68.08 mg/g	Chen et al. 2019
5.	Plantain peel	Plantain peel biochar (PPB)	NA	Pb (II)	Dose of adsorbent – 10 g/L Equilibrium time – 150 mins	Pore size – 8.79 cc/g Surface area – 16.69 m^2/g	4 mg/g	Nworie et al. 2020

6.	Walnut shell	Walnut shell biochar (WSBA)	NA	Ni (II)	Dose of adsorbent – 3 g/L Concentration of adsorbate – 150 mg/L pH – 6.5 Temperature –303 K Equilibrium time – 120 mins	425 m^2/g	13.92 mg/g	Georgieva et al. 2020
7.	Rice husk	Iron oxide rice husk hybrid biochar (RHIOB)	$FeCl_3$ impregnated Pyrolysis – 600°C, 1 hour	As (III)	Dose of adsorbent – 2 g/L pH – 7.5 Temperature – 25°C Equilibrium time – 24 hours	300 m^2/g	85 %	Singh et al. 2020
	Wheat husk	Iron oxide wheat husk hybrid biochar (WHIOB)				399 m^2/g	90 %	

Table 3: Performance evaluation of different biochar sources to remove dyes from wastewater

S. No.	Source of Biochar	Adsorbent	Chemical used/ Pyrolysis	Dyes	Operating Conditions	Adsorbent Properties	Batch Results	Reference
						Adsorbent Surface Area (m^2/g)	Maximum Capacity (q_{max}) (mg/g)/% Removed	
1.	Cattle manure	Cattle manure-low temperature biochar (CMB)	Pyrolysis – 200°C, 30 mins	Methylene Blue	Dose of adsorbent – 1.25 g/L Concentration of adsorbate – 200 mg/L pH – 3 - 4 Temperature – 5 to 25°C Equilibrium time – 24 hours	Surface area – 0.3276 m^2/g Pore volume – 10.0061 cm^3/g	241.9 mg/g	Zhu et al. 2018
2.	Municipal waste	MMDM biochar (MMDM consists of paper, cardboard, food scraps, and garden waste)	Pyrolysis – 300°C, 12 hours	Methylene Blue	Dose of adsorbent – 5 g/L Concentration of adsorbate – 100 mg/L pH – 5 Equilibrium time – 360 mins	NA	7.2 mg/g	Hoslett et al. 2020
3.	Rice husk	Rice husk biochar	Pyrolysis – 400°C to 600°C, 1 hour	Malachite Green	Dose of adsorbent – 2 g/L Concentration of adsorbate – 5 mg/L pH – 6.2 Equilibrium time – 2 hours	NA	99.8%	Ganguly et al. 2020
4.	Cactus	Activated biochar	Activation NaOH / 400°C, 1 hour	Malachite Green	Dose of adsorbent – 0.6 g/L Concentration of adsorbate – 100 mg/L pH – 6	32.07 m^2/g	1341 mg/g	Choudhary et al. 2020

					Temperature – 30°C Equilibrium time – 210 mins			
5.	Date palm	Date palm petioles biochar (DPP – biochar)	HCl / 700°C	Crystal violet	Dose of adsorbent – 1 g/L Concentration of adsorbate – 5 to 100 mg/L pH – 7 Temperature – 30°C Equilibrium time – 24 hours	640 m^2/g	209 mg/g	Chahinez et al. 2020
6.	Animal waste	Bovine Bones (BB) biochar	Pyrolysis – 800°C 60 mins	Basic red 9	Dose of adsorbent – 2.5 g/L Concentration of adsorbate – 50 mg/L pH – 2 to 9	90 m^2/g	49.5 mg/g	Cortes et al. 2019
		Fish scales (FS) biochar			Temperature – 25°C Equilibrium time – 2 hours		52.3 mg/g	
7.	Pine nut shells	Cetyl trimethyl ammonium bromide modified magnetic biochar (CTAB-MC)	$FeCl_3$, Cetyl trimethyl ammonium bromide (CTAB) Pyrolysis – 700°C	Acid chrome Blue	Dose of adsorbent – 0.4 g/L Concentration of adsorbate – 40 mg/L pH – 3 Temperature – 303 K Equilibrium time – 600 mins	NA	27.24 mg/g	Wang et al. 2020

and garden waste appeared to be heterogeneous, with different-sized holes (Hoslett et al. 2020).

Rice husk-derived biochar was reported by Ganguly et al. via pyrolysis at 400–600°C to remove malachite green from polluted water. Under optimum conditions at pH 6.2, a dose of adsorbent of 2 g/L, concentration of adsorbate as 5 mg/L, with a duration time of 2 hours. The pyrolysed biochar showed a 99.8% removal rate for dye from wastewater (Ganguly et al. 2020).

Choudhary et al. reported cladodes of *Opuntia ficus-indica* (OFI) cactus-derived activated biochar via pyrolysis at 400°C for 1 hour, followed by chemical activation with NaOH solution. The engineered biochar was subsequently used to eliminate malachite green from polluted water. The adsorption potential was exhibited as 1341 mg/g for the fabricated adsorbent under optimum conditions at pH 6, dose of adsorbent as 0.6 g/L, 100 mg/L concentration of adsorbate, with a duration time of 210 mins at 30°C temperature. BET studies demonstrated a 32.07 m^2/g surface area (Choudhary et al. 2020).

Furthermore, Chahinez et al. synthesised biochar derived from date palm petioles via direct pyrolysis at 700°C using an HCl solution. The adsorptive capability was shown as 209 mg/g for the fabricated adsorbent. Under optimum conditions at pH 7, dose of adsorbent was 1 g/L, and the concentration of adsorbate was 5–100 mg/L for 24 hours at 30°C temperature. BET studies demonstrated a 640 m^2/g surface area (Chahinez et al. 2020).

Cortes et al. used bovine bones (BB) and fish scales (FS) as precursors for the production of biochar via pyrolysis at 800°C for 60 mins to remove basic red 9 dye from aqueous solution. Under optimum conditions at pH 2-9, dose of adsorbent 2.5 g/L, concentration of adsorbate as 50 mg/L with duration of time as 2 hours at temperature 298K, synthesized fish scale biochar exhibited more adsorption capability at 52.3 mg/g than the bovine bones biochar at 49.5 mg/g. BET studies demonstrated a 90 m^2/g surface area (Cortes et al. 2019).

Furthermore, in 2020, Wang et al. fabricated pine nut shell-derived chemically modified magnetic biochar via pyrolysis at 700°C and modified it with cetyl trimethyl bromide and $FeCl_3$ to remove acid chrome blue from polluted water. The adsorptive capability was shown as 27.24 mg/g for the fabricated adsorbent. Under optimum conditions at a temperature of 303 K, the dose of adsorbent was 0.4 g/L, 40 mg/L as the initial concentration of adsorbate with a duration time of 10 hours at pH 3 (Wang et al. 2020).

Removal of Pharmaceuticals

Table 4 shows the performance assessment of different biochar for pharmaceuticals from wastewater.

In 2016, Shang et al. reported an *Astragalus mongholicus*-derived carbonised biochar via thermal activation at 800°C for 5 hours to remove ciprofloxacin from contaminated water. The adsorptive capability was exhibited as 40.11 mg/g for the fabricated adsorbent. Under optimum conditions at pH 3–10, the dose of adsorbent was 2 g/L, 200 mg/L as the initial concentration of adsorbate, with a duration time of 12 hours. BET studies demonstrated a 176.3 m^2/g surface area (Shang et al. 2016).

Table 4: Performance evaluation of different biochar sources to remove pharmaceuticals from wastewater

S. No.	Source of Biochar	Adsorbent	Chemical used/ Pyrolysis	Pharmaceuticals	Operating Conditions	Adsorbent properties	Batch Results	Reference
						Adsorbent Surface Area (m^2/g)	Maximum Capacity (q_{max}) (mg/g)/ % Removed	
1.	*Astra-galusmongholicus*	*Astra-galusmongholicus* biochar (AMBC)	Pyrolysis – 800°C	Cipro-floxacin (CIP)	Dose of adsorbent – 2 g/L Concentration of adsorbate – 200 mg/L pH – 3 to 10 Equilibrium time – 12 hours	176.3 m^2/g	40.11 mg/g	Shang et al. 2016
2.	Pine wood	Wood biochar $A_{1/5}C800$ (1/5 = air-gas ratio and 800 = pyrolysis temperature)	Pyrolysis – 600°C to 800°C	Tetra-cycline Sulfa-diazine	Dose of adsorbent – 0.5 g/L Adsorbate dosage – 6 to 48 mg/L pH – 5 Temperature – 298K Equilibrium time – 48 hours	738 m^2/g	163 mg/g 261 mg/g	Li et al. 2019
3.	Bull manure	Bull manure with sawdust (BM600)	Pyrolysis – 600°C	Linco-mycin	Dose of adsorbent – 0.5 g/L	237 m^2/g	20 microM	Liu et al. 2016

(Contd.)

(*Contd.*)

S. No.	Source of Biochar	Adsorbent	Chemical used/ Pyrolysis	Pharmaceuticals	Operating Conditions	Adsorbent properties	Batch Results	Reference
						Adsorbent Surface Area (m^2/g)	Maximum Capacity (q_{max}) (mg/g)/ % Removed	
		Daily manure with sawdust (M600)	Pyrolysis – 600°C		pH – 10 Equilibrium time – 168 hours	221 m^2/g		
		Poultry manure with sawdust (PM600)	Pyrolysis – 600°C			47 m^2/g		
		Anaerobically digested dairy manure (AM600)	Pyrolysis – 600°C			187 m^2/g		
4.	Rice straw	Alkali-acid modified magnetic biochar (MSABC)	Pyrolysis – 300°C	Tetra-cycline	Dose of adsorbent – 1.2 g/L Concentration of adsorbate – 10 to 200 mg/L pH – 8 to 9 Temperature – 25°C Equilibrium time – 120 mins	140.78 m^2/g	98.33 mg/g	Dai et al. 2020

5.	Sludge	Sludge biochar and carbon nanotube composite (CNT-SBC)	Modified by carbon nanotube (CNT)	Sulfa-methoxa-zole (SMX)	Dose of adsorbent – 0.5 g/L Equilibrium time – 120 mins	119 m^2/g	$2.35*10^4$ µg/g 83.51%	Ma et al. 2020
6.	Rape stalk	Rape stalk biochar (RB300)	H_2O_2 modified Pyrolysis – 300°C	Tetracycline	Dose of adsorbent – 0.05 g/L pH – 9 Equilibrium time – 48 hours	3.87 m^2/g	35.9 mg/g	Tan et al. 2019
7.	Hair waste	Hair waste biochar	KOH and K_2CO_3 / Pyrolysis – 450°C	Amoxicillin (AMOX) Diclofenac (DFC)	Dose of adsorbent – 200 mg/L Equilibrium time – 30 mins	430.61 m^2/g	90% 80%	Rodriguez et al. 2020

Li et al. synthesised pine wood biochar (A1/5C800) via pyrolysis in an inert gas environment for 3 hours using an air:gas ratio of 1:5. The adsorptive capability at pH 5 showed 163 mg/g for tetracycline and 261 mg/g for sulfadiazine, respectively. The dose of adsorbent was 0.5 g/L, and the concentration of adsorbate was 6–48 mg/L, with a duration of 48 hours at 25°C. BET studies demonstrated a 738 m^2/g surface area (Li et al. 2019).

Liu et al. fabricated biochar from bull manure (bull manure with sawdust BM600, dairy manure with rice hulls DM600, poultry manure with sawdust PM600, and anaerobically digested dairy manure AM600) via pyrolysis at 600°C to remove lincomycin from polluted water. The adsorptive capability was reported as 20 μM for the fabricated adsorbent. Under optimum conditions at pH 10, the adsorbent dose was 500 mg/L with a duration of 168 hours. BET studies revealed a higher surface area of bull manure with sawdust (BM600 at 237 m^2/g) than that of the other three adsorbents (DM600 at 221 m^2/g, PM600 at 47 m^2/g, and AM600 at 187 m^2/g) (Liu et al. 2016).

Furthermore, the study reported by Dai et al. used rice straw-derived alkali-acid modified magnetic biochar, fabricated via heating at 300°C and modified with acid, to remove tetracycline from polluted water. The adsorptive capability was reported as 98.3 mg/g for the fabricated adsorbent. Under optimum conditions at a temperature of 25°C, the dose of adsorbent was 1.2 g/L, with an adsorbate concentration of 10–200 mg/L and a duration time of 1 hour. BET studies demonstrated a 140.78 m^2/g surface area. SEM images revealed many pores on the surface of the modified biochar (Dai et al. 2020).

A study was cited for synthesising sludge biochar with a carbon nanotube composite (CNT-SBC) to remove sulfamethoxazole (SMX) from a contaminated aqueous solution. Under optimum conditions, the dose of adsorbent was 500 mg/L, and the equilibrium time was 120 mins. The maximum adsorptive capacity was 2.35×10^4 μg/g. BET studies demonstrated a 119 m^2/g surface area for CNT-SBC. Furthermore, the removal rate was observed as 83.51% (Ma et al. 2020).

Tan et al., in 2019, used H_2O_2 for the modification of rape stalk, with pyrolysis at 300°C, and synthesised rape stalk biochar (RB300) to remove the pharmaceutical (tetracycline) from polluted water. The adsorptive capability of the RB300 at pH 9 was reported as 35.9 mg/g with a duration of 48 hours, and the dose of adsorbent was 50 mg/L. BET studies demonstrated a 3.87 m^2/g surface area (Tan et al. 2019).

A study was cited for the fabrication of hair waste biochar via chemical activation with KOH and K_2SO_3 and pyrolysis at 450°C to remove amoxicillin (AMOX) and diclofenac (DFC) from a contaminated aqueous solution. Under optimum conditions, the dose of adsorbent was 200 mg/L with a duration time of half an hour. BET studies demonstrated a 430.61 m^2/g surface area. Removal rates were observed at 90% and 80% for AMOX and DFC, respectively (Rodriguez et al. 2020).

Conclusion and Future Perspectives

Looking towards the future, numerous exciting prospects emerge. More research is widely needed to fully understand the technical and economic aspects of biochar.

Studies have already been conducted on the effectiveness of biochar for treating wastewater, but greater consideration towards the desorption process for the removal of contaminants is required. The reusability of biochar also needs to be addressed for wastewater remediation.

This book chapter summarises the sources of biochar, the fabrication of biochar, the physicochemical characteristics of biochar, modified or engineered biochar, and the application of biochar, providing a sustainable and environmentally responsible answer to the problems posed by agricultural waste. The chapter has highlighted the significant impact that different activation techniques have on the structural and performance features of biochar. The potency of biochar for removing different pollutants is discussed in this book chapter. Forthcoming scientists will benefit from the direction of this book chapter and gain an excellent understanding of sustainable technologies. Therefore, this book chapter addresses the promising role of biochar as an eco-friendly, green-synthesised material in the field of aqueous toxin elimination.

References

Achieng, G.O. and Shikuku, V.O. 2020. Adsorption of copper ions from water onto fish scales derived biochar: Isothermal perspectives. Journal of Materials and Environmental Science, 11(11): 1816-1827.

Angın, D., Altintig, E. and Köse, T.E. 2013. Influence of process parameters on the surface and chemical properties of activated carbon obtained from biochar by chemical activation. Bioresource Technology, 148: 542-549.

Barakat, M.A. 2011. New trends in removing heavy metals from industrial wastewater. Arabian Journal of Chemistry, 4(4): 361-377.

Bulut, Y. and Aydın, H. 2006. A kinetics and thermodynamics study of methylene blue adsorption on wheat shells. Desalination, 194(1-3): 259-267.

Cha, J.S., Park, S.H., Jung, S.C., Ryu, C., Jeon, J.K., Shin, M.C. and Park, Y.K. 2016. Production and utilization of biochar: A review. Journal of Industrial and Engineering Chemistry, 40: 1-15.

Chahinez, H.O., Abdelkader, O., Leila, Y. and Tran, H.N. 2020. One-stage preparation of palm petiole-derived biochar: Characterization and application for adsorption of crystal violet dye in water. Environmental Technology & Innovation, 19, 100872.

Chen, X., Chen, G., Chen, L., Chen, Y., Lehmann, J., McBride, M.B. and Hay, A.G. 2011. Adsorption of copper and zinc by biochars produced from pyrolysis of hardwood and corn straw in aqueous solution. Bioresource Technology, 102(19): 8877-8884.

Chen, Z.L., Zhang, J.Q., Huang, L., Yuan, Z.H., Li, Z.J. and Liu, M.C. 2019. Removal of Cd and Pb with biochar made from dairy manure at low temperature. Journal of Integrative Agriculture, 18(1): 201-210.

Cho, D.W., Yoon, K., Kwon, E.E., Biswas, J.K. and Song, H. 2017. Fabrication of magnetic biochar as a treatment medium for As (V) via pyrolysis of $FeCl_3$-pretreated spent coffee ground. Environmental Pollution, 229: 942-949.

Choudhary, M., Kumar, R. and Neogi, S. 2020. Activated biochar derived from Opuntia ficus-indica for the efficient adsorption of malachite green dye, Cu^{+2} and Ni^{+2} from water. Journal of Hazardous Materials, 392: 122441.

Côrtes, L.N., Druzian, S.P., Streit, A.F.M., Godinho, M., Perondi, D., Collazzo, G.C. et al. 2019. Biochars from animal wastes as alternative materials to treat colored effluents containing basic red 9. Journal of Environmental Chemical Engineering, 7(6): 103446.

Crini, G., Lichtfouse, E., Wilson, L.D. and Morin-Crini, N. 2019. Conventional and non-conventional adsorbents for wastewater treatment. Environmental Chemistry Letters, 17: 195-213.

Dai, J., Meng, X., Zhang, Y. and Huang, Y. 2020. Effects of modification and magnetization of rice straw derived biochar on adsorption of tetracycline from water. Bioresource Technology, 311: 123455.

Dalahmeh, S., Ahrens, L., Gros, M., Wiberg, K. and Pell, M. 2018. Potential of biochar filters for onsite sewage treatment: Adsorption and biological degradation of pharmaceuticals in laboratory filters with active, inactive and no biofilm. Science of the Total Environment, 612: 192-201.

Danish, M.S.S., Estrella, L.L., Alemaida, I.M.A., Lisin, A., Moiseev, N., Ahmadi, M. et al. 2021. Photocatalytic applications of metal oxides for sustainable environmental remediation. Metals, 11(1): 80.

Darban, Z., Shahabuddin, S., Gaur, R., Ahmad, I. and Sridewi, N. 2022. Hydrogel-based adsorbent material for the effective removal of heavy metals from wastewater: A comprehensive review. Gels, 8(5): 263.

Enders, A., Hanley, K., Whitman, T., Joseph, S. and Lehmann, J. 2012. Characterization of biochars to evaluate recalcitrance and agronomic performance. Bioresource Technology, 114, 644-653.

Escudero, L.B., Quintas, P.Y., Wuilloud, R.G. and Dotto, G.L. 2019. Recent advances on elemental biosorption. Environmental Chemistry Letters, 17: 409-427.

Frankel, M.L., Bhuiyan, T.I., Veksha, A., Demeter, M.A., Layzell, D.B., Helleur, R.J. et al. 2016. Removal and biodegradation of naphthenic acids by biochar and attached environmental biofilms in the presence of co-contaminating metals. Bioresource Technology, 216: 352-361.

Ganguly, P., Sarkhel, R. and Das, P. 2020. Synthesis of pyrolyzed biochar and its application for dye removal: Batch, kinetic and isotherm with linear and non-linear mathematical analysis. Surfaces and Interfaces, 20: 100616.

Georgieva, V.G., Gonsalvesh, L. and Tavlieva, M.P. 2020. Thermodynamics and kinetics of the removal of nickel (II) ions from aqueous solutions by biochar adsorbent made from agro-waste walnut shells. Journal of Molecular Liquids, 312: 112788.

Hoslett, J., Ghazal, H., Mohamad, N. and Jouhara, H. 2020. Removal of methylene blue from aqueous solutions by biochar prepared from the pyrolysis of mixed municipal discarded material. Science of the Total Environment, 714: 136832.

Jin, J., Li, S., Peng, X., Liu, W., Zhang, C., Yang, Y. et al. 2018. HNO_3 modified biochars for uranium (VI) removal from aqueous solution. Bioresource Technology, 256: 247-253.

Jindo, K., Mizumoto, H., Sawada, Y., Sanchez-Monedero, M.A. and Sonoki, T. 2014. Physical and chemical characterization of biochars derived from different agricultural residues. Biogeosciences, 11(23): 6613-6621.

Lee, J.W., Kidder, M., Evans, B.R., Paik, S., Buchanan Iii, A.C., Garten, C.T. and Brown, R.C. 2010. Characterization of biochars produced from cornstovers for soil amendment. Environmental Science & Technology, 44(20): 7970-7974.

Li, C., Zhu, X., He, H., Fang, Y., Dong, H., Lü, J. et al. 2019. Adsorption of two antibiotics on biochar prepared in air-containing atmosphere: Influence of biochar porosity and molecular size of antibiotics. Journal of Molecular Liquids, 274: 353-361.

Liu, C.H., Chuang, Y.H., Li, H., Teppen, B.J., Boyd, S.A., Gonzalez, J.M. et al. 2016. Sorption of lincomycin by manure-derived biochars from water. Journal of Environmental Quality, 45(2): 519-527.

Ma, Y., Yang, L., Wu, L., Li, P., Qi, X., He, L. et al. 2020. Carbon nanotube supported sludge biochar as an efficient adsorbent for low concentrations of sulfamethoxazole removal. Science of The Total Environment, 718: 137299.

Mo, J., Yang, Q., Zhang, N., Zhang, W., Zheng, Y. and Zhang, Z. 2018. A review on agro-industrial waste (AIW) derived adsorbents for water and wastewater treatment. Journal of Environmental Management, 227: 395-405.

Mukome, F.N., Zhang, X., Silva, L.C., Six, J. and Parikh, S.J. 2013. Use of chemical and physical characteristics to investigate trends in biochar feedstocks. Journal of Agricultural and Food Chemistry, 61(9): 2196-2204.

Novak, J.M., Lima, I., Xing, B., Gaskin, J.W., Steiner, C., Das, K.C. et al. 2009. Characterization of designer biochar produced at different temperatures and their effects on a loamy sand. Annals of Environmental Science, 3: 195-206.

Nworie, F.S., Oroke, E.C., Ikelle, I.I. and Nworu, J.S. 2020. Equilibrium and kinetic studies for the adsorptive removal of lead (II) ions from aqueous solution using activated plantain peel biochar. Acta Chemica Malaysia, 4(1): 9-16.

Pawar, A., Panwar, N.L. and Salvi, B.L. 2020. Comprehensive review on pyrolytic oil production, upgrading and its utilization. Journal of Material Cycles and Waste Management, 22(6): 1712-1722.

Prabhahar, R.S.S., Nagaraj, P. and Jeyasubramanian, K. 2019. Enhanced recovery of H_2 gas from rice husk and its char enabled with nano catalytic pyrolysis/gasification. Microchemical Journal, 146: 922-930.

Prabhahar, R.S.S., Nagaraj, P. and Jeyasubramanian, K. 2020. Promotion of bio oil, H_2 gas from the pyrolysis of rice husk assisted with nano silver catalyst and utilization of bio oil blend in CI engine. International Journal of Hydrogen Energy, 45(33): 16355-16371.

Qin, Y., Zhu, X., Su, Q., Anumah, A., Gao, B., Lyu, W. et al. 2020. Enhanced removal of ammonium from water by ball-milled biochar. Environmental Geochemistry and Health, 42: 1579-1587.

Rajapaksha, A.U., Vithanage, M., Ahmad, M., Seo, D.C., Cho, J.S., Lee, S. E. et al. 2015. Enhanced sulfamethazine removal by steam-activated invasive plant-derived biochar. Journal of Hazardous Materials, 290: 43-50.

Rashed, M.N. 2013. Adsorption technique for the removal of organic pollutants from water and wastewater. Organic Pollutants-monitoring, Risk and Treatment, 7: 167-194.

Rashidi, N.A. and Yusup, S. 2020. A mini review of biochar synthesis, characterization, and related standardization and legislation. Applications of Biochar for Environmental Safety, 11.

Read, D.J., Leake, J.R. and Perez-Moreno, J. 2004. Mycorrhizal fungi as drivers of ecosystem processes in heathland and boreal forest biomes. Canadian Journal of Botany, 82(8): 1243-1263.

Rodríguez, F., Montoya-Ruiz, C., Estiati, I. and Saldarriaga, J.F. 2020. Removal of drugs in polluted waters with char obtained by pyrolysis of hair waste from the tannery process. ACS Omega, 5(38): 24389-24402.

Shang, J.G., Kong, X.R., He, L.L., Li, W.H. and Liao, Q.J.H. 2016. Low-cost biochar derived from herbal residue: Characterization and application for ciprofloxacin adsorption. International Journal of Environmental Science and Technology, 13: 2449-2458.

Singh, P., Sarswat, A., Pittman Jr, C.U., Mlsna, T. and Mohan, D. 2020. Sustainable low-concentration arsenite [As (III)] removal in single and multicomponent systems using hybrid iron oxide–biochar nanocomposite adsorbents—A mechanistic study. ACS Omega, 5(6): 2575-2593.

Sodha, V., Shahabuddin, S., Gaur, R., Ahmad, I., Bandyopadhyay, R. and Sridewi, N. 2022. Comprehensive review on zeolite-based nanocomposites for treatment of effluents from wastewater. Nanomaterials, 12(18): 3199.

Tan, Z., Zhang, X., Wang, L., Gao, B., Luo, J., Fang, R. et al. 2019. Sorption of tetracycline on H_2O_2-modified biochar derived from rape stalk. Environmental Pollutants and Bioavailability, 31(1): 198-207.

Torres-Pérez, J., Soria-Serna, L.A., Solache-Ríos, M. and McKay, G. 2015. One step carbonization/activation process for carbonaceous material preparation from pecan shells for tartrazine removal and regeneration after saturation. Adsorption Science & Technology, 33(10): 895-913.

Tsang, D.C., Zhou, F., Zhang, W. and Qiu, R. 2016. Stabilization of cationic and anionic metal species in contaminated soils using sludge-derived biochar. Chemosphere, 149: 263-271.

Wang, H., Wang, S. and Gao, Y. 2020. Cetyl trimethyl ammonium bromide modified magnetic biochar from pine nut shells for efficient removal of acid chrome blue K. Bioresource Technology, 312: 123564.

Wang, Y. and Liu, R. 2018. H_2O_2 treatment enhanced the heavy metals removal by manure biochar in aqueous solutions. Science of The Total Environment, 628: 1139-1148.

Yan, M., Zhang, S., Wibowo, H., Grisdanurak, N., Cai, Y., Zhou, X. et al. 2020. Biochar and pyrolytic gas properties from pyrolysis of simulated municipal solid waste (SMSW) under pyrolytic gas atmosphere. Waste Disposal & Sustainable Energy, 2: 37-46.

Zhu, Y., Yi, B., Yuan, Q., Wu, Y., Wang, M. and Yan, S. 2018. Removal of methylene blue from aqueous solution by cattle manure-derived low temperature biochar. RSC Advances, 8(36): 19917-19929.

CHAPTER
5

Role of Constructed Wetlands in Wastewater Treatment and Mitigation of Emerging Contaminants

Sweta Upadhyay[1], Pradeep Kumar Sharma[1*], Rachan Karmakar[1], Vijay Tripathi[2], Varij Panwar[3] and Amit Mittal[4]

[1] Department of Environmental Science, Graphic Era Deemed to be University, Dehradun, Uttarakhand

[2] Department of Molecular and Cellular Engineering, Jacob Institute of Biotechnology and Bioengineering, Sam Higginbottom University of Agriculture Technology and Sciences, Prayagraj, Uttar Pradesh

[3] Department of Electronics and Communication Engineering, Graphic Era Deemed to be University, Dehradun, Uttarakhand

[4] Department of Environmental Science, Graphic Era Hill University, Bhimtal Campus, Uttarakhand

Introduction

Wastewater treatment is a critical environmental management issue worldwide. It involves removing contaminants from water discharged by households, industries, various types of institutions, and agriculture before it re-enters the natural water cycle. The depletion of freshwater resources and insufficient water treatment infrastructure in low-income Global South countries, including India, has led to challenges in providing clean water and sanitation, with only 37% of sewage treated and high costs associated with centralized systems (Ravichandran et al. 2023). Wastewater from domestic, industrial, agricultural, hospital, and municipal establishments has been treated with the help of traditional wastewater treatment methods on a primary basis before discharging into the environment. These methods usually consist of primary treatment, in which processes like sedimentation are used to settle out solids; secondary treatment, where biological processes such as activated sludge and biofilm reactors are used to degrade and eliminate the pollutants; and tertiary treatment, where advanced techniques like filtration and chemical disinfection are used to further treat

*Corresponding author: pksharma@geu.ac.in

DOI: 10.1201/9781003527350-5

the water. Traditional treatment methods, such as activated sludge processes and chemical treatments, have proven effective at removing many pollutants. However, they often require high energy consumption, specialized infrastructure, and extensive maintenance, making them less suitable for rural or low-resource settings. The conventional techniques are not suitable for remote regions and areas with lower populations due to their requirement of area, cost, skilled labor, and maintenance. Since conventional wastewater treatment techniques cannot be fully operational in such situations, the need for alternative treatment systems is required—ones that are efficient yet economical, easy to operate, and environmentally friendly. Furthermore, these conventional techniques may not be fully effective in removing "emerging contaminants," the recently introduced pollutants that pose new or increasingly recognized risks. The growing concern over emerging contaminants has prompted the search for innovative and sustainable treatment technologies that can supplement or replace traditional methods.

Researchers (Vymazal et al. 2021) have defined "emerging pollutants" as those contaminants that, although not yet subject to regulations requiring public reporting or monitoring of their presence in water sources or wastewater discharges, have the potential to enter the environment and have unfavorable ecological and/or human health effects. Emerging contaminants include a wide range of substances such as pharmaceuticals, personal care products, pesticides, microplastics, endocrine-disrupting compounds (EDCs), and industrial chemicals. Since their detection and high concentration in aquatic environments offer a rising hazard to aquatic organisms and human health, EDCs and antibiotics have drawn the most attention among them (Christofilopoulos et al. 2019). These contaminants are found in wastewater at trace levels but can have significant adverse effects on ecosystems and human health even at low concentrations. For instance, pharmaceutical residues can disrupt aquatic life, while endocrine-disrupting compounds can interfere with hormone systems in animals and humans (Ahmed et al. 2017, Chen et al. 2022). Additionally, the evolving global trend of wastewater recycling and reuse has brought attention to the need for sustainable treatment systems that can eliminate a variety of contaminants, including ECs, from biomedical waste and hospital wastewater (Parida et al. 2022). Hospital wastewater released from hospitals and healthcare institutions is frequently contaminated with medications, disinfectants, and other toxic substances (Parashar et al. 2022).

Conventional treatment processes are not always sufficient to remove the full spectrum of ECs, making it necessary to explore hybrid systems such as membrane bioreactors (MBRs) and constructed wetlands integrated with advanced oxidation processes that combine biological, chemical, and physical treatment processes.

Constructed wetlands (CWs) offer a promising solution to the challenges associated with conventional wastewater treatment. These engineered systems mimic the natural processes occurring in natural wetlands to treat contaminated water through physical, chemical, and biological mechanisms. Constructed wetlands are designed to remove a range of pollutants, including nutrients, heavy metals, pathogens, and organic matter. With the capacity to address emerging contaminants, CWs have gained recognition as a sustainable and cost-effective wastewater treatment technology. Recent studies have demonstrated the potential of CWs to mitigate the impacts of industrial wastewater, as they are able to filter and degrade pollutants through complex biogeochemical processes (Overton et al. 2023, Chaturvedi et al. 2021). A variety of organic contaminants, such as pesticides (Boucher-Carrier et al. 2022), heavy metals (Hu et al. 2019), and pharmaceuticals (Youssef et al. 2023), have been effectively removed by constructed wetlands under different operational conditions. Thus, the potential of CWs to treat wastewater and their cost-effectiveness make them a promising alternative for the sustainable management of wastewater in areas with limited access to centralized treatment facilities. Both natural and constructed wetlands, as complex biogeochemical systems, utilize mechanisms such as sorption, photodegradation, microbial biodegradation, and phytoremediation to remove these contaminants, with effectiveness influenced by contaminant properties and wetland design (Overton et al. 2023).

Compared to traditional treatment facilities, constructed wetlands require less energy, minimal chemical inputs, and lower operational costs. Additionally, their aesthetic and ecological benefits, such as habitat creation and landscape enhancement, make them an attractive option for urban and rural settings alike. The integration of natural processes with engineering design allows for adaptable solutions that can be customized for specific wastewater treatment needs, including domestic, industrial, hospital, and agricultural effluents. Even though the benefits of CWs in wastewater treatment are widely known, adapting these systems for large-scale applications is still challenging (Vymazal et al. 2021). Their efficiency in some conditions has been challenged because of factors such as climate variability, the requirement for land space, and seasonal variations in pollutant loads, etc. (Tan et al. 2020, Misiti et al. 2011). Research is being conducted to improve CW designs and integrate them with supplementary treatment methods to increase treatment efficiency and reliability in order to overcome these issues. Countries like China are focusing on CW systems and providing funds, and developing countries like India should implement similar steps (Ji and Zhao 2023).

The objective of this chapter is to provide a comprehensive overview of the role of constructed wetlands in the treatment of wastewater, with a particular focus on mitigating emerging contaminants. This includes exploring the mechanisms of contaminant removal in CWs, the design and operational factors that influence treatment efficiency, and real-world applications across different settings. The chapter will also review case studies on the removal of specific emerging contaminants and highlight innovations that enhance the performance of constructed wetlands. This chapter highlights the performance, removal mechanisms, and successful practical applications of CWs, emphasizing their potential for addressing water security and environmental protection in resource-constrained regions. This chapter aims to bridge the gap between traditional wastewater treatment approaches and emerging sustainable practices. By examining both the strengths and limitations of constructed wetlands, this chapter seeks to inform the development of more effective wastewater management strategies. It will also address the current policy and regulatory landscape, outlining challenges and future directions for integrating natural treatment systems in water management frameworks.

Emerging Contaminants and their Impact on Environment

Emerging contaminants (ECs) are a broad category of anthropogenic substances that have drawn greater attention recently because of their environmental durability and potential threats to human health and aquatic ecosystems. Pharmaceuticals, personal care items, endocrine-disrupting chemicals, pesticides, industrial chemicals, nanomaterials, microplastics, and antibiotic resistance genes (ARGs) are some examples of these pollutants, which come from a range of man-made and natural sources. Each kind of EC poses different challenges to the environment and requires distinct mitigating strategies. Endocrine-disrupting pollutants (EDPs), including pharmaceuticals, personal care products, and surfactants, are exogenous chemicals found in municipal wastewater at concentrations of 1 ng/L to 500 µg/L, causing health issues like neurotoxicity and reproductive dysfunction (Ghosh and Biswas 2023). Pharmaceuticals are among the most researched ECs, including hormones, analgesics, antibiotics, and antiepileptics, all playing a major role in the environment. These chemicals find their way into the environment through agricultural runoff from animal farms, incorrect disposal of unwanted pharmaceuticals, and household wastewater. Antibiotics that persist in aquatic environments, such as ciprofloxacin and tetracycline, for instance, affect microbial diversity and promote the transmission of antibiotic resistance genes (ARGs) (Nõlvak et al. 2013). Commonly found in contraceptives, hormonal medications like ethinylestradiol can interfere with aquatic species' endocrine systems, feminizing male fish and lowering fertility (Hamilton et al. 2022, Nasri et al. 2021).

The most commonly found ECs in the environment are personal care products (PCPs). PCPs consist of substances including preservatives, perfumes, and ultraviolet (UV) filters that are frequently found in sunscreen, soaps, and cosmetics. For instance, UV filters used in sunscreens such as oxybenzone and octinoxate are associated with coral bleaching and bioaccumulation in marine creatures (Miller et al. 2021, de Miranda et al. 2021). These substances can linger in aquatic environments and are frequently resistant to standard wastewater treatment techniques. It has been demonstrated that the antibiotic triclosan used in toothpaste and soaps disturbs microbial ecosystems and is harmful to algae and invertebrates (Noriega-Rico et al. 2021).

Endocrine disrupting chemicals (EDCs) are another kind of emerging contaminants posing threat to the environment and human health. EDCs such as Bisphenol A (BPA) and phthalates are examples of EDCs that have been shown to disrupt human and animal hormone signaling pathways (Akanbi et al. 2025). Industrial and consumer products such as plastics, detergents, and packaging materials, frequently leak these substances into the environment. In fish and amphibians, exposure to EDCs in aquatic environments has been associated with behavioral changes, reproductive dysfunction, and developmental abnormalities. Because of their ability to act at extremely low concentrations, EDCs are especially concerning. Industrial chemicals such as polychlorinated biphenyls (PCBs), dioxins, and PFAS are highly persistent and pose significant challenges due to their longevity and bioaccumulative nature. PFAS, also referred to as "forever chemicals," are frequently found in waterproof textiles, non-stick cookware, and firefighting foams (Zhao et al. 2025). They have been found in drinking water and are linked to thyroid disturbance, liver damage, and an elevated risk of cancer (Savvidou et al. 2024). For efficient removal, these compounds frequently need advanced treatment techniques such as adsorption and enhanced oxidation.

Another category of ECs includes microplastics and nanomaterials, which are of growing concern in the current scenario. The degradation of various plastic-based materials gives rise to microplastics (MPs) in the environment. Microplastics are fragments smaller than 5 mm in size. They have emerged as a significant environmental concern, having been found in almost every ecosystem on Earth as well as in atmospheric fallout (Upadhyay et al. 2024). Microplastics, present in freshwater, marine, and terrestrial habitats, are consumed by species of all trophic levels and can cause chemical, physical, and biological damage (R.R. Kumar et al. 2021). They serve as vectors for other ECs, including persistent organic pollutants (POPs) and heavy metals, aggravating their environmental impact (Hildebrandt et al. 2021). Similar to MPs, nanoparticles are also very small in size and capable of posing threats to the environment. The usage of nanoparticles, such as titanium dioxide, in consumer products, technology, and the medical field is growing. They are reactive and able to pass through biological membranes due to their large surface area and nanoscale size, which causes cytotoxicity in aquatic creatures (Sarma et al. 2022). Furthermore, nanomaterials may change the toxicity and environmental fate of other pollutants through their interactions. Studies indicate that nanomaterials can interfere with wastewater treatment processes by inhibiting microbial activity (López et al. 2022).

Pesticides are another type of EC that includes herbicides, insecticides, and fungicides, which can leach into water bodies through runoff or atmospheric deposition. DDT and other persistent organochlorine pesticides have long-term impacts on living beings, such as endocrine disruption and bioaccumulation in food chains (Campanale et al. 2020). Despite their rapid-breakdown design, modern pesticides frequently produce toxic metabolites that can endanger non-target organisms. For example, neonicotinoids have been linked to toxicity in aquatic invertebrates and pollinator decline (Morrissey et al. 2015). Microplastics are known to directly or indirectly affect the growth and overall health of aquatic organisms. Exposure to the highest concentration of 50 nm carboxyl-coated polystyrene spheres (10,000 particles per spermatozoon) reduced sperm motility, delayed offspring hatching, and resulted in smaller body size and impaired swimming ability in European whitefish (*Coregonus lavaretus*) (Yaripour et al. 2021). These findings suggest that microplastic/nanoplastic pollution can decrease male fertilization potential and have significant transgenerational impacts on offspring phenotype and performance, with potential ecological and evolutionary consequences.

They are often eliminated from wastewater treatment systems and released into the environment. As a result, ECs have been detected in a wide range of water bodies, including rivers, lakes, and groundwater, raising concerns about their long-term impacts on both environmental and human health. The chemical stability, low biodegradability, and intricate molecular structures of ECs are major factors responsible for their persistence in aquatic settings (Dhangar and Kumar 2020,

Table 1: Types of ECs, their sources and impacts on the environment

S. No	Emerging Contaminant	Sources	Impacts	References
1.	Pharmaceuticals (e.g. drugs, antidepressants)	Human excretion, wastewater effluents, hospital discharges	Disruption of endocrine systems in aquatic organisms, resistance in bacteria, accumulation in the food chain, impact on biodiversity, behavioral changes in fish	(Song et al. 2018, Zhou et al. 2022, Hazra et al. 2022, Rampuria et al. 2021)
2.	Antibiotics (e.g. Tetracycline, Ciprofloxacin, Penicillin)	Agriculture, aquaculture, wastewater, hospital effluents	Antibiotic resistance, changes in microbial communities, potential transfer of resistance genes to humans via the food chain, endocrine disruption	(Du et al. 2020)
3.	Personal Care Products (PCPs) (e.g. Triclosan, Parabens)	Household waste, bathing, showering, cosmetics, and personal care products	Hormonal disruptions (endocrine disruption), reproductive toxicity in aquatic organisms, bioaccumulation, developmental toxicity	(Ahmed et al. 2017, Dey Chowdhury et al. 2022, Talib and Randhir 2016)
4.	Microplastics (e.g. Polyurethane, Polystyrene, Polyethylene)	Urban runoff, plastic waste, breakdown of larger plastic debris	Physical harm to aquatic organisms, ingestion by marine life, disruption of ecosystems, chemical leaching, potential entry into the food chain	(Yuan et al. 2022, Issac and Kandasubramanian 2021)
5.	Surfactants (e.g. Nonylphenol, Alkylphenols)	Household products, detergents, industrial effluents	Toxicity to aquatic life, developmental effects in aquatic organisms, endocrine disruption, persistence in water bodies	(Ghosh and Biswas 2023, Oliveira et al. 2020, Gross et al. 2004)
6.	Pesticides (e.g. Atrazine, Glyphosate, Malathion)	Agricultural runoff, irrigation systems, industrial discharges, leaching from soils	Acute and chronic toxicity in aquatic species, disruption of reproductive cycles, biodiversity loss, accumulation in sediments	(Morrissey et al. 2015, Boucher-Carrier et al. 2022, Mukherjee et al. 2022)

7.	Heavy Metals (e.g. Lead, Mercury, Arsenic)	Industrial effluents, mining activities, atmospheric deposition, wastewater discharges	Neurotoxicity, developmental and behavioral disorders in aquatic species, bioaccumulation through the food web, impacts on ecosystems, potential carcinogens in humans	(R. Kumar et al. 2021, Sonam et al. 2021, Premarathna et al. 2023)
8.	Endocrine Disrupting Chemicals (EDCs) (e.g. Bisphenol A, Phthalates)	Industrial processes, plastic products, landfill leaching, packaging materials	Altered reproductive health, behavioral changes, reduced fertility, carcinogenic effects, developmental issues in humans and wildlife	(Chen et al. 2022, Ahmed et al. 2017, Tesfamariam et al. 2022)
9.	Plasticizers (e.g. DEHP, DBP)	Plastics production, household products, food packaging, personal care items	Hormonal disruption, developmental and reproductive toxicity, carcinogenic potential, environmental persistence, impacts on biodiversity	(Sol et al. 2020, Nas et al. 2022, Matamoros et al. 2016)
10.	Fluorinated Chemicals (PFAS)	Industrial production, firefighting foams, food packaging, consumer products	Persistence in the environment, carcinogenic effects, liver and kidney toxicity, immunotoxicity, bioaccumulation in the food chain	(Zhao et al. 2025, Dey Chowdhury et al. 2022, Wang et al. 2024, Savvidou et al. 2024)
11.	Nanomaterials (e.g. Silver nanoparticles, Titanium dioxide)	Consumer goods (cosmetics, sunscreens), wastewater, industrial waste	Ecotoxicity, disruption of cellular functions in aquatic organisms, risk of bioaccumulation in food chains, long-term environmental persistence	(Sarma et al. 2022, Anawar et al. 2019)

Chaturvedi et al. 2021). For instance, pharmaceuticals like antibiotics, analgesics, and antidepressants are commonly found in wastewater effluents, frequently at concentrations that exceed their original background levels (Andrade et al. 2024, Chaves-Barquero et al. 2018). The impact of ECs extends beyond environmental concerns. Due to their persistence in surface water and wastewater, ECs have the potential to deteriorate drinking water supplies, increasing the risk of human exposure through ingestion of contaminated water or aquatic sources of food. These substances can disrupt the endocrine systems of aquatic organisms, leading to altered reproductive and growth characteristics (Talib and Randhir 2016, Chen et al. 2022). According to previous studies, personal care items such as sunscreen chemicals, detergents, and fragrances have also been found to be important sources of ECs in water bodies (de Miranda et al. 2021, Matamoros et al. 2016). These compounds have the ability to accumulate in the tissues of aquatic organisms, potentially entering the food chain and affecting biodiversity (Jena et al. 2023, Noriega-Rico et al. 2021).

The emergence of antibiotic-resistant bacteria, which is becoming a major global public health concern, has also been associated with the presence of ECs in water bodies (García et al. 2020). Antibiotics released from hospitals and wastewater treatment plants can contribute to the development of antibiotic-resistant microorganisms in the environment, posing significant challenges for the treatment of infectious diseases. Similarly, antibiotic-resistant genes (ARGs) represent a unique category of ECs linked to the misuse and overuse of antibiotics. COVID-19 has also caused a spike in the concentration of ECs in various compartments of the environment. Significant increases in medicines, insecticides, and stimulants during the pandemic compared to pre-pandemic levels demonstrate the effect of the COVID-19 pandemic on emerging contaminant (EC) concentrations in sewage. Surfactants, primarily introduced into graywater from soaps, detergents, and personal care products, have increased significantly due to heightened cleaning during the COVID-19 pandemic, posing environmental toxicity and treatment challenges (Jena et al. 2023). The pandemic intensified ecological risks for aquatic organisms by 40%, with substances like clarithromycin and sulfamethoxazole shifting to high-risk categories and caffeine showing particularly severe ecological impacts ($RQ > 2500$) (Andrade et al. 2024). These genetic elements tend to exist in bacteria that are resistant to antibiotics, can spread throughout environmental microbial populations, and increase resistance. ARGs are commonly found in wastewater treatment plant discharges, agricultural runoff, and hospital effluents. Since they reduce the efficiency of antibiotics, their proliferation in aquatic systems presents a serious risk to the health of both humans and animals.

The survival and dispersion of resistant bacterial strains and antibiotic-resistant genes (ARGs) are facilitated by the selective action of ECs, especially antibiotics (Helt 2012). ARGs may be transferred among microbial populations as a result of selective pressure exerted by antibiotics, even at trace levels. In addition to mitigating ECs from wastewater, constructed wetlands (CWs) may help in the removal of ARB and ARGs. The persistence of compounds like ciprofloxacin, sulfamethoxazole, and tetracyclines in treated wastewater has been directly linked to elevated ARG concentrations in natural water bodies, posing significant challenges to public health (Karungamye et al. 2023, Liu et al. 2019). Many of these compounds are associated with human health risks, including potential carcinogenic effects, hormone disruption, and neurological impairments (Saravanakumar et al. 2022). EDCs, such as bisphenol A (BPA) and phthalates, interfere with hormonal systems in aquatic organisms, leading to developmental and reproductive abnormalities (Fang et al. 2020). Many pesticides, such as insecticides and herbicides, are used extensively in agriculture and frequently leach into water systems, where they can exhibit acute and chronic toxicity to aquatic organisms (Chen et al. 2022). Industrial chemicals such as perfluoroalkyl and polyfluoroalkyl substances (PFAS) are highly persistent and have the ability to bioaccumulate in ecosystems, posing risks to aquatic and terrestrial food chains (Savvidou et al. 2024, Zhao et al. 2025). Because of their high reactivity and tiny size, nanomaterials like silver nanoparticles and titanium dioxide, which are being utilized more and more in consumer goods, contribute to ecotoxicity. These various ECs show the complex and multifaceted nature of contamination in modern ecosystems. They are capable of transferring to different trophic levels of the aquatic food chain and may reach the human body as well. The wide range of ECs demonstrates how intricately they interact with their surroundings and affect it. Understanding their sources, pathways, and mechanisms will help in developing targeted mitigation strategies for efficient treatment.

Treatment Methods for Emerging Contaminants in Wastewater

The persistence, chemical diversity, and range of environmental effects of emerging contaminants make their removal from wastewater difficult. Since traditional wastewater treatment facilities are unable to completely eliminate ECs, research is being carried out regarding better treatment techniques. The most common methods currently used for the removal of ECs are physical, biological, chemical, and hybrid systems. Advanced treatment technologies, such as aerated reactors with biomembranes and planted wetlands, achieve over 90% removal of certain ECs, offering cost-effective solutions for sewage management (Andrade et al. 2024). Bioremediation techniques such as algal bioreactors, constructed wetlands, and aerobic granular sludge can achieve up to 80% removal efficiency for EDPs, but challenges like slow biodegradation and the need for sophisticated detection hinder effective treatment (Ghosh and Biswas 2023). Some emerging, promising green solutions are also being explored for this purpose. These methods and techniques are described briefly as follows:

Traditional Treatment Methods

Municipal and industrial wastewater treatment frequently uses biological treatments, such as membrane bioreactors (MBRs) and activated sludge processes. To eliminate organic pollutants, these techniques rely on biosorption, microbial degradation, and other naturally occurring processes (Hassan et al. 2021, Parida et al. 2022). However, research shows that although these methods are successful in lowering traditional pollutants, they only eliminate 50–70% of ECs because of their low biodegradability and resistance (Parida et al. 2022). Traditional WWTPs, designed primarily for the removal of conventional pollutants like organic matter and nutrients, often fail to effectively remove ECs. In order to break down or capture these pollutants, processes like primary sedimentation and secondary biological treatment are not specific enough. For example, compounds such as EDCs often persist through these systems and eventually get discharged into recipient water bodies. When it comes to eliminating ECs that are resistant to biological degradation, physical-chemical techniques are essential. These methods include adsorption, membrane filtration, and ion exchange. Activated carbon and biochar are widely used as adsorbents for ECs. Due to their large surface area and pore volume, they are able to adsorb a wide range of ECs, such as surfactants, personal care products, pharmaceuticals, endocrine-disrupting compounds (EDCs), and pesticides. It is evident from previous studies that modified biochar and nanocomposites tend to increase removal efficiencies and can be used as sustainable alternatives to activated carbon (Dhangar and Kumar 2020, Dey Chowdhury et al. 2022). Techniques such as ultrafiltration (UF), nanofiltration (NF), and reverse osmosis (RO) are effective in physically separating contaminants from wastewater (Sonam et al. 2021, Talib and Randhir 2016). RO, in particular, is known for high removal rates of ECs, including pharmaceuticals and pesticides. However, issues like membrane fouling and concentrate disposal can arise and limit their results (Li et al. 2021). Nanomaterials like TiO_2 and ZnO exhibit high photocatalytic activity under UV light, leading to effective degradation of ECs. In rural areas where resources are limited, these materials can show great potential. Biochar, produced from agricultural waste, is effective for adsorbing ECs like personal care products, microplastics, pharmaceuticals, pesticides, and EDCs (Hanif et al. 2023). Modified biochar, enhanced with metal ions or nanocomposites, shows greater efficiency and environmental sustainability. New research is focusing on biosorbents derived from plant or microbial biomass, such as chitosan and cellulose (Julisa Burbano Benavides Tutor and Lenin Javier Ramírez Cando 2022). These materials are renewable, biodegradable, and cost-effective for removing ECs, including pharmaceuticals and industrial chemicals. Despite their high efficiency in EC removal, high operational costs and complex maintenance present limitations to their widespread application.

Advanced Techniques

Advanced technologies like advanced oxidation processes (AOPs), photocatalysis, UV irradiation, nanofiltration, the use of nanomaterials, and the Fenton process are being explored to provide better treatment of wastewater, especially for eliminating ECs. Various other removal methods include adsorption with advanced materials like hydrophilic silica nanoparticles, membrane filtration, electrocoagulation, and microbial fuel cells that degrade surfactants while generating power. Advanced oxidation processes (AOPs) and integrated bio-electrochemical systems (BESs) are explored for their efficiency, cost-effectiveness, and environmental compatibility. Advanced oxidation processes (AOPs) are characterized by the production of highly reactive hydroxyl radicals, which can degrade even the most resilient organic contaminants. Key AOPs include ozone, UV, Fenton's reagent, and photocatalysis, often applied in combination to achieve greater efficiency. Ozone is highly effective for degrading various classes of ECs, especially those with complex chemical structures, such as pharmaceuticals, EDCs, and personal care products (Overton et al. 2023). Hybrid AOPs and bio-electrochemical systems (BESs) show significant promise for EC degradation (Bashir et al. 2023). However, potential byproducts and high operational costs remain significant challenges (Mohapatra and Kirpalani 2019). AOPs are effective but costly, leading to the exploration of integrated approaches like combining ozonation with activated carbon, as well as low-cost alternatives like photo-Fenton and solar radiation-based methods, and constructed wetlands. UV photolysis alone shows limited EC degradation, but when combined with H_2O_2, hydroxyl radicals are generated, enhancing removal efficiency (Ahmed et al. 2017). UV-C irradiation has shown promise in removing pharmaceutical residues from hospital wastewater, with H_2O_2 addition further increasing degradation rates (Thakur et al. 2022). These processes involve the reaction of hydrogen peroxide with ferrous ions to produce hydroxyl radicals (Yu et al. 2020). They are effective in degrading ECs like pesticides and pharmaceuticals but can result in sludge generation, which requires proper disposal (Engida et al. 2020). Activated carbon adsorption followed by the Fenton process offers effective removal of persistent ECs, particularly pharmaceuticals and personal care products. Studies show that coupling adsorption with Fenton or electro-Fenton systems enhances removal rates and improves process efficiency (Yu et al. 2020). Similarly, over 95% removal efficiency for ECs may be achieved by using MBRs followed by nanofiltration or reverse osmosis. While energy demand and fouling may be problems, this combination works well for a range of ECs, such as beta-blockers, insecticides, and EDCs (Sonam et al. 2021, Ahmed et al. 2017). Anaerobic baffled reactors efficiently convert organic carbon into biogas and achieve over 90% COD removal under mesophilic conditions but require post-treatment to meet discharge standards. They also show potential for degrading emerging contaminants like perchlorate and antibiotics, though an additional aerobic step may be needed for complete oxidation of complex organics (Lin Lau and Trzcinski 2024). A study (Matamoros et al. 2016) compared the removal efficiencies of 25 emerging contaminants (ECs) in four wastewater treatment systems in north-eastern Spain: an activated sludge (AS) system, a rotating biological contactor (RBC), a constructed wetland (CW), and a waste stabilization pond (WSP). The removal efficiencies were 42% for the CW, 62% for AS, 63% for RBC, and 82% for the WSP, with seasonal variability observed in all systems except WSP. Ecotoxicological assessments showed that while all technologies reduced aquatic risk, only the WSP system eliminated the risk in both seasons. Different traditional techniques are being aided with these advanced technologies for better removal of ECs and sustainable treatment of wastewater.

Green Technologies

Emerging contaminants (ECs) are becoming more common in wastewater, which has made the development of environmentally friendly and sustainable treatment systems necessary. While maintaining high pollution removal efficiency, these green technologies place a strong priority on cost-effectiveness, low energy consumption, and environmental sustainability. Membrane technologies, phytoremediation, advanced biological systems, and constructed wetlands (CWs) have become important environmentally friendly options for reducing the negative effects of ECs on aquatic ecosystems. Phytoremediation uses the capacity of plants to absorb, degrade, and sequester pollutants to provide an affordable and aesthetically pleasing alternative. Many studies have been conducted on the effectiveness of some plant species in eliminating ECs, heavy metals, and nutrients, such as *Canna indica*, *Phragmites australis*, and *Eichhornia crassipes* (Liu et al. 2020, (Youssef et al. 2023). Another popular green technology currently being used is membrane bioreactors (MBRs). MBRs achieve substantial removal efficiency for ECs and other pollutants by combining membrane filtration with biological treatment. This hybrid method removes emerging contaminants and improves biodegradation simultaneously. Various types of wastewater, such as industrial and municipal wastewater, may be treated with MBRs. Studies reveal that MBRs remove >90% of pharmaceuticals, pesticides, and endocrine-disrupting compounds (Ahmed et al. 2017). Advanced configurations, such as MBRs coupled with ultrafiltration, further enhance contaminant removal (Dhangar and Kumar 2020). However, operational expenses and maintenance demands limit their use in underdeveloped nations. Another green technology being used for this purpose is the algal-based treatment system. Algal-based treatment systems use sorption processes and photosynthetic activities of microalgae to absorb and transform ECs. Microalgae have shown potential in removing a significant amount of nutrients and pharmaceuticals like diclofenac and sulfamethoxazole (Mehariya et al. 2024). Since these systems clean wastewater and produce biofuel as a byproduct, they are becoming increasingly popular. Different types of biochars are also being used to remove ECs and other contaminants from wastewater. For instance, a study demonstrated that biochar derived from sugarcane bagasse effectively adsorbs naphthalene, achieving 100% removal within 75 minutes. Using low-frequency ultrasonication (40 kHz), over 80% of the adsorbed naphthalene was desorbed in 45 minutes, highlighting its potential as an efficient treatment for trace organic pollutants (Shruthi et al. 2022). Another study used palm kernel shell and coconut shell biochar as filter media to remove microplastics extracted from facial cleanser. The study achieved a maximum MP removal efficiency of 96.65% with palm kernel shell biochar at an optimal particle size of 0.6–1.18 mm and a column size of 20 mm (Hanif et al. 2023).

Among these, the most popular green technology being explored worldwide for the treatment of ECs from various types of wastewater is constructed wetlands. Constructed wetlands (CWs) are artificial wastewater treatment systems that imitate natural wetland processes. These systems provide an environmentally beneficial, economical, and sustainable way to manage a range of water pollutants, such as organic matter, nutrients, pathogens, and emerging contaminants (ECs). CWs are extremely adaptable in a variety of environmental conditions because they use the combined action of vegetation, substrate, and microbial populations to break down, change, or sequester pollutants. Constructed wetlands, through their complex microbial ecosystems and diverse redox conditions, offer a promising approach for the removal of antibiotics and the attenuation of ARGs. These systems can limit the spread of antibiotic resistance in the environment by supporting microbial communities that can break down pollutants and compete with resistant strains. Their design offers an environmentally sustainable substitute for traditional wastewater treatment technologies by combining efficiency and convenient operation. Furthermore, CWs are an appealing alternative for resolving wastewater issues in both developed and developing countries due to their cost-effectiveness, low energy needs, and reliance on natural processes.

Constructed Wetlands and their Role in Removal of ECs

Through processes including adsorption, biodegradation, sedimentation, and phytoremediation, these systems combine biological, chemical, and physical mechanisms to eliminate contaminants. CWs have become a viable wastewater management option due to their low costs, minimal operating requirements, and applicability to various types of wastewater. Because of the increasing emphasis on sustainable development across the world, CWs are now being recognized as a cutting-edge technique for wastewater treatment. Constructed wetlands effectively remove pharmaceutical compounds from hospital wastewater through various mechanisms, but optimizing system design and performance is essential for addressing emerging contaminants sustainably (Mumtaj et al. 2024). Where traditional wastewater treatment methods are either unfeasible or too expensive, CWs are especially beneficial in decentralized and rural environments. They can be categorized based on the characteristics of the direction of the water flow. The two main types are subsurface flow (SF) and free water surface flow (FWS), each offering distinct advantages for different applications. Subsurface flow systems (SF) can be further divided into vertical flow (VF) and horizontal subsurface flow (HSF).

Researchers (Decamp and Warren 2000) found in comparative studies on pilot-scale and laboratory-constructed wetlands that the *E. coli* removal rates were higher in pilot-scale systems (96.6–98.9%) compared to microcosms (41–72%), likely due to longer retention times and more efficient microbial communities in the larger systems. *E. coli* removal kinetics showed that the majority of removal occurred in the first third of the system, with planted gravel beds achieving faster removal than unplanted soil beds, and a hybrid system of planted soil followed by planted gravel performing better than the reverse configuration. A study focused on the treatment of municipal wastewater from a village using a horizontal flow constructed wetland (CW) system, monitored over a year. The CW system efficiently removed organic pollutants, including COD (92.3%), BOD_5 (93%), TSS (96%), turbidity (96.4%), total phosphate, and total nitrogen (74%). The treatment also successfully reduced thermotolerant coliforms. The study found that the system was both economical and effective, with higher hydraulic retention times leading to improved pollutant removal efficiency. Another study (Benny and Chakraborty 2020) investigated the treatment of phenol, thiocyanate, and ammonia in wastewater using a horizontal subsurface flow constructed wetland (CW) with *Typha angustifolia* planted in three out of four zones. The system achieved high removal efficiencies, with phenol and COD removals at 99% and 93%, respectively, while thiocyanate removal reached 91%, primarily through a combination of aerobic and anaerobic processes. Although ammonia and total nitrogen removal were less efficient (17–30% and 7–28%, respectively), plant uptake contributed to the removal, and *Typha angustifolia* grew well without toxicity effects (Benny and Chakraborty 2020).

The VF systems, which are unsaturated, are designed to encourage the flow of water through a filter bed from top to bottom, creating conditions for both biological and physical treatment processes. On the other hand, HSF systems, also known as saturated systems, allow the water to flow horizontally, providing an ideal environment for microbial processes that degrade pollutants. Another classification of CW systems is based on plant species diversity, which plays a crucial role in pollutant removal. These systems include emergent macrophyte CWs, submerged macrophyte CWs, and floating treatment wetlands (FTWs). Among these, emergent macrophyte-based systems are the most commonly employed due to their ability to provide significant treatment for a range of wastewater types (Omondi Donde et al. 2020). These systems utilize plants that grow above the water surface, contributing to water purification through processes like filtration, adsorption, and nutrient uptake. The choice of CW system depends on several factors, including the type of wastewater, the treatment goals, and the specific environmental conditions of the location.

A study (Karabelnik et al. 2008) evaluated methods to enhance oxygen supply in constructed wetlands (CWs) using a pilot-scale lightweight aggregate filter system for municipal wastewater treatment. The system achieved high purification efficiencies for BOD_7 (96%) and total nitrogen (51%) under optimal conditions, with organic matter removal rates double that of typical horizontal-flow CWs, though its aeration capacity remained lower than advanced vertical-flow CWs. Another study (Zheng et al. 2021) investigated the impact of sulfamethoxazole (SMX) on nitrogen removal processes in integrated vertical-flow constructed wetlands (IVCW), finding a moderate reduction (circa 10%) in total nitrogen removal efficiency due to chronic exposure to SMX. High-throughput sequencing revealed that SMX exposure significantly affected functional bacteria, including ammonium-oxidizing bacteria (AOB) and ANAMMOX, with NO_3^--N accumulation and carbon deficiency observed in the system. Despite these challenges, IVCWs demonstrated a high and stable removal efficiency (circa 95%) for sulfamethoxazole, with biodegradation—primarily in the first stage—being the main removal pathway, and microbial enrichment of *Pseudomonas*, *Bradyrhizobium*, *Sphingomonas*, and *Luteimonas* indicating potential for SMX biodegradation.

In addition to these traditional designs, hybrid or combined systems have gained popularity. These systems integrate multiple types of CWs, such as combining surface and subsurface flow systems to take advantage of the benefits of both. Such hybrid systems are increasingly used to improve treatment efficiency, especially in cases involving complex or heavily polluted wastewater. A study (Sgroi et al. 2017) evaluated a hybrid CW system comprising vertical subsurface flow (VF), partially saturated VF (VF sat), horizontal flow (HF), and free water surface (FWS) wetlands for removing conventional pollutants, emerging organic contaminants (EOCs), and fluorescing organic matter. Results indicated that aerobic VF wetlands were most effective for BOD_5 and COD removal, while anoxic HF wetlands excelled in nitrogen removal. Partially saturated VF wetlands showed superior performance in reducing nitrogen and biodegradable EOCs, with fluorescence indices proving valuable as surrogate indicators of water treatment efficiency and EOC monitoring (Sgroi et al. 2017). Similarly, a study (Koukoura et al. 2024) assessed the performance of four pilot-scale vertical sub-surface flow treatment wetlands (TWs) for removing pollutants and organic micropollutants from domestic wastewater, with three systems filled with electroconductive material and operated under saturated (MET SAT), unsaturated (MET UNSAT), and unsaturated-saturated (MET HYBRID) conditions, while the fourth system (AEW) was filled with gravel. The MET-TWs achieved significant COD removal (>78%) and TSS removal (>84%) at the maximum organic loading rate (OLR) of 249 g COD m^{-3} d^{-1}. The MET UNSAT system removed 25 g NH_4-N m^{-3} d^{-1}, while AEW achieved a higher removal rate of 35 g NH_4-N m^{-3} d^{-1}, with MET SAT showing phosphorus removal via precipitation in anaerobic pretreated wastewater. Organic micropollutant removal ranged from 70 ± 18% (MET UNSAT) to 91 ± 4% (AEW), and the MET HYBRID system achieved lower operational costs (0.02 € m^{-3}) and a smaller area requirement (0.51 m^2 per person equivalent, PE) compared to conventional TWs. Another study (Nõlvak et al. 2013) examined the dynamics of seven antibiotic resistance genes (ARGs) in a hybrid constructed wetland treating municipal wastewater, focusing on their abundance in influent, effluent, and wetland media biofilm. The results showed that while ARG concentrations fluctuated initially, they stabilized in the wetland biofilm, with notable reductions in gene abundance, especially sulfonamide resistance genes (sul1), which were more effectively removed compared to traditional wastewater treatments. The presence and removal of ARGs were influenced by operational parameters, including time and temperature, and were correlated with the removal of nitrogen compounds and organic matter in the system. In a study a full-scale hybrid constructed wetland (CW) system combining vertical subsurface flow (VF), horizontal subsurface flow (HF), and free water surface (FWS) wetlands demonstrated robust performance for treating and reusing domestic wastewater in small communities. The system achieved 98–99% removal efficiency for conventional pollutants (TSS, BOD_5, NH_4-N) and over 80% removal for various emerging contaminants, meeting Spanish reuse standards. The high efficiency was attributed to synergistic abiotic and biotic removal mechanisms, including biodegradation, sorption, hydrolysis, and photodegradation, facilitated by the distinct conditions in each CW stage.

Researchers (Verma et al. 2024) developed a multistage Constructed Wetland-Microbial Fuel Cell (CW-MFC) system using the halophyte *Juncus ridigus* to treat brackish sewage, showing significantly higher removal efficiencies for COD, BOD, PO_4^{3-}-P, NH_4^+-N, NO_2^--N, and NO_3^--N compared to a conventional CW and a control system. The CW-MFC also achieved a 40.6% decrease in salinity, with plant absorption and electrode adsorption identified as key mechanisms for removing salt-contributing ions. While conventional WWTPs are not designed to address AR, constructed wetlands (CWs) have shown promising performance in reducing antibiotic-resistant bacteria and genes, as well as general pollutants, with removal efficiencies ranging from 2 to 4 logs (Hazra and Durso 2022).

Pollutant Removal Mechanism of CWs

The capacity of CWs to mimic and enhance natural purifying processes accounts for their effectiveness. In CWs, sedimentation, adsorption, microbial degradation, plant uptake, and chemical precipitation work together to remove pollutants. Because each mechanism targets distinct contaminants, CWs are adaptable systems that can manage intricate wastewater matrices. Filtration and sedimentation are the main mechanisms used by CWs to eliminate suspended particles. The suspended particles in wastewater settle out as it moves through the substrate under the influence of gravity, or the matrix of the filter media traps them. The binding of dissolved contaminants, including heavy metals and medications, onto substrate surfaces is known as adsorption, and it is another important process. Sand, gravel, and charcoal are among the materials frequently utilized to improve this process. The efficiency of a CW is directly related to the filter media, as it can significantly affect the processes responsible for the treatment of wastewater. Pharmaceutical active compounds (PhACs) in aquatic environments, such as carbamazepine, diclofenac, and sulfamethoxazole, pose risks to human and ecosystem health. Constructed wetlands (CWs) have demonstrated removal efficiencies of up to 80–95% for certain PhACs, employing mechanisms like photodegradation, biodegradation, phytoremediation, and sorption (Zhang et al. 2023). A study suggested that the removal of NH_4-N becomes higher when we increase the size of filter media (Minakshi et al. 2022). Development of suitable matrix or filter media is one of the crucial factors to achieve better removal efficiency in CWs. Another study developed a unique lanthanum/aluminum (La/Al)-modified attapulgite/biochar composite (SAlLa@AB) as a substrate for constructed wetlands (CWs), with a high capacity for phosphorus (P) sorption. During a 120-day column trial with an influent concentration of 10 mg P/L and an 8-hour hydraulic retention period, the composite showed a maximum P sorption capacity of 21.3 mg/g and over 95% P removal efficiency. Since the removal of phosphorus was mainly accomplished by the creation of calcium-bound and Al-bound P fractions, which were verified by XPS analysis displaying La–O–P and Al–O–P complexes, it was confirmed that SAlLa@AB can be used as a promising substrate for CWs (Yin et al. 2022). Various ECs are removed by the adsorption process of the CW system. For instance, substrate adsorption is identified as the primary mechanism for the removal of PFAS, while plant uptake, harvesting, and microbial degradation further contribute to long-term treatment efficiency. The average removal efficiencies reported were 64% in free water surface (FWS), 46% in horizontal subsurface flow (HF), and 0% in vertical flow (VF) wetlands, with PFAS removal primarily attributed to adsorption by the substrate or plant root/rhizosphere. Studies show that wetland vegetation effectively adsorbs long-chain PFAS and transfers them from roots to aboveground tissues, as evidenced by bioaccumulation factors (BCF) and translocation factors (TF). Microbial communities in TWs play a crucial role, with bacteria capable of degrading PFAS through processes like decarboxylation, desulfonation, and reductive defluorination, contributing significantly to PFAS removal (Zhao et al. 2025).

The metabolism of organic contaminants and nutrients by microorganisms makes microbial degradation essential to CW operation. For instance, bacteria-mediated nitrification and denitrification processes facilitate the removal of nitrogen compounds. A parallel function is played by plant absorption, whereby nutrients such as phosphorus and nitrogen are absorbed by wetland vegetation and then integrated into plant biomass. Furthermore, through the release of oxygen via their roots, plants create microenvironments that facilitate aerobic microbial activity. Another example of the plant uptake process is a study where a CW system used to treat laundry wastewater was planted with *Trachelospermum jasminoides*, and the vegetation showed remarkable growth in harsh conditions (Stefanatou et al. 2024). This suggests the promising nature of planted CWs for onsite wastewater treatment. Plants also support the microbial community within the CW system and enhance the degradation of ECs and other pollutants through plant–microbe interaction processes. For instance, a study revealed that Phragmites australis and Typha domingensis in constructed wetlands harbor distinct bacterial communities. The P. australis rhizospheres showed significantly higher levels of ARGs like ermB, blaTEM, and tetA, as well as the intl1 integrase gene, highlighting the importance of plant species and microbiome interactions in optimizing pollutant and ARG removal in wastewater treatment systems (Riva et al. 2024). Further research is needed to optimize wetland systems and deepen understanding of their removal processes to better address this environmental challenge.

Factors Affecting the Efficiency of CWs

Physical, chemical, biological, and operational factors altogether play a part in determining the efficiency of constructed wetlands (CWs) to remove emerging contaminants (ECs). Physical elements are important, including the type, design, and filter material of the wetland. For instance, hybrid CWs that combine vertical flow (VF) and horizontal flow (HF) systems show higher removal efficiencies due to their ability to support a wide range of redox conditions. In VF systems, higher levels of dissolved oxygen (DO) promote aerobic degradation, while in HF wetlands, anaerobic zones aid in processes like denitrification, which are crucial for eliminating nitrogenous chemicals. Research shows that anaerobic pathways help in the degradation of persistent pollutants like triclosan and sulfamethoxazole, whereas aerobic conditions are efficient at eliminating non-steroidal anti-inflammatory drugs (NSAIDs) (Alsubih et al. 2022). Furthermore, photoreactive ECs like carbamazepine and ciprofloxacin may be degraded by solar radiation in free water surface wetlands, making these systems appropriate for areas with high sunlight exposure (Sánchez et al. 2022). Environmental conditions also play an important role in the removal efficiency of CWs. Microbial degradation rates and wetland vegetation-mediated processes typically decrease in colder climates, leading to reduced removal efficiencies. However, strategies like hybrid system designs and insulated substrates can mitigate these impacts. For example, in a study conducted under cold climatic conditions, hybrid CWs still maintained a removal efficiency of over 80% for pharmaceuticals like ibuprofen and carbamazepine by optimizing HRT and substrate design (Alsubih et al. 2022). The depth of the filter matrix significantly influences its treatment efficiency by affecting factors such as hydraulic retention time, oxygen transfer, and microbial activity. Filter media layers with higher depth are generally associated with improved removal of organic matter, suspended solids, and nutrients due to enhanced contact between the wastewater and the media, promoting biodegradation and adsorption processes. However, excessive depths can lead to challenges such as clogging and restricted oxygen penetration, which can inhibit aerobic microbial processes. Thus, an optimal filter media depth is essential to balance efficient pollutant removal with sustainable operation, particularly in multistage CWs designed for agricultural irrigation in regions where resources are limited. Depth of filter media also plays a major role in the removal of different pollutants. A study (Ren et al. 2011) evaluated how different substrate depths (0.1 m, 0.3 m, and 0.6 m) in constructed wetlands (CWs) affect wastewater treatment efficiency. The results showed that the 0.1 m substrate depth was best for removing ammonium nitrogen (NH_4^+-N) and total phosphorus (TP), while the 0.6 m depth was most effective for removing chemical oxygen demand (COD). The 0.1 m depth also improved dissolved oxygen levels and root growth (Ren et al. 2011). Along with CW design, the choice of substrate is also a crucial aspect. Sand, gravel, and charcoal are examples of substrates or filter media that affect the sorption of hydrophobic ECs and offer surfaces for microbial biofilms. A study in France explored the effects of filter media type and depth, finding that 30 cm of pea gravel significantly outperformed crushed gravel for COD (83% vs. 76%), TSS (87% vs. 81%), and TKN (64% vs. 57%) removal, likely due to better biofilm attachment and improved hydrodynamics (Ruiz-Ocampo et al. 2021). Tracer tests and core sampling confirmed that the shape and size of filter media influence performance. Substrates like biochar, coke, and charcoal are showing great promise in enhancing the efficacy of CW systems. For example, biochar-enriched CWs have shown excellent adsorption capacities, eliminating over 90% of pollutants such as ibuprofen and diclofenac under optimal conditions (Wu et al. 2022).

In CWs designed for the removal of ECs, such as pharmaceuticals and personal care products, and other pollutants, hydraulic retention time (HRT) has a substantial influence on removal efficiency. HRT refers to the average time that water retains within the wetland system. Longer hydraulic retention times (HRTs) ensure that pollutants have enough time to come into contact with microbial populations and substrates, which maximizes the removal process. Studies have shown that longer HRTs, typically between 5 to 7 days, enhance contact time, promoting adsorption and microbial degradation. In vertical-flow constructed wetlands (VFCWs), the removal efficiency of helminth eggs ranged from 82% to 100%, with hydraulic retention times (HRT) of 7 days and 14 days leading to higher removal rates compared to free flow conditions. The presence of plants further enhanced the efficiency, with longer HRTs and planted units showing improved helminth egg removal in the leachates (Nzouebet et al. 2022). According to a study, the antibiotic tetracycline was significantly removed at HRTs of 2–7 days, and a substantial decrease in the amount of antibiotic-resistant genes was shown at longer HRTs (Thanh et al. 2017). Efficient removal of heavy metals like Cd and Pb was reported at higher HRTs (Si et al. 2019). Optimal HRT depends on the type of CW and the nature of the influent. Multistage systems combining VF-CWs and HF-CWs often operate with combined HRTs of 5 to 15 days, achieving robust removal efficiencies for both conventional pollutants and ECs. A pilot-scale study in Senegal demonstrated that a 3-stage CW system with a total HRT of 7 days removed over 95% of BOD, SS, and nutrients while reducing microbiological indicators to safe levels for agricultural reuse (Torrens et al. 2020). While longer HRTs generally result in better treatment outcomes, they can increase land area requirements and reduce system throughput. However, in areas with water scarcity, the trade-off is often acceptable given the potential for high-quality effluent reuse. Research on CW-MFC systems indicates that even moderate HRTs of 2 to 3 days can achieve significant removal efficiencies while enabling energy recovery (Mu et al. 2020, Han et al. 2023).

Hydraulic loading rate (HLR) directly influences the efficacy of a CW system. Opposite to HRT, higher HLR can lead to poor performance of the CW system. Excessively high HLRs can lead to reduced residence times, compromising the system's efficacy. Increasing the HLR can reduce pollutant removal efficiency as it decreases the contact time and leads to hydraulic overloading, affecting the wetland's overall treatment capabilities. For example, an experimental CW system treating tetracycline-laden wastewater achieved over 98% removal when the HRT was maintained at 7 days, compared to 85% at shorter retention times. Additionally, influent composition, particularly the concentration and type of ECs, dictates removal efficiencies. High concentrations of hydrophobic contaminants tend to saturate adsorption sites, while hydrophilic compounds may escape treatment under certain conditions (Jena et al. 2023). Studies also suggest that optimizing the balance between HRT and HLR is crucial for maintaining treatment efficiency without risking clogging or insufficient wastewater processing. Pilot studies have indicated that HLRs ranging between 0.5–1.5 m/day with corresponding HRTs of 3–5 days are ideal for achieving high removal efficiencies of contaminants like COD, BOD, and nitrogen compounds in most CW configurations (Sharma et al. 2021, Hu et al. 2019).

Biological processes associated with vegetation and microbial communities present in the CW also play a major role in enhancing the efficiency. A study (Thanh Hien et al. 2017) evaluated fill-and-drain constructed wetlands (CWs) for the removal of tetracycline (TC) and tetracycline resistance genes (tet genes) from municipal wastewater. Planted CWs with common reed (Phragmites australis) achieved higher TC removal (95.4%) compared to non-planted CWs (87.2%) within one day, with complete removal observed for both systems at longer treatment times. The planted CWs also demonstrated superior long-term performance, completely removing tet genes and maintaining a 3-log reduction, highlighting their effectiveness for mitigating antibiotics and resistance genes in wastewater (Thanh Hien et al. 2017). Another study (Mancini et al. 2023) investigated the effectiveness of constructed wetlands (CWs) with aquatic vegetation *Juncus maritimus* and interspersed lagoons in retaining microplastics (MPs) through experimental simulations. Results showed that vegetated patches enhanced particle dispersion and sedimentation, achieving high retention rates of 88 ± 3% for PA fragments and 99 ± 1.5% for PET fibers. The type of vegetation is also a crucial factor in determining the efficiency of CWs in the removal of ECs. For instance, a study (Hamad 2023) evaluated the effectiveness of macrophytes *Typha domingensis* and *Cyperus papyrus* in removing heavy metals and antibiotics (roxithromycin and levofloxacin) from wastewater for agricultural reuse. *Cyperus papyrus* outperformed *T. domingensis* in removing biochemical oxygen demand (BOD), chemical oxygen demand (COD), ammonium (NH_4^+), and solid pollutants, while *T. domingensis* excelled in heavy metal removal. Constructed

Table 2: Factors affecting the efficiency of CW

Category	Factors	Impacts	References
Physical	System design	Impacts oxygen availability and degradation pathways.	(Caselles-Osorio and García 2006)
	Type of the filter matrix	Sorption capacity and microbial attachment sites enhance removal.	(Minakshi et al. 2022)
Biological	Vegetation	Promotes oxygenation, nutrient uptake, and microbial habitats.	(Saz et al. 2018)
	Microbial community	Microorganisms facilitate the biodegradation of contaminants.	(Rajan et al. 2019)
	Plant-microbe interactions	Root exudates of CW vegetation enhance microbial degradation.	(Meng et al. 2023)
Chemical	pH	Affects microbial activity and sorption processes.	(Patyal et al. 2021)
	Dissolved oxygen	Influences aerobic microbial activity.	(Karabelnik et al. 2008)
	Redox conditions	Affects the degradation mechanism of contaminants.	(Cui et al. 2023)
Environmental	Temperature	Directly affects microbial activity and plant growth.	(López et al. 2019)
	Seasonal variations	Influences removal rates and affects vegetation and microbial community.	(Nuel et al. 2018)
	Solar radiation	Facilitates photodegradation of photoreactive contaminants.	(Zhang et al. 2023)
Operational	Inlet load	Directly influences efficiency as high loads may saturate removal mechanisms.	(Sharma et al. 2023)
	Hydraulic retention time (HRT)	Longer HRT improves contact time for microbial and chemical processes.	(Merino-Solís et al. 2015)
	Flow rate	Slower flow enhances contaminant-substrate contact.	(Teoh et al. 2024)
	Depth of filter media	Provides surface area for microbial growth, pollutant removal through adsorption, ensures proper flow distribution.	(Ruiz-Ocampo et al. 2021, Ren et al. 2011)

wetlands with zeolite and *C. papyrus* significantly enhanced treatment capacity, achieving high removal efficiencies for bacterial indicators (96.9–99.8%) and pollutants, with roxithromycin showing higher elimination rates than levofloxacin. A study (Riva et al. 2024) investigated the role of plant-microbe interactions in constructed wetlands (CWs) using *Phragmites australis* and *Typha domingensis* to understand bacterial community structures and their potential for pollutant removal. Significant differences in microbiome composition and antibiotic resistance gene (ARG) abundance were observed between the two species, with P. australis showing higher levels of ARGs like ermB, blaTEM, and tetA in its rhizosphere. A study (Youssef et al. 2023) evaluated the performance of a laboratory-scale integrated constructed wetland-microbial fuel cell (CW-MFC) system seeded with *Eichhornia crassipes* in removing ibuprofen, generating electricity, and altering bacterial community structures, compared to sediment MFC (S-MFC) and abiotic S-MFC (AS-MFC). In comparison to S-MFC (29–42%) and AS-MFC (20–22%), the CW-MFC system demonstrated a much greater ibuprofen elimination efficiency (49–62%) and electricity production, with plants aiding in the alleviation of ibuprofen-induced inhibition on electroactive microbial populations. High-throughput sequencing revealed that the superior performance of CW-MFC was linked to increased relative abundances of microbial groups involved in anode respiration and organic matter fermentation, showcasing its potential for wastewater treatment and sustainable energy generation (Youssef et al. 2023). Another study stated that bioaugmented vertical flow constructed wetlands improved the removal of certain contaminants of emerging concern and shifted antibiotic resistance gene compositions (Tadić et al. 2024).

Efficiency of CWs in EC Removal

Factors like substrate properties, hydraulic retention time, hydraulic loading rate, type of vegetation, microbial community, oxygenation, and redox conditions influence the efficiency of CWs. Their efficiency depends on processes like sedimentation, precipitation, adsorption, microbial degradation, and plant uptake. For example, CWs often outperform conventional wastewater treatment plants (WWTPs) in removing pharmaceutical active compounds (PhACs), with factors like hydraulic retention time, vegetation type, and substrate composition playing critical roles in achieving high removal rates (Zhang et al. 2023).

A hybrid constructed wetland system (40 m^2) comprising vertical-horizontal-vertical beds and planted with tropical species (*Arundo donax L.*, *Hibiscus esculentus L.*, and *Solanum melongena L.*) effectively treated dairy farm wastewater, achieving significant reductions in TSS (92.2%), BOD_3 (95%), total N (83.6%), and P (86.1%). This system demonstrates potential for improving groundwater quality and health, with future research suggested on integrating anaerobic reactors for enhanced treatment of high-strength wastewater in tropical regions (Sharma et al. 2021). A study (Nyieku et al. 2022) evaluated the contaminant removal efficiency of three indigenous wetland plants *Typha latifolia*, *Ruellia simplex*, and *Alternanthera philoxeroides* used in a free water surface flow constructed wetland to polish effluent from a conventional oily wastewater treatment plant. The results showed that planted wetlands, especially with *A. philoxeroides*, significantly outperformed unplanted wetlands in removing oil and grease, BOD, COD, and TDS, while *T. latifolia* was most effective for nitrate and phosphorus removal, and *R. simplex* excelled in removing coliform bacteria. This study highlights the potential of using native plants in constructed wetlands as a sustainable, low-cost solution to improve wastewater treatment and mitigate environmental and health impacts from oilfield contaminants. Another study (Stefanatou et al. 2024) evaluated the effectiveness of vertical flow constructed wetlands (VFCWs), planted with *Trachelospermum jasminoides*, for treating real laundry wastewater, focusing on conventional pollutants and benzotriazoles (BTRs). Both planted and unplanted systems showed high removal efficiencies (92–98%) for turbidity, TSS, and COD, and excellent removal rates for CBTR and XTR. According to the results of this study, *T. jasminoides* thrived under harsh wastewater conditions, highlighting VFCWs as a promising solution for on-site treatment while providing a green facade for buildings. Another study (Falahi et al. 2021) evaluated the effectiveness of vegetated constructed wetlands (CWs) with *Scirpus grossus* and a control system for removing ibuprofen and paracetamol from domestic wastewater at hydraulic retention times (HRT) of 3, 4, and 5 days. The highest removal efficiencies of 99% for ibuprofen and 94% for paracetamol were achieved in vegetated CWs at an HRT of 5 days, compared to 54% and 63%, respectively, in non-vegetated systems. Enhanced biodegradation was attributed to the symbiotic effects of plants and bacteria, as systems with higher bacterial colony-forming units (CFUs) demonstrated significantly greater contaminant removal. A similar study investigated a pilot-scale vertical subsurface flow constructed wetland (VSSFCW) with *Scirpus grossus* and aeration to remove ibuprofen, COD, and nutrients (NH_3-N, NO_3-N, PO_4-P) from domestic wastewater. The system achieved maximum removal efficiencies under 2 L/min aeration and a 5-day hydraulic retention time (HRT), with 99.3% for ibuprofen, 88.2% for COD, 99.1% for ammonia, 72.9% for nitrate, and 83.2% for orthophosphate, meeting Malaysian effluent standards. Filtration, adsorption, and longer HRT were crucial for ibuprofen removal, while nutrient removal was less influenced by HRT (Falahi et al. 2022).

A study (Tian et al. 2023) evaluated the removal efficiency of oxytetracycline (OTC) and sulfamethoxazole (SMZ) and the occurrence of antibiotic resistance genes (ARGs) in vertical down-up flow constructed wetlands (VDUF-CWs). VDUF-CWs effectively removed OTC (84.3%) and SMZ (77.7%), with microbial degradation as the primary pathway contributing 58.52% and 65.13%, respectively, followed by photodegradation, plant absorption, and substrate adsorption. However, the detection and increased abundance of ARGs (e.g. tetA, tetC, tetW, sulI, and sulII) over time highlight the potential risk of ARG spread, emphasizing the need for strategies to mitigate this threat in CWs. A study (Nguyen et al. 2019) evaluated a bench-scale constructed wetland (CW) for removing organic matter (OM) and tetracycline (TC) from water, assessing the effects of initial concentrations and hydraulic retention times (HRTs). The CW achieved OM removal efficiencies of 55.2–80.8% for UV-254, 28.1–71.9% for dissolved organic carbon (DOC), and 72.1–79.7% for soluble chemical oxygen demand (sCOD) under a 1-day HRT, with higher DOC concentrations enhancing removal efficiency.

CWs filled with electroconductive media (coke) showed improved ARG retention compared to conventional substrates. For instance, ARG concentrations were 10,690 copies/g in coke-filled CWs compared to 8,576 copies/g in quartz sand systems, demonstrating the influence of substrate on ARG dynamics (Bai et al. 2022). Constructed wetland–microbial fuel cells (CW-MFCs) achieved 89.29% removal of ibuprofen (IBP) and 84.10% of diclofenac (DCF) over 60 days with a 3-day hydraulic retention time, with glucose as the most effective co-substrate for degradation and sodium acetate enriching electroactive bacteria (Qin et al. 2023).

Modification of CWs for Enhanced Efficiency

Constructed wetlands (CWs) incorporating biochar as a substrate have shown promise in wastewater treatment, leveraging biochar's properties such as high specific surface area and large pore volume for pollutant removal. The efficiency of biochar depends on its preparation conditions (e.g. pyrolysis temperature, feedstock type) and its placement within CWs (El Barkaoui et al. 2023). A study used biochar made from wetland plants to improve the elimination of diclofenac in vertical subsurface flow constructed wetlands (VSF-CWs). With an initial dose of 0.75–1 mg/L, pH 6.5–7.5, and a hydraulic retention period of 7 days, optimal conditions for DCF removal (>90%) were reached. The addition of BC drastically altered the architecture of the microbial communities and increased the effectiveness of the CW (Wu et al. 2022). Another study (Asaad et al. 2022) evaluated the use of reed plants (*Phragmites australis*) and their biochar for enhancing wastewater reclamation in wetlands within 72 hours, focusing on water with low irrigation quality and contamination from BOD_5, fecal coliform, and metals such as Cu, Mn, and Cd. The treatments tested included a sand and gravel bed, reed plants grown on a sand and gravel bed, and biochar with a sand and gravel bed, all of which significantly reduced BOD_5, fecal coliform, cations, anions, and toxic metals, with biochar treatment showing superior results. While the wastewater quality improved to a medium-grade level, further protocols are needed for safe handling and sustainable use of reclaimed water in wetlands (Asaad et al. 2022). Placement of biochar in the CW matrix is also an important factor that should not be overlooked. For instance, a study (Sha et al. 2024) investigated the removal of cadmium (Cd) and lead (Pb) in constructed wetlands (CWs) with biochar placed in different layers, showing that the biochar at the bottom of the CW system (BCW) achieved maximum removal efficiencies of 96.6–98.6% for Cd and 97.2–98.9% for Pb. The BCW system also significantly altered the microbial community, increasing the relative abundance of various phyla and genera, including *Proteobacteria*, *Firmicutes*, and *Basidiomycota*. Furthermore, Cd and Pb concentrations were positively correlated with the abundance of specific microbial taxa, such as *Verrucomicrobiota* and *Basidiomycota*, suggesting that BCW improves both metal removal and microbial community structure (Sha et al. 2024). In order to prevent clogging or flotation, the best placement of biochar is suggested to be between two layers of inert matrix materials (El Barkaoui et al. 2023). Along with biochar, different types of suitable vegetation can be used to further enhance the efficiency of CWs. A study (Bai et al. 2022) examined the effectiveness of constructed wetlands (CWs) filled with electroconductive coke media for removing nutrients, antibiotics, and antibiotic resistance genes (ARGs) from domestic sewage. The CWs with coke achieved higher antibiotics removal efficiencies (13–100%) compared to quartz sand (7–100%), with wetland plants further enhancing nutrient and tetracycline removal. However, coke-filled CWs retained more ARGs (10,690 copies/g), highlighting the potential environmental risk of ARG release and emphasizing the need for careful substrate and plant selection for sustainable wastewater treatment. Ecological restoration using wetland plants, particularly through root iron plaque (IP) formation, offers an environmentally friendly approach to treating secondary effluent wastewater by influencing the bioavailability and transformation of key elements like carbon, nitrogen, and phosphorus in constructed wetlands (CWs). Root IP, regulated by rhizosphere redox conditions and microbial activity, enhances pollutant removal while interacting with biogeochemical processes and mitigating emerging contaminants and heavy metals (Fan et al. 2023).

The addition of specific kinds of microorganisms that are capable of biodegrading emerging contaminants and other related pollutants is also becoming popular to enhance the overall efficiency of CWs. For instance, a study (Tadić et al. 2024) investigated the removal of contaminants of emerging concern (CECs) and antibiotic resistance genes (ARGs) in a pilot-scale vertical flow constructed wetland, with or without bioaugmentation using *Trichoderma asperellum* (strain T34). Bioaugmentation was effective, with *Trichoderma asperellum* thriving under competitive conditions and showing improved removal efficiencies for certain compounds like benzotriazole and diclofenac, though polar compounds like melamine were harder to remove. A correlation was observed between removal efficiency, biodegradability, and sorption capacity, with optimal plant uptake seen for compounds with a log Dow between 0 and 3. Transformation products were also identified, and while bioaugmentation shifted ARG compositions, no significant improvement in ARG removal rates was observed. A combined microalgal-assisted CW approach is proposed to leverage their complementary strengths, improving MP removal efficiency, creating biodegradable conditions, and enhancing effluent quality. This integrated strategy offers a sustainable and synergistic solution for managing MP pollution and co-pollutants in wastewater treatment (Ding et al. 2024).

Constructed wetlands, along with natural coagulation-flocculation and microbial fuel cells, are highlighted as eco-friendly and sustainable solutions for surfactant removal from wastewater. Constructed wetlands can be associated with microbial fuel cells for sustainable treatment of wastewater along with the production of energy. Constructed wetlands (CWs) coupled with microbial fuel cells (MFCs) offer a promising low-energy and low-carbon wastewater treatment solution, enabling simultaneous nutrient removal and bioenergy recovery. Integrated CW-MFC systems enhance wastewater treatment efficiency through phytoremediation, increase power generation by utilizing wetland plants, and reduce greenhouse gas emissions due to electrogenic bacteria (Ebrahimi et al. 2021). The efficiency of integrated CW-MFC systems depends on various factors, including organic loading rate, hydraulic retention time, system design, plant species, substrate type, electrode materials, and dissolved oxygen levels. Rapid development of CW-MFC has been seen in the past decade. A study (Niu et al. 2023) investigated two types of constructed wetland-microbial fuel cells (CW-MFC), a closed-circuit system (CCW-MFC) and an open-circuit system (OCW-MFC), using petroleum coke as both the electrode and filter matrix. The CCW-MFC system outperformed traditional CWs, achieving high removal efficiencies of COD, NH_4^+-N, and TN at 94.49%, 94.99%, and 84.67%, respectively, when optimal conditions of aeration rate (0.4 L/min), COD concentration (300 mg/L), and hydraulic retention time (3 days) were maintained. Additionally, the CCW-MFC system demonstrated a higher denitrification ability, likely due to the increased abundance of *Thiothrix*, which could stimulate quorum sensing and enhance the participation of other denitrifying bacteria (Niu et al. 2023). Another study (Qin et al. 2023) evaluated the efficiency of small-scale constructed wetland–microbial fuel cells (CW-MFCs) in removing ibuprofen (IBP) and diclofenac (DCF) from wastewater over 60 days with a hydraulic retention time of 3 days, achieving removal rates of 89.29% for IBP and 84.10% for DCF. Co-substrates like glucose, sucrose, and sodium acetate influenced bacterial community diversity, with glucose proving most effective for degrading IBP and DCF, and sodium acetate favoring electroactive bacterial enrichment. While these systems show significant potential, further research is required to clarify the interactive mechanisms between CW and MFC components and to optimize their performance for sustainable wastewater management. While CW-MFC systems outperform standalone CWs and MFCs, larger-scale trials and long-term studies are needed to confirm their viability for widespread domestic or metropolitan wastewater applications.

According to a study, the inclusion of photocatalysis (TiO_2/α-Al_2O_3 with UVA irradiation) as post-treatment improved effluent quality, especially for OM removal, highlighting the CW's potential for cost-effective treatment of OM and TC in water. Currently, up-flow anaerobic sludge blanket (UASB) reactors integrated with constructed wetlands (CW) are being used for the treatment of domestic and industrial wastewater in developing countries. The combined UASB-CW system demonstrated high removal efficiencies, achieving 79.2–93.9% for COD, 89.2–92.9% for BOD, 87.2–96.3% for TSS, and 97.9–99.99% for fecal coliforms, meeting global effluent standards (Engida et al. 2020). These systems are cost-effective, easy to operate and maintain, making them suitable for resource-scarce regions. The implementation of UASB-CW systems can address the indiscriminate disposal of wastewater and mitigate its environmental impacts, particularly in developing countries.

Limitations

Although constructed wetlands (CWs) are acknowledged as an economical and environmentally friendly method of treating wastewater, they do have several drawbacks. The incomplete removal of emerging contaminants (ECs), such as microplastics, pesticides, personal care items, and medications, is one of the main problems. The efficiency of CWs in removing ECs largely depends on the chemical structure and properties of the contaminants, as well as the removal mechanisms, such as sorption, biodegradation, and photodegradation. For instance, while hydrophobic compounds may be effectively adsorbed onto the substrate, more hydrophilic or persistent contaminants can bypass treatment processes. Studies indicate that certain ECs, such as antibiotics and flame retardants, often exhibit lower removal rates compared to conventional pollutants like biochemical oxygen demand (BOD) and chemical oxygen demand (COD).

The incomplete removal and retention of antibiotic resistance genes (ARGs) is another area of concern. Nevertheless, research has demonstrated that whereas CWs can lower the amount of ARGs by means of adsorption and biodegradation, they may also serve as ARB reservoirs in suboptimal conditions (Krzeminski et al. 2019). Using advanced materials like biochar and coke may enhance the removal efficiency of CWs, but they too come with issues. Even though the coke-filled CWs removed antibiotics more efficiently, they also accumulated higher levels of ARGs, suggesting that the use of electroconductive media could potentially increase the environmental risk of ARGs being released back into water systems (Bai et al. 2022). This indicates the urgent demand for more effective treatment methods that may eliminate ECs and other pollutants from wastewater before it is discharged into natural water systems. Results of another study showed that they were efficient in the removal of ECs like ARGs, but issues like variability in performance and challenges to completely remove the polar compounds remained (Tadić et al. 2024).

The efficiency of CWs is significantly affected by seasonal variations, particularly in areas with extreme weather conditions. During colder months, microbial activity and plant-mediated processes decrease, leading to reduced removal efficiencies for nutrients and ECs. Processes like nitrification and denitrification, which are vital for nitrogen removal, slow down at lower temperatures, further compromising treatment efficiency. As a result, the wide applicability of CWs is limited due to their seasonal dependence. The efficiency of CWs is closely tied to the health and functionality of macrophytes and microbial communities. Stressors like high contaminant concentrations, substrate clogging, or unfavorable environmental conditions can impair plant growth and microbial activity, reducing the system's pollutant removal capacity. Certain plants may also exhibit limited tolerance to specific contaminants, which can further affect the system's resilience and performance over time.

The performance of CWs is further impacted by hydraulic loading rate and HRT. High hydraulic loading rates or insufficient retention times can reduce the interaction between wastewater and wetland media, leading to suboptimal removal of both ECs and conventional pollutants. Moreover, clogging of substrates, especially in HF systems, can impair hydraulic performance, necessitating frequent maintenance. Such issues are particularly challenging in decentralized areas where resources for regular upkeep may be limited. Variability in treatment efficiency due to factors like hydraulic retention time, loading rate, pollutant concentration, hydrology, vegetation type, climatic conditions, microbiome, plant-microbe interaction, etc., makes them not widely applicable. Challenges like clogging and flotation also arise during operation. While CWs present an environmentally friendly and efficient solution for wastewater treatment, addressing their limitations is crucial for optimizing performance. Innovations such as enhanced aeration, substrate modifications, and advanced hybrid designs hold promise for overcoming these challenges and improving the removal efficiency of pollutants and emerging contaminants.

Conclusion

Constructed wetlands provide a viable pathway for sustainable wastewater management and environmental protection. Their potential to mitigate ECs while offering energy-efficient and eco-friendly solutions makes them a critical tool in advancing global water security. They are capable of eliminating ECs, which are challenging to address through conventional wastewater treatment technologies. Research highlights the robust capacity of CWs in removing ECs, including pharmaceuticals and personal care products. For example, a study (Chowdhury et al. 2022) emphasized the combined efficiency of CWs and vermifiltration in reducing EC toxicity in ecosystems. Another study (Overton et al. 2023) provided valuable insights into the biogeochemical pathways responsible for EC removal, underscoring the importance of substrate, hydrology, and vegetation in enhancing these processes. However, certain challenges like variability in treatment efficiency, clogging, and flotation still persist. Addressing these limitations will require an interdisciplinary approach to refine CW designs and incorporate emerging technologies that improve their effectiveness. Furthermore, research gaps remain in understanding the fate of a broad range of emerging contaminants in CWs, identifying specific functional microorganisms, elucidating pharmaceutical transformation pathways, and evaluating the ecotoxicity of CW effluents to optimize system performance. By integrating emerging technologies and fostering interdisciplinary research, CWs can be further developed to meet the growing demands of wastewater treatment and the mitigation of emerging contaminants in an environmentally friendly and sustainable manner.

References

A.L. Falahi, O.A., Abdullah, S.R.S., Hasan, H.A. Othman, A.R. Ewadh, H.M. Al-Baldawi, I.A. et al. 2021. Simultaneous removal of ibuprofen, organic material, and nutrients from domestic wastewater through a pilot-scale vertical sub-surface flow constructed wetland with aeration system. J Water Process Eng, 43. doi:10.1016/j.jwpe.2021.102214.

A.L. Falahi, O.A., Abdullah, S.R.S., Hasan, H.A., Othman, A.R., Ewadh, H.M., Al-Baldawi, I.A. et al. 2022. Elimination of mixed ibuprofen and paracetamol from spiked domestic wastewater via a pilot continuous aerated sub-surface constructed wetland system. J Water Process Eng, 50. doi:10.1016/j.jwpe.2022.103308.

Ahmed, M.B., Zhou, J.L., Ngo, H.H., Guo, W., Thomaidis, N.S. and Xu, J. 2017. Progress in the biological and chemical treatment technologies for emerging contaminant removal from wastewater: A critical review. J Hazard Mater, 323: 274-298. doi:10.1016/j.jhazmat.2016.04.045.

Akanbi, C.A., Rotimi, D.E., Ojo, A.B. and Ojo, O.A. 2025. Endocrine-disrupting chemicals (EDCs) and epigenetic regulation in embryonic development: Mechanisms, impacts, and emerging trends. Toxicol. Rep, 14: 101885. doi: https://doi.org/10.1016/J.TOXREP.2024.101885.

Alsubih, M., El Morabet, R., Khan, R.A., Khan, N.A., Khan, A.R., Khan, S. et al. 2022. Performance evaluation of constructed wetland for removal of pharmaceutical compounds from hospital wastewater: Seasonal perspective. Arab J Chem, 15(12). doi:10.1016/j.arabjc.2022.104344.

Anawar, H.M., Ahmed, G. and Strezov, V. 2019. Long-term performance and feasibility of using constructed wetlands for treatment of emerging and nanomaterial contaminants in municipal and industrial wastewater. pp. 63–81. *In*: A.K. Mishra, H.M. Anawar, N. Drouiche [eds.]. Emerging and Nanomaterial Contaminants in Wastewater: Advanced Treatment Technologies. Elsevier.

Andrade, H.N.D., Oliveira, J.F.D., Siniscalchi, L.A.B., Costa, J.D.D. and Fia, R. 2024. Global insight into the occurrence, treatment technologies and ecological risk of emerging contaminants in sanitary sewers: Effects of the SARS-CoV-2 coronavirus pandemic. Sci Total Environ, 921. doi:10.1016/j.scitotenv.2024.171075.

Asaad, A.A., El-Hawary, A.M., Abbas, M.H.H., Mohamed, I., Abdelhafez, A.A. and Bassouny, M.A. 2022. Reclamation of wastewater in wetlands using reed plants and biochar. Sci Rep, 12(1): 1-12. doi:10.1038/s41598-022-24078-9. https://doi.org/10.1038/s41598-022-24078-9.

Ávila, C., Bayona, J.M., Martín, I., Salas, J.J. and García, J. 2015. Emerging organic contaminant removal in a full-scale hybrid constructed wetland system for wastewater treatment and reuse. Ecol Eng, 80: 108-116. doi:10.1016/j.ecoleng.2014.07.056.

Bai, S., Wang, X., Zhang, Y., Liu, F., Shi, L., Ding, Y. et al. 2022. Constructed wetlands as nature-based solutions for the removal of antibiotics: Performance, microbial response, and emergence of antimicrobial resistance (AMR). Sustain, 14(22). doi:10.3390/su142214989.

Bashir, Y., Raj, R., Ghangrekar, M.M., Nema, A.K. and Das, S. 2023. Critical assessment of advanced oxidation processes and bio-electrochemical integrated systems for removing emerging contaminants from wastewater. RSC Sustain, 1(8): 1912-1931. doi:10.1039/d3su00112a.

Benny, C. and Chakraborty, S. 2020. Continuous removals of phenol, organics, thiocyanate and nitrogen in horizontal subsurface flow constructed wetland. J Water Process Eng, 33. doi:10.1016/j.jwpe.2019.101099.

Boucher-Carrier, O., Brisson, J., Abas, K., Duy, S.V., Sauvé, S. and Kõiv-Vainik, M. 2022. Effects of macrophyte species and biochar on the performance of treatment wetlands for the removal of glyphosate from agricultural runoff. Sci Total Environ, 838. doi:10.1016/j.scitotenv.2022.156061.

Campanale, C., Dierkes, G., Massarelli, C., Bagnuolo, G. and Uricchio, V.F. 2020. A relevant screening of organic contaminants present on freshwater and pre-production microplastics. Toxics, 8(4): 1-21. doi:10.3390/toxics8040100.

Caselles-Osorio, A. and García, J. 2006. Performance of experimental horizontal subsurface flow constructed wetlands fed with dissolved or particulate organic matter. Water Res, 40(19): 3603-3611. doi:10.1016/j.watres.2006.05.038.

Chaturvedi, M., Mishra, A., Sharma, K., Sharma, G., Saxena, G. and Singh, A.K. 2021. Emerging contaminants in wastewater: Sources of contamination, toxicity, and removal approaches. Emerg Treat Technol Waste Manag, 103-132. doi:10.1007/978-981-16-2015-7_5.

Chaves-Barquero, L.G., Luong, K.H., Rudy, M.D., Frank, R.A., Hanson, M.L. and Wong, C.S. 2018. Attenuation of pharmaceuticals, nutrients and toxicity in a rural sewage lagoon system integrated with a subsurface filtration technology. Chemosphere, 209: 767-775. doi:10.1016/j.chemosphere.2018.06.119.

Chen, Y., Yang, J., Yao, B., Zhi, D., Luo, L. and Zhou, Y. 2022. Endocrine disrupting chemicals in the environment: Environmental sources, biological effects, remediation techniques, and perspective. Environ Pollut, 310. doi:10.1016/j.envpol.2022.119918.

Chowdhury, S.D., Surampalli, R.Y. and Bhunia, P. 2022. Potential of the constructed wetlands and the earthworm-based treatment technologies to remove the emerging contaminants: A review. J Hazardous, Toxic, Radioact Waste, 26(2). doi:10.1061/(ASCE)HZ.2153-5515.0000668.

Christofilopoulos, S., Kaliakatsos, A., Triantafyllou, K., Gounaki, I., Venieri, D. and Kalogerakis, N. 2019. Evaluation of a constructed wetland for wastewater treatment: Addressing emerging organic contaminants and antibiotic resistant bacteria. N Biotechnol, 52(May): 94-103. doi:10.1016/j.nbt.2019.05.006.

Cui, E., Zhou, Z., Gao, F., Chen, H. and Li, J. 2023. Roles of substrates in removing antibiotics and antibiotic resistance genes in constructed wetlands: A review. Sci Total Environ, 859. doi:10.1016/j.scitotenv.2022.160257.

de Miranda, L.L.R., Harvey, K.E., Ahmed, A. and Harvey., S.C. 2021. UV-filter pollution: Current concerns and future prospects. Environ Monit Assess, 193(12). doi:10.1007/S10661-021-09626-6.

Decamp, O. and Warren, A. 2000. Investigation of *Escherichia coli* removal in various designs of subsurface flow wetlands used for wastewater treatment. Ecol Eng, 14(3): 293-299. doi:10.1016/S0925-8574(99)00007-5.

Dey Chowdhury, S., Bhunia, P., Surampalli, R.Y. and Zhang, T.C. 2022. Nature and characteristics of emerging contaminants as a triggering factor for selection of different configurations and combinations of constructed wetlands: A review. J Environ Eng (United States), 148(8). doi:10.1061/(ASCE)EE.1943-7870.0002012.

Dhangar, K. and Kumar, M. 2020. Tricks and tracks in removal of emerging contaminants from the wastewater through hybrid treatment systems: A review. Sci Total Environ, 738. doi:10.1016/j.scitotenv.2020.140320.

Ding, S., Gu, X., Sun, S. and He, S. 2024. Optimization of microplastic removal based on the complementarity of constructed wetland and microalgal-based system. Sci Total Environ, 912. doi:10.1016/j.scitotenv.2023.169081.

Du, L., Zhao, Y., Wang, C., Zhang, H., Chen, Q., Zhang, X. et al. 2020. Removal performance of antibiotics and antibiotic resistance genes in swine wastewater by integrated vertical-flow constructed wetlands with zeolite substrate. Sci Total Environ, 721: 1-10. doi:10.1016/j.scitotenv.2020.137765.

Ebrahimi, A., Sivakumar, M., McLauchlan, C., Ansari, A. and Vishwanathan, A.S. 2021. A critical review of the symbiotic relationship between constructed wetland and microbial fuel cell for enhancing pollutant removal and energy generation. J Environ Chem Eng, 9(1): 105011. doi:10.1016/j.jece.2020.105011.

El Barkaoui, S., Mandi, L., Aziz, F., Del Bubba, M. and Ouazzani, N. 2023. A critical review on using biochar as constructed wetland substrate: Characteristics, feedstock, design and pollutants removal mechanisms. Ecol Eng, 190. doi:10.1016/j.ecoleng.2023.106927.

Engida, T.M., Wu, J.M., Xu, D. and Wu, Z.B. 2020. Review paper on treatment of industrial and domestic wastewaters using uasb reactors integrated into constructed wetlands for sustainable reuse. Appl Ecol Environ Res, 18(2): 3101-3129. doi:10.15666/aeer/1802_31013129.

Hien, P.T. 2018. Performance of common reed constructed wetland on removal of tetracycline antibiotic and resistance genes. Ph.D. Thesis, University of Yamanashi, Japan.

Fan, Y., Sun, S. and He, S. 2023. Iron plaque formation and its effect on key elements cycling in constructed wetlands: Functions and outlooks. Water Res, 235. doi:10.1016/j.watres.2023.119837.

Fang, Zheng, Gao, Y., Wu, X., Xu, X., Sarmah, A.K., Bolan, N., Gao, B. et al. 2020. A critical review on remediation of bisphenol S (BPS) contaminated water: Efficacy and mechanisms. Crit Rev Environ Sci Technol, 50(5): 476-522. doi:10.1080/10643389.2019.1629802.

García, J., Jesús García-Galán, M., Day, J.W., Boopathy, R., Wallace, S. and Hunter, R.G. 2020. Antibiotics, antibiotic resistant bacteria (ARB), and antibiotic resistance genes (ARGs): Increasing removal with wetlands and reducing environmental impacts. 1-41. doi: 10.1016/j.biortech.2020.123228.

Ghosh, S. and Biswas, A. 2023. Emerging contaminants in municipal wastewater: Occurrence, characteristics, and bioremediation. pp. 153-178. *In*: I. Haq, A. Kalamdhad, A. Pandey [eds.]. Current Developments in Biotechnology and Bioengineering. Elsevier.

Gross, B., Montgomery-Brown, J., Naumann, A. and Reinhard, M. 2004. Occurrence and fate of pharmaceuticals and alkylphenol ethoxylate metabolites in an effluent-dominated river and wetland. Environ Toxicol Chem, 23(9): 2074-2083. doi:10.1897/03-606.

Hamad, M.T.M.H. 2023. Comparing the performance of Cyperus papyrus and Typha domingensis for the removal of heavy metals, roxithromycin, levofloxacin and pathogenic bacteria from wastewater. Environ Sci Eur, 35(1). doi:10.1186/s12302-023-00748-x. https://doi.org/10.1186/s12302-023-00748-x.

Hamilton, P.B., Baynes, A., Nicol, E., Harris, G., Uren Webster, T.M., Beresford, N. et al. 2022. Feminizing effects of ethinylestradiol in roach (Rutilus rutilus) populations with different estrogenic pollution exposure histories. Aquat Toxicol, 249: 106229. doi:10.1016/J.AQUATOX.2022.106229.

Han, J., Zhao, J., Wang, Y., Shu, L. and Tang, J. 2023. Performance optimization of two-stage constructed wetland-microbial fuel cell system for the treatment of high-concentration wastewater. Environ Sci Pollut Res, 30(23): 63620-63630. doi:10.1007/s11356-023-26488-2.

Hanif, M.A., Ibrahim, N., Dahalan, F.A., Umi, U.F., Hasan, M., Azhari, A.W. and Jalil, A.A. 2023. Microplastics in facial cleanser: Extraction, identification, potential toxicity, and continuous-flow removal using agricultural waste–based biochar. Environ Sci Pollut Res, doi:10.1007/S11356-023-26741-8.

Hassan, I., Chowdhury, S.R., Prihartato, P.K. and Razzak, S.A. 2021. Wastewater treatment using constructed wetland: Current trends and future potential. Processes, 9(11): 1-27. doi:10.3390/pr9111917.

Hazra, M., Joshi, H., Williams, J.B. and Watts, J.E.M. 2022. Performance efficiency of conventional treatment plants and constructed wetlands towards reduction of antibiotic resistance. Antibiotics, 11(1). doi:10.3390/antibiotics11010114.

Hazra, M. and Durso, L.M. 2022. Antibiotics and antibiotic resistant bacteria/genes in urban wastewater: A comparison of their fate in conventional treatment systems and constructed wetlands. Chemosphere, 303. doi:10.1016/j.chemosphere.2022.135148.

Helt, C. 2012. Occurrence, Fate, and Mobility of Antibiotic Resistant Genes among Microbial Communities Exposed to Alternative Wastewater Treatment Systems. Ph.D. Thesis. University of Waterloo, Waterloo, Ontario, Canada.

Hildebrandt, L., Nack, F.L., Zimmermann, T. and Pröfrock, D. 2021. Microplastics as a Trojan horse for trace metals. J Hazard Mater Lett, 2: 100035. doi:10.1016/J.HAZL.2021.100035.

Hu, S., Chen, Z., Lv, Z., Chen, K., Huang, L., Zuo, X. et al. 2019. Purification of leachate from sludge treatment beds by subsurface flow constructed wetlands: Effects of plants and hydraulic retention time. Environ Sci Pollut Res, 26(6): 5769-5781. doi:10.1007/s11356-018-4006-7.

Issac, M.N. and Kandasubramanian, B. 2021. Effect of microplastics in water and aquatic systems. Environ Sci Pollut Res, 28(16): 19544-19562. doi:10.1007/s11356-021-13184-2.

Jena, G., Dutta, K. and Daverey, A. 2023. Surfactants in water and wastewater (greywater): Environmental toxicity and treatment options. Chemosphere, 341. doi:10.1016/j.chemosphere.2023.140082.

Ji, B. and Zhao, Y. 2023. Trend and status of the NSFC funded projects for constructed wetlands in the last decade in China. Ecol Eng, 194. doi:10.1016/j.ecoleng.2023.107057.

Julisa Burbano Benavides Tutor, C. and Lenin Javier Ramírez Cando, P.D. 2022. Application of an aquaculture byproduct-based biosorbent for the removal of pollutants in aquatic systems. https://repositorio.yachaytech.edu.ec/handle/123456789/570.

Karabelnik, K., Noorvee, A., Põldvere, E. and Mander, Ü. 2008. Batch-operation as a method to enhance oxygen supply in a constructed wetland. pp. 131-142. *In*: U. Mander, C.A. Brebbia and J.F. Martin-Duque [eds.]. WIT Transactions on the Built Environment. Vol. 100. WIT Press, U.K.

Karungamye, P., Rugaika, A., Mtei, K. and Machunda, R. 2023. The removal of ciprofloxacin from synthetic wastewater in constructed wetland. HydroResearch, 6: 138-146. doi:10.1016/j.hydres.2023.04.001.

Koukoura, A., Seintos, T., Statiris, E., Barka, E., Gatidou, G., Noutsopoulos, C. et al. 2024. Comparing the performance of microbial electrochemical assisted and aerated treatment wetlands in pilot-scale: Removal of major pollutants and organic micropollutants. Sci Total Environ, 951. doi:10.1016/j.scitotenv.2024.175550.

Krzeminski, P., Tomei, M.C., Karaolia, P., Langenhoff, A., Almeida, C.M.R., Felis, E. et al. 2019. Performance of secondary wastewater treatment methods for the removal of contaminants of emerging concern implicated in crop uptake and antibiotic resistance spread: A review. Sci Total Environ, 648: 1052-1081. doi:10.1016/j.scitotenv.2018.08.130.

Kumar, R., Sharma, P., Manna, C. and Jain, M. 2021. Abundance, interaction, ingestion, ecological concerns, and mitigation policies of microplastic pollution in riverine ecosystem: A review. Sci Total Environ, 782: 146695. doi:10.1016/j.scitotenv.2021.146695. https://doi.org/10.1016/j.scitotenv.2021.146695.

Kumar, R.R., Verma, A., Shome, A., Sinha, R., Sinha, S., Jha, P.K. et al. 2021. Impacts of plastic pollution on ecosystem services, sustainable development goals, and need to focus on circular economy and policy interventions. Sustainability, 13(17): 9963. doi:10.3390/su13179963.

Li, J., Wang, B., Chen, Z., Ma, B. and Chen, J.P. 2021. Ultrafiltration membrane fouling by microplastics with raw water: Behaviors and alleviation methods. Chem Eng J, 410: 128174. doi:10.1016/J.CEJ.2020.128174.

Lin Lau, P. and Trzcinski, A.P. 2024. A review of modified and hybrid anaerobic baffled reactors for municipal wastewater treatment with a focus on emerging contaminants. Environ Sci Water Res Technol, 10(6): 1335-1354. doi:10.1039/d3ew00822c.

Liu, F., Sun, L., Wan, J., Shen, L., Yu, Y., Hu, L. and Zhou, Y. 2020. Performance of different macrophytes in the decontamination of and electricity generation from swine wastewater via an integrated constructed wetland-microbial fuel cell process. J Environ Sci (China), 89: 252-263. doi:10.1016/j.jes.2019.08.015. https://doi.org/10.1016/j.jes.2019.08.015.

Liu, L., Li, J., Fan, H., Huang, X., Wei, L. and Liu, C. 2019. Fate of antibiotics from swine wastewater in constructed wetlands with different flow configurations. Int Biodeterior Biodegrad, 140(March): 119-125. doi:10.1016/j.ibiod.2019.04.002. https://doi.org/10.1016/j.ibiod.2019.04.002.

López, A.D.F.F., Fabiani, M., Lassalle, V.L., Spetter, C.V. and Severini, M.D.F.F. 2022. Critical review of the characteristics, interactions, and toxicity of micro/nanomaterials pollutants in aquatic environments. Mar Pollut Bull, 174(January): 113276. doi:10.1016/j.marpolbul.2021.113276.

López, D., Leiva, A.M., Arismendi, W. and Vidal, G. 2019. Influence of design and operational parameters on the pathogens reduction in constructed wetland under the climate change scenario. Rev Environ Sci Biotechnol, 18(1): 101-125. doi:10.1007/s11157-019-09493-1.

Mancini, M., Solari, L., Colomer, J. and Serra, T. 2023. Retention of microplastics by interspersed lagoons in both natural and constructed wetlands. J Water Process Eng, 56. doi:10.1016/j.jwpe.2023.104559.

Matamoros, V., Rodríguez, Y. and Albaigés, J. 2016. A comparative assessment of intensive and extensive wastewater treatment technologies for removing emerging contaminants in small communities. Water Res, 88: 777-785. doi:10.1016/j.watres.2015.10.058.

Mehariya, S., Das, P., Thaher, M.I., Abdul Quadir, M., Khan, S., Sayadi, S. et al. 2024. Microalgae: A potential bioagent for treatment of emerging contaminants from domestic wastewater. Chemosphere, 351: 141245. doi:10.1016/J.CHEMOSPHERE.2024.141245.

Meng, X., Jin, M., Feng, Q., Sha, A., Bai, S. and Zhao, X. 2023. Resource and energy utilization of swine wastewater treatment: Recent progress and future directions. Separations, 10(12). doi:10.3390/separations10120591.

Merino-Solís, M.L., Villegas, E., de Anda, J. and López-López, A. 2015. The effect of the hydraulic retention time on the performance of an ecological wastewater treatment system: An anaerobic filter with a constructed wetland. Water (Switzerland), 7(3): 1149-1163. doi:10.3390/W7031149.

Miller, I.B., Pawlowski, S., Kellermann, M.Y., Petersen-Thiery, M., Moeller, M., Nietzer, S. and Schupp, P.J. 2021. Toxic effects of UV filters from sunscreens on coral reefs revisited: Regulatory aspects for "reef safe" products. Environ Sci Eur, 33(1): 1-13. doi:10.1186/S12302-021-00515-W/TABLES/3. https://enveurope.springeropen.com/articles/10.1186/s12302-021-00515-w.

Minakshi, D., Sharma, P.K. and Rani, A. 2022. Effect of filter media and hydraulic retention time on the performance of vertical constructed wetland system treating dairy farm wastewater. Environ Eng Res, 27(1). doi:10.4491/eer.2020.436.

Misiti, T.M., Hajaya, M.G. and Pavlostathis, S.G. 2011. Nitrate reduction in a simulated free-water surface wetland system. Water Res, 45(17): 5587-5598. doi:10.1016/j.watres.2011.08.019.

Mohapatra, D.P. and Kirpalani, D.M. 2019. Advancement in treatment of wastewater: Fate of emerging contaminants. Can J Chem Eng, 97(10): 2621-2631. doi:10.1002/cjce.23533.

Morrissey, C.A., Mineau, P., Devries, J.H., Sanchez-Bayo, F., Liess, M., Cavallaro, M.C. and Liber, K. 2015. Neonicotinoid contamination of global surface waters and associated risk to aquatic invertebrates: A review. Environ Int, 74: 291-303. doi:10.1016/J.ENVINT.2014.10.024.

Mu, C., Wang, L. and Wang, L. 2020. Performance of lab-scale microbial fuel cell coupled with unplanted constructed wetland for hexavalent chromium removal and electricity production. Environ Sci Pollut Res, 27(20): 25140-25148. doi:10.1007/s11356-020-08982-z.

Mukherjee, A.G., Wanjari, U.R., Eladl, M.A., El-Sherbiny, M., Elsherbini, D.M.A., Sukumar, A. et al. 2022. Mixed contaminants: Occurrence, interactions, toxicity, detection, and remediation. Molecules, 27(8): 1-35. doi:10.3390/molecules27082577 [accessed 2022 Nov 15].

Mumtaj, Z.A., Khan, A.R., Alsubih, M., Aleya, L., Khan, R.A. and Khan, S. 2024. Removal of pharmaceutical contaminants from hospital wastewater using constructed wetlands: A review. Environ Sci Pollut Res, 31(9): 12856-12870. doi:10.1007/s11356-024-32022-9.

Nas, B., Ateş, H., Dolu, T., Yel, E., Argun, M.E., Koyuncu, S. et al. 2022. Evaluation of occurrence, fate and removal of priority phthalate esters (PAEs) in wastewater and sewage sludge by advanced biological treatment, waste stabilization pond and constructed wetland. Chemosphere, 295. doi:10.1016/j.chemosphere.2022.133864.

Nasri, A., Mezni, A., Lafon, P.A., Wahbi, A., Cubedo, N., Clair, P. et al. 2021. Ethinylestradiol (EE2) residues from birth control pills impair nervous system development and swimming behavior of zebrafish larvae. Sci Total Environ, 770: 145272. doi:10.1016/J.SCITOTENV.2021.145272.

Nguyen, H.T.T., Chao, H.R. and Chen, K.C. 2019. Treatment of organic matter and tetracycline inwater by using constructed wetlands and photocatalysis. Appl Sci, 9(13). doi:10.3390/app9132680.

Niu, Y., Qu, M., Du, J., Wang, X., Yuan, S., Zhang, L. et al. 2023. Effects of multiple key factors on the performance of petroleum coke-based constructed wetland-microbial fuel cell. Chemosphere, 315. doi:10.1016/j.chemosphere.2023.137780.

Nõlvak, H., Truu, M., Tiirik, K., Oopkaup, K., Sildvee, T., Kaasik, A. et al. 2013. Dynamics of antibiotic resistance genes and their relationships with system treatment efficiency in a horizontal subsurface flow constructed wetland. Sci Total Environ, 461-462: 636-644. doi:10.1016/j.scitotenv.2013.05.052.

Noriega-Rico, E.A., Caselles-Osorio, A., Ortega Herrera, A. and Cerro Medina, J. 2021. Uptake and accumulation of triclosan in cyperus articulatus L. planted in a constructed wetland for the treatment of domestic wastewater. Water Air Soil Pollut, 232(11). doi:10.1007/S11270-021-05413-8.

Nuel, M., Laurent, J., Bois, P., Heintz, D. and Wanko, A. 2018. Seasonal and ageing effect on the behaviour of 86 drugs in a full-scale surface treatment wetland: Removal efficiencies and distribution in plants and sediments. Sci Total Environ, 615: 1099-1109. doi:10.1016/j.scitotenv.2017.10.061.

Nyieku, F.E., Essandoh, H.M.K., Armah, F.A. and Awuah, E. 2022. Oilfield wastewater contaminants removal efficiencies of three indigenous plants species in a free water surface flow constructed wetland. Sustain Environ, 8(1). doi:10.1080/27658511.2022.2076361.

Nzouebet, W.A.L., Kengne, E.S., Rechenburg, A., Ndangang, J.J.N. and Noumsi, I.M.K. 2022. Fecal sludge treatment in a pilot-scale vertical-flow constructed wetland for the removal of helminth eggs. J Water Sanit Hyg Dev, 12(9): 634-646. doi:10.2166/washdev.2022.090.

Oliveira, J.F.D., Fia, R., Rodrigues, F.N., Fia, F.R.L., Matos, M.P.D., Siniscalchi, L.A.B. and Sanson, A.L. 2020. Quantification, removal and potential ecological risk of emerging contaminants in different organic loads of swine wastewater treated by integrated biological reactors. Chemosphere, 260. doi:10.1016/j.chemosphere.2020.127516.

Omondi Donde, O., Makindi, S.M., Tian, C., Tian, Y., Hong, P., Cai, Q. et al. 2020. A novel integrative performance evaluation of constructed wetland on removal of viable bacterial cells and related pathogenic, virulent and multi-drug resistant genes from wastewater systems. J Water Process Eng, 33(August 2019): 101060. doi:10.1016/j.jwpe.2019.101060. https://doi.org/10.1016/j.jwpe.2019.101060.

Overton, O.C., Olson, L.H., Majumder, S.D., Shwiyyat, H., Foltz, M.E. and Nairn, R.W. 2023. Wetland removal mechanisms for emerging contaminants. Land, 12(2). doi:10.3390/land12020472.

Parashar, V., Singh, S., Purohit, M.R., Tamhankar, A.J., Singh, D., Kalyanasundaram, M. et al. 2022. Utility of constructed wetlands for treatment of hospital effluent and antibiotic resistant bacteria in resource limited settings: A case study in Ujjain, India. Water Environ Res, 94(9): 1-10. doi:10.1002/wer.10783.

Parida, V.K., Sikarwar, D., Majumder, A. and Gupta, A.K. 2022. An assessment of hospital wastewater and biomedical waste generation, existing legislations, risk assessment, treatment processes, and scenario during COVID-19. J Environ Manage, 308. doi:10.1016/j.jenvman.2022.114609.

Patyal, V., Jaspal, D. and Khare, K. 2021. Materials in constructed wetlands for wastewater remediation: A review. Water Environ Res, 93(12): 2853-2872. doi:10.1002/wer.1648.

Premarathna, K.S.D., Mohan, D., Biswas, J.K., Wijesekara, H., Jayasanka, J. and Vithanage, M. 2023. Arsenic interaction with microplastics: Implications for soil-water-food nexus. Curr Opin Environ Sci Heal, 34: 100482. doi:10.1016/J.COESH.2023.100482.

Qin, G., Feng, H., Yu, R., Zheng, F., Jiang, X., Xia, L. and An, S. 2023. Study on the removal characteristics of IBP and DCF in wastewater by CW-MFC with different co-substrates. Water (Switzerland), 15(21). doi:10.3390/w15213862.

Rajan, R.J., Sudarsan, J.S. and Nithiyanantham, S. 2019. Microbial population dynamics in constructed wetlands: Review of recent advancements for wastewater treatment. Environ Eng Res, 24(2): 181-190. doi:10.4491/EER.2018.127.

Rampuria, A., Gupta, A.B., Kulshreshtha, N.M. and Brighu, U. 2021. Microbiological analysis of two deep constructed wetlands with special emphasis on the removal of pathogens and antibiotic-resistant bacteria. Water Air Soil Pollut, 232(5). doi:10.1007/s11270-021-05121-3.

Ravichandran, M.K., Thayyil, M.I. and Philip, L. 2023. Application of engineered natural treatment systems for pollution abatement. *In*: L. Philip, T. Pradeep and S.M. Bhallamudi [eds.]. Technological Solutions for Water Sustainability: Challenges and Prospects: Towards a Water-secure India. IWA Publishing, London, U.K.

Ren, Y.-X., Zhang, H., Wang, C., Yang, Y.-Z., Qin, Z. and Ma, Y. 2011. Effects of the substrate depth on purification performance of a hybrid constructed wetland treating domestic sewage. J Environ Sci Heal – Part A Toxic/Hazardous Subst Environ Eng, 46(7): 777-782. doi:10.1080/13632469.2011.571960.

Riva, V., Vergani, L., Rashed, A.A., El Saadi, A., Sabatino, R., Di Cesare, A. et al. 2024. Plant species influences the composition of root system microbiome and its antibiotic resistance profile in a constructed wetland receiving primary treated wastewater. Front Microbiol, 15. doi:10.3389/fmicb.2024.1436122.

Ruiz-Ocampo, H., Tondera, K., Katusic, V., Paing, J., Molle, P. and Chazarenc, F. 2021. Effect of filter media and depth on hydrodynamics and treatment performances of single-stage French vertical flow treatment wetlands treating domestic effluent. Water Air Soil Pollut, 232(7). doi:10.1007/s11270-021-05228-7.

Sánchez, M., Ruiz, I. and Soto, M. 2022. The potential of constructed wetland systems and photodegradation processes for the removal of emerging contaminants—A review. Environ – MDPI. 9(9). doi:10.3390/environments9090116.

Saravanakumar, K., Siva Santosh, S., Sathiyaseelan, A., Naveen, K.V., Afaan Ahamed, M.A., Zhang, X. et al. 2022. Unraveling the hazardous impact of diverse contaminants in the marine environment: Detection and remedial approach through nanomaterials and nano-biosensors. J Hazard Mater, 433. doi:10.1016/j.jhazmat.2022.128720.

Sarma, H. 2022. Understanding emerging contaminants in soil and water: Current perspectives on integrated remediation approaches. pp. 1-38. *In:* H. Sarma, D. C. Dominguez and W. Y. Lee [eds.]. Emerging Contaminants in the Environment Challenges and Sustainable Practices. Elsevier.

Savvidou, P., Dotro, G., Campo, P., Coulon, F. and Lyu, T. 2024. Constructed wetlands as nature-based solutions in managing per- and poly-fluoroalkyl substances (PFAS): Evidence, mechanisms, and modelling. Sci Total Environ, 934. doi:10.1016/j.scitotenv.2024.173237.

Saz, Ç., Türe, C., Türker, O.C. and Yakar, A. 2018. Effect of vegetation type on treatment performance and bioelectric production of constructed wetland modules combined with microbial fuel cell (CW-MFC) treating synthetic wastewater. Environ Sci Pollut Res, 25(9): 8777-8792. doi:10.1007/s11356-018-1208-y.

Sgroi, M., Pelissari, C., Ávila, C., Sezerino, P.H., Vagliasindi, F.G.A., García, J. et al. 2017. Removal of conventional water quality parameters, emerging contaminants and fluorescing organic matter in a hybrid constructed wetland system. pp. 313-317. *In:* Mannina, G. [eds.]. Frontiers in Wastewater Treatment and Modelling. FICWTM 2017. Lecture Notes in Civil Engineering , vol 4. Springer, Cham. https://doi.org/10.1007/978-3-319-58421-8_50.

Sha, H., Song, X., Abdullah Al-Dhabi, N., Zeng, T., Mao, Y., Fu, Y. et al. 2024. Effects of biochar layer position on treatment performance and microbial community in subsurface flow constructed wetlands for removal of cadmium and lead. Bioresour Technol, 394. doi:10.1016/j.biortech.2023.130194.

Sharma, P.K., Rausa, K., Rani, A., Mukherjee, S. and Kumar, M. 2021. Biopurification of dairy farm wastewater through hybrid constructed wetland system: Groundwater quality and health implications. Environ Res, 200: 111426. doi:10.1016/j.envres.2021.111426. https://doi.org/10.1016/j.envres.2021.111426.

Sharma, P.K., Rani, A., Minakshi, D. and Malaviya, P. 2023. Effect of influent load fluctuation on the efficiency of vertical constructed wetlands treating dairy farm wastewater. Adv Environ Technol., 9(3): 227-241. doi:10.22104/aet.2023.6224.1716.

Shruthi, R., Gaonkar, O., Kumar, R.M., Mazumdar, D., Karthik, P., Kumar, S. et al. 2022. Effective removal of naphthalene adsorbed on biochar by low-frequency ultrasonication in a customized reactor. pp. 323-337. *In:* P. Chakraborty and D. Snow [eds.]. Legacy and Emerging Contaminants in Water and Wastewater. Emerging Contaminants and Associated Treatment Technologies. Springer, Cham. https://doi.org/10.1007/978-3-030-95443-7_14.

Si, Z., Wang, Y., Song, X., Cao, X., Zhang, X. and Sand, W. 2019. Mechanism and performance of trace metal removal by continuous-flow constructed wetlands coupled with a micro-electric field. Water Res, 164. doi:10.1016/j.watres.2019.114937.

Sol, D., Laca, A., Laca, A. and Díaz, M. 2020. Approaching the environmental problem of microplastics: Importance of WWTP treatments. Sci Total Environ, 740. doi:10.1016/j.scitotenv.2020.140016.

Song, H.L., Zhang, S., Guo, J., Yang, Y.L., Zhang, L.M., Li, H. et al. 2018. Vertical up-flow constructed wetlands exhibited efficient antibiotic removal but induced antibiotic resistance genes in effluent. Chemosphere, 203: 434-441. doi:10.1016/j.chemosphere.2018.04.006. https://doi.org/10.1016/j.chemosphere.2018.04.006.

Stefanatou, A., Vouzi, L., Petousi, I., Koukoura, A., Gatidou, G., Stasinakis, A.S. and Fountoulakis, M.S. 2024. Treatment of real laundry wastewater using vertical flow constructed wetland planted with the ornamental climbing plant Trachelospermum jasminoides: Assessing the removal of conventional pollutants and benzotriazoles. Environ Sci Pollut Res, 31(30): 43281-43291. doi:10.1007/s11356-024-34035-w.

Tadić, Đ., Sauvêtre, A., Cerqueira, F., Lestremau, F., Ait-Mouheb, N. and Chiron, S. 2024. Partially saturated vertical surface flow constructed wetland for emerging contaminants and antibiotic resistance genes removal from wastewater: The effect of bioaugmentation with Trichoderma. J Environ Chem Eng, 12(2). doi:10.1016/j.jece.2024.112128.

Talib, A. and Randhir, T.O. 2016. Managing emerging contaminants: Status, impacts, and watershed-wide strategies. Expo Heal, 8(1): 143-158. doi:10.1007/s12403-015-0192-4.

Tan, Y.Y., Tang, F.E., Saptoro, A. and Khor, E.H. 2020. Effect of loading rate and sludge deposit on a VFCW treating septage. J Appl Water Eng Res, 8(1): 1-14. doi:10.1080/23249676.2020.1719217.

Teoh, T.-P., Ong, S.-A., Ho, L.-N., Wong, Y.-S., Lutpi, N.A., Tan, S.-M. et al. 2024. Discerning the effect of operating conditions on the improvement of up-flow constructed wetland-microbial fuel cell performance in treating mixed azo dyes wastewater and bioelectricity generation. Energy, Ecol Environ, doi:10.1007/s40974-023-00314-4.

Tesfamariam, E.H., Badza, T. and Cogger, C. 2022. Impact of processing technology on the chemical contaminants occurrence in end products. *In*: S. Thakur et al. [eds.]. Environmental Toxicity, Health Hazards, and Bioremediation Strategies for Removal of Microplastics from Wastewater. John Wiley & Sons Ltd., U.K.

Thanh H.P., Tadashi, T. and Kazuhiro, M. 2017. Removal of tetracycline and tetracycline resistance genes from municipal wastewater in microcosm fill-and-drain constructed wetlands. Japanese J Water Treat Biol, 53(1): 11-21. doi:10.2521/JSWTB.53.11.

Tian, Y., Wang, X., Liang, Y., Li, B., Liu, Y., Feng, L. et al. 2023. Removal of oxytetracycline and sulfamethoxazole in vertical down-up flow constructed wetlands: Efficiency, removal pathways, and occurrence of antibiotic resistance genes. J Water Process Eng, 55(August). doi:10.1016/j.jwpe.2023.104142.

Tiwari, S., Kushwaha, R., Tripathi, P., Singh, R.S. and Mohan, D. 2021. Emerging pollutants: Occurrence, fate, toxicological impacts and removal technologies. *In*: M.P. Shah, G.Y. Tore and S.P. Bera [eds.]. An Innovative Approach of Advanced Oxidation Processes in Wastewater Treatment. Nova Science Publishers, Inc., New York, USA.

Torrens, A., de la Varga, D., Ndiaye, A.K., Folch, M. and Coly, A. 2020. Innovative multistage constructed wetland for municipal wastewater treatment and reuse for agriculture in Senegal. Water (Switzerland), 12(11): 1-12. doi:10.3390/w12113139.

Upadhyay, S., Sharma, P.K., Dogra, K., Bhattacharya, P., Kumar, M., Tripathi, V. and Karmakar, R. 2024. Microplastics in freshwater: Unveiling sources, fate, and removal strategies. Groundw Sustain Dev, 26(August 2023): 101185. doi:10.1016/j.gsd.2024.101185. https://doi.org/10.1016/j.gsd.2024.101185.

Verma, P., Rajgor, D., Satasiya, G. and Ray, S. 2024. In-depth performance evaluation of an innovative halophyte-based multistage constructed wetland-microbial fuel cell for the treatment of brackish sewage. Chem Eng J, 498. doi:10.1016/j.cej.2024.155262.

Vymazal, J., Zhao, Y. and Mander, Ü. 2021. Recent research challenges in constructed wetlands for wastewater treatment: A review. Ecol Eng, 169: 106318. doi:10.1016/J.ECOLENG.2021.106318.

Wang, L., Ma, L., Wang, J., Zhao, X., Jing, Y., Liu, C. et al. 2024. Research progress on the removal of contaminants from wastewater by constructed wetland substrate: A review. Water (Switzerland), 16(13). doi:10.3390/w16131848.

Wu, B., Xu, D., Wang, H., Xu, R., Qin, N. and Han, J. 2022. Wetland plant-derived biochar enhances the diclofenac treatment performance in vertical subsurface flow constructed wetlands. Environ Res, 215. doi:10.1016/j.envres.2022.114326.

Yaripour, S., Huuskonen, H., Rahman, T., Kekäläinen, J., Akkanen, J., Magris, M. et al. 2021. Pre-fertilization exposure of sperm to nano-sized plastic particles decreases offspring size and swimming performance in the European whitefish (Coregonus lavaretus). Environ Pollut, 291: 118196. doi:10.1016/j.envpol.2021.118196. [accessed 2023 Oct 16]. http://creativecommons.org/licenses/by/4.0/.

Yin, H., Zhang, M., Huo, L. and Yang, P. 2022. Efficient removal of phosphorus from constructed wetlands using solidified lanthanum/aluminum amended attapulgite/biochar composite as a novel phosphorus filter. Sci Total Environ, 833. doi:10.1016/j.scitotenv.2022.155233.

Youssef, Y.A., Abuarab, M.E., Mahrous, A. and Mahmoud, M. 2023. Enhanced degradation of ibuprofen in an integrated constructed wetland-microbial fuel cell: Treatment efficiency, electrochemical characterization, and microbial community dynamics. RSC Adv, 13(43): 29809-29818. doi:10.1039/d3ra05729a.

Yu, Y.H., Su, J.F., Shih, Y., Wang, J., Wang, P.Y. and Huang, C.P. 2020. Hazardous wastes treatment technologies. Water Environ Res, 92(10): 1833-1860. doi:10.1002/WER.1447. [accessed 2023 Jun 14]. https://onlinelibrary.wiley.com/doi/full/10.1002/wer.1447.

Yuan, Z., Nag, R. and Cummins, E. 2022. Human health concerns regarding microplastics in the aquatic environment – From marine to food systems. Sci Total Environ, 823: 153730. doi:10.1016/j.scitotenv.2022.153730. https://doi.org/10.1016/j.scitotenv.2022.153730.

Zhang, H., Wang, X.C., Zheng, Y. and Dzakpasu, M. 2023. Removal of pharmaceutical active compounds in wastewater by constructed wetlands: Performance and mechanisms. J Environ Manage, 325. doi:10.1016/j.jenvman.2022.116478.

Zhao, C., Liu, H., Cheng, D., Wang, Y., Hu, Z., Wu, H. et al. 2025. Insights into poly- and perfluoroalkyl substances (PFAS) removal in treatment wetlands: Emphasizing the roles of wetland plants and microorganisms. Water Res, 268. doi:10.1016/j.watres.2024.122702.

Zheng, Y., Liu, Y., Qu, M., Hao, M., Yang, D., Yang, Q. et al. 2021. Fate of an antibiotic and its effects on nitrogen transformation functional bacteria in integrated vertical flow constructed wetlands. Chem Eng J, 417(February): 129272. doi:10.1016/j.cej.2021.129272. https://doi.org/10.1016/j.cej.2021.129272.

Zhou, Y., Li, W.-B., Kumar, V., Necibi, M.C., Mu, Y.-J., Shi, C.-Z. et al. 2022. Synthetic organic antibiotics residues as emerging contaminants waste-to-resources processing for a circular economy in China: Challenges and perspective. Environ Res, 211. doi:10.1016/j.envres.2022.113075.

CHAPTER

6

Microalgae-based Wastewater Treatment and Production of Value Added Products

Sonal Singh[1], Kuldeep Dwivedi[1], Rwitabrata Mallick[1], Shashank Gupta[2], Kumar Shrestha[3], Vijay Tripathi[4,5] and Nidhi Shukla[1*]

[1] Department of Environmental Science, Amity University, Gwalior, M.P., India

[2] Department of Civil Engineering, ITM University, Gwalior, M.P., India

[3] National Centre for Aquatic Animal Health, Cochin University, Kerala - 682016, India

[4] Department of Molecular and Cellular Engineering, Jacob Institute of Biotechnology & Bioengineering, Sam Higginbottom University of Agriculture, Technology & Sciences, Prayagraj, U.P. - 211007

[5] Department of Microbiology, Graphic Era Deemed to be University, Clement Town, Dehradun, U.K. - 248002

Introduction

Microalgae, such as cyanobacteria and eukaryotic algae, are a sustainable and environmentally beneficial substitute for the energy-intensive, traditional biological treatment processes still in use today (Singh et al. 2015). Microalgae treatment for wastewater treatment is a viable and affordable way to bio-fix CO_2 and serve as a renewable biomass supply (Almomani 2020). Mixotrophic microalgae are used to remediate wastewater because they can use the inorganic carbon, organic carbon, inorganic nitrogen and phosphorus found in wastewater to grow, which reduces the pollution level in the wastewater (Mohsenpour et al. 2021). The main benefit of using microalgae to clean wastewater is that they produce oxygen through photosynthesis, which is required for heterotrophic bacteria to break down carbonaceous pollutants (Abdelfattah et al. 2023).

Many studies have shown that algal formation can promote nutrient removal in wastewater, although it is challenging to compare the effect of algal growth in wastewater treatment (Chawla et al. 2020, Magalhães et al. 2024, Bang Truong et al. 2024).

*Corresponding author: nidthesaviour@gmail.com; nshukla@gwa.amity.edu

DOI: 10.1201/9781003527350-6

The employment of algae granules in synthetic wastewater is particularly effective at removing phosphorus and recovering it for later usage from the resulting P-rich algae biomass (Cai et al. 2019). Microalgae have demonstrated their efficiency in removing nutrients and CO_2 from wastewater and their potential as a source of energy generation (Arun et al. 2020).

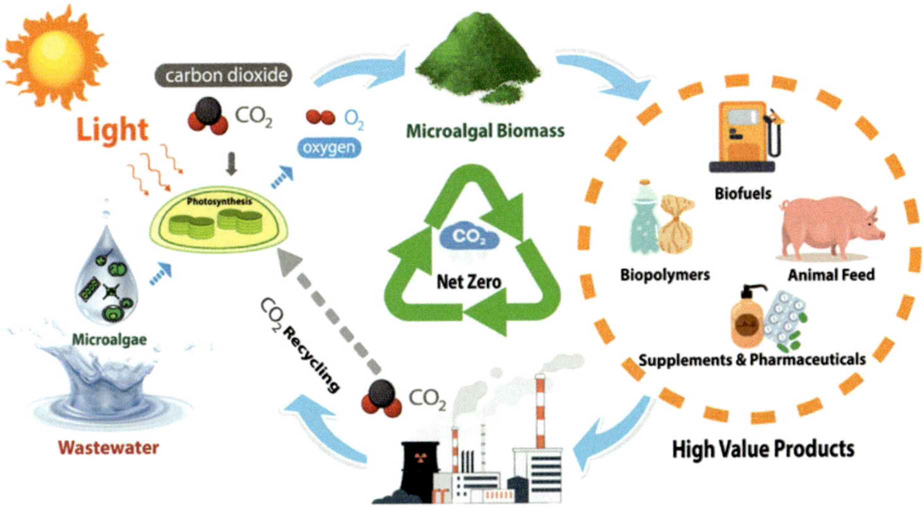

Figure 1: Microalgae-based wastewater treatment for a circular green economy (Srimongkol et al. 2022).

Both organic nitrogen (urea) and inorganic nitrogen (ammonium/ammonia), as well as nitrite and nitrates, can be used by microalgae (Ross et al. 2018). The environmental circumstances in which N-removal occurs cause the emission of N_2O throughout the wastewater treatment process (Arun et al. 2020). Moreover, wastewater treated by an algal-bacterial co-culture requires only a single treatment step, lowering the complexity and energy of the treatment process by facilitating the removal of inorganic N and P without switching between different operating environments (Gouveia et al. 2016). This is because microalgae directly and jointly assimilate phosphate (PO_4) and ammonia (NH_3) for metabolic processes and cell growth (Falkowski and Raven 2007, Borowitzka 2016). The microalgae absorb most nitrogen rather than transforming it into nitrogen oxides. Numerous research works have documented the small amount of N_2O released from the combination of microalgae and related microbes throughout the wastewater treatment process (Guieysse et al. 2013, Fagerstone et al. 2011). An emission factor of 0.0047% g N_2O-N g^{-1} N-input is determined for a microalgae wastewater treatment process, according to Alcantara et al. 2015. Providing dissolved oxygen to wastewater via microalgae photosynthesis is reliable for saving substantial energy and reducing related greenhouse gas emissions.

Emerging Pollutants in Wastewater

Pharmaceutical and Personal Care Products (PPCPs)

A broad range of substances found in PPCPs are among the micropollutants classified as "emerging contaminants" or "emerging pollutants". They are a sizable category of pollutants from sources directly connected to human activities, including pharmaceuticals, cosmetics, personal cleanliness products, industrial and agricultural run-off, and hospital discharges (Sangion and Gramatica 2016). Numerous PPCP chemicals are unintentionally and uncontrollably released into the environment from direct (factories, refineries, waste treatment plants) and indirect sources (agricultural run-off, acid rain, etc.). Previous studies have observed and reported unregulated sales and misuse of pharmaceutical items, mainly due to the considerable benefits received from their use in limiting the spread of human and veterinary diseases (Miyazaki et al. 2020, Rees et al. 2021). Using detection techniques like Gas Chromatography-Mass Spectrometry (GC-MS), clofibric acid (a pharmaceutical metabolite) and salicylic acid (which can be dangerous to aquatic life in high concentrations) are found in sewage and river effluent (Garrison et al. 1976, Hignite and Azarnoff 1977). More than 50 medications have been found in surface waters across 139 rivers spread over 30 states in the United States (Kolpin et al. 2002). Sewage treatment facilities in southern India contain PPCPs (Subedi et al. 2015). Amphetamine, saccharin, cyclamate, and sucralose were found by these authors at 4300 ng/l, 303,000 ng/l, 3460 ng/l, and 1460 ng/l, respectively. Subedi and colleagues assessed the regular release of amphetamine and saccharin at 6.93 kg and 2.52 kg, respectively, based on a normal daily sewage flow rate of 20.7 million litres received from a population of 325,000 (Subedi et al. 2015). Because rivers are essential to anthropogenic and socioeconomic activity, the prevalence of PPCPs in river systems is extensively researched (Sharma et al. 2019). Kumar et al. (2019) stated that PPCPs are mainly released into major water bodies like rivers and lakes by wastewater drainage systems and precipitation run-off.

Rivers are the primary sources of irrigation and drinking water in India. It's also not surprising that multiple studies regarding the existence of PPCPs in some major rivers of India, like the Yamuna (Mutiyar et al. 2018), Ganga (Sharma et al. 2019), and Brahmaputra (Kumar et al. 2019), were conducted (Balakrishna et al. 2017), given that India is ranked 13th in the world in terms of pharmaceutical product consumption (Mutiyar et al. 2018). Singh and Suthar investigated the presence of triclosan, caffeine, acetaminophen, and tetracycline in the Ganga River, and the overall concentration of PPCP compounds was 1104.84 mg/l below the detection limit (Singh and Suthar 2021). The presence of PPCPs in the river system can affect the river microbiome and induce several illnesses that impact human body organs, including the heart, kidneys, liver, and others (Figure 2).

Agrochemicals

Agrochemicals, fertilisers, and pesticides are necessary to increase crop yields and have become an essential component of modern agriculture (Alix and Capri 2018). In addition, agrochemicals improve soil health and crop growth (Simonne

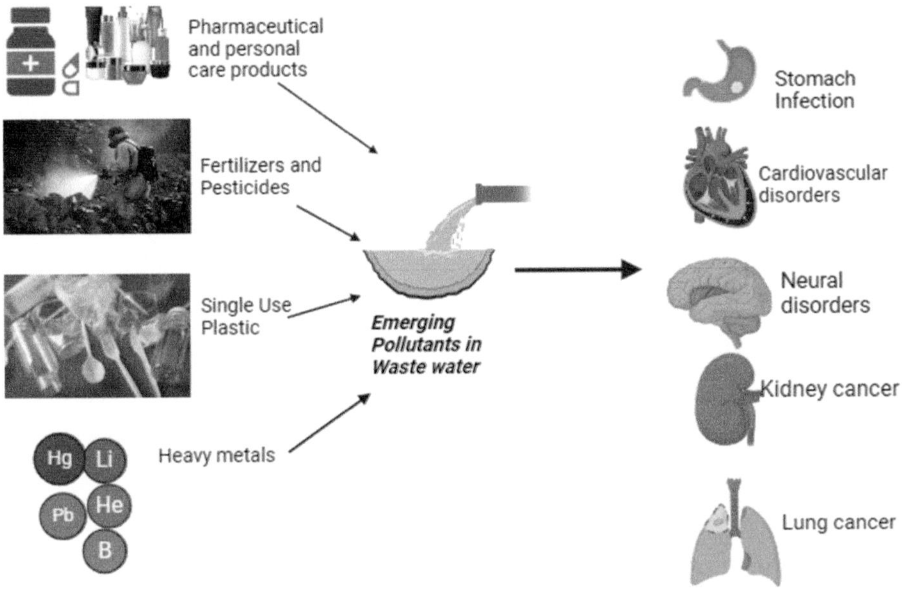

Figure 2: Sources of emerging pollutants in wastewater and diseases caused by wastewater.

et al. 2017). According to statistics, fertilisers (nitrogen (N), phosphorus (P), and potassium (K)) are applied to produce the majority of the world's food in the modern era. By 2050, fertiliser consumption levels are expected to rise by 150%, 172%, and 175%, respectively, as reported by Khan et al. (2018). Because fertilisers can also contain certain heavy metals during manufacturing, the indiscriminate use of fertilisers has lowered the soil's hydrogen ion concentration (pH), promoting the availability of heavy metals in soil (Alengebawy et al. 2021). Pesticides are typically used to eliminate weeds, insects, and other pests from agricultural fields. In very high concentrations, pesticides are considered the most common pollutants in ecological systems, as Alvarez et al. (2017) reported. Generally, DDT (Dichloro Diphenyl Trichloroethane), HCH (Hexa Chlorocyclohexane), Aldrin, Dieldrin, Endosulfan, Chlordane, Endrin, Heptachlor, Mirex, Hexachlorobenzene, Toxaphene, Methoxychlor, and Metolachlor are classified as organochlorine-based insecticides used to protect both human and crop health (Yadav et al. 2015).

According to estimates from the Food and Agriculture Organization (FAO/UNEP/OECD/SIB 2001), approximately 5 lakh tonnes of pesticides are improperly utilised in agricultural fields, posing a significant risk to human and environmental health. Due to pesticide and fertiliser residues in food products, many developed and developing nations are currently experiencing health issues in their populations (Singh et al. 2006, Agoramoorthy 2008). According to Yadav et al. (2015), the toxic nature of pesticides causes about 3 million cases of severe poisoning each year. Every environmental component, including the air (Lammel et al. 2007), food (Bajpai et al. 2007), aquatic animal like fishes (Malik et al. 2007), milk (Zhao et al. 2007), soil

(Fuentes et al. 2010), water (Kumari et al. 2007), human plasma and adipose tissues (Ridolfi et al. 2014), and water (Malik et al. 2007), is reported to contain pesticides.

The researchers found that there are two categories for the severity of health problems caused by pesticide poisoning in humans: mild or moderate, such as headache, faintness, stomach issues, numbness, hyperirritability, flu, skin inflammation, blurred vision, and neural issues; and severe, such as paraplegia, loss of eyesight, and occasionally loss of life (Yadav et al. 2015). A surplus of fertilisers enters water bodies through leakage, irrigation, rain, soil erosion, or other means and affects the physicochemical characteristics of the water. Groundwater is also impacted by leaching of contaminated water (Namdev et al. 2006).

Single-use Plastic

Over 300 forms of plastic are created, with about 60 being the most widely used. Based on their intended use, plastics can be separated into engineering and general plastics (Yuan 2009). The most common polymers found in everyday plastic products, particularly disposable ones like plastic packaging and water bottles, are polyethylene (PE), polypropylene (PP), polyvinyl chloride (PVC), polystyrene (PS), polyurethane (PU), and phenolic resin (Giacovelli et al. 2018, Kankanige and Babel 2020). These polymers are called single-use plastics (SUPs) since they are meant to be used once.

Additionally, SUPs have demonstrated a slow rate of degradation and fragmentation into microscopic particles, known as secondary microplastics (MPs), when exposed to heat, mechanical stress, or UV radiation (Schopel and Stamminger 2019). Plasticisers like phthalates or phthalic acid esters (henceforth called "PAEs") are frequently used in SUP products and found in food containers, plastic bags, and tableware. Their purpose is to increase the plasticity of resin molecules, making the plastic more flexible and easier to process (Huo et al. 2017). Plastic-encased PAEs have the potential to leak into packaged foods or the environment, where they could subsequently find their way into the public food chain.

Four widely consumed drinks—soda, lemonade, cola, and mineral water—sold in Turkish local markets in plastic bottles have been shown to contain PAEs (Ustun et al. 2014). In Saudi markets, plastic water bottles kept at varying temperatures were found to contain the following chemicals: dimethyl phthalate (DMP), diethyl phthalate (DEP), benzyl butyl phthalate (BBP), and diethylhexyl phthalate (DEHP) (Alshehri et al. 2022). Several documented instances exist of animals being injured or dying after ingesting plastics, both on land and in water (Cartraud et al. 2019). Plastic trash, such as the residue from mulching films, impedes the flow of air and water through soils in agricultural areas. Because of this, cropland production may decrease, and plants may be unable to absorb nutrients (Yan et al. 2014).

SUP waste has caused significant pollution to lakes, rivers, and seas (Boucher et al. 2019). It is commonly reported that dolphins and turtles consume plastic materials, especially plastic bags, mistaking them for food (Cartraud et al. 2019). Medical supplies, including sanitary pads and newborn diapers, have been found to contain high concentrations of PAEs (Gimeno et al. 2014), as have extension lines, drip chambers, and bags (Park et al. 2019).

According to estimates from the Federation of Indian Chambers of Commerce and Industry (2017), India consumes 11 kg of plastic per person annually, and 661 kg of plastic garbage was produced in 2017–2018 (CPCB 2018). According to Padgelwar et al. (2019), the main sources of household plastic trash include plastic packaging (which makes up 37% of garbage), bags, wraps, and sacks (29%), PET bottles (16%), and other plastic containers (13%).

Heavy Metals

Any living thing requires water to survive, notably humans, who comprise 70% of the planet's water. 785 million people lack access to drinking water, a basic necessity for living, and 144 million depend on surface water. Water pollution from many sources is causing problems for people worldwide (Verma et al. 2020). According to a recent estimate, at least 2 billion people consume tainted water (World Health Organization 2019). Numerous industrial processes, including mining, the production of iron, steel, and non-ferrous metals; glass manufacturing; tanneries; the production and consumption of energy and fuel; pipe corrosion; agro-industrial processes; municipal waste processing facilities and WWTPs, release heavy metals into the environment (EEA 2019). Human activities, vehicles, and industrial pollutants are responsible for the emission of environmental pollutants like heavy metals (HMs). Lead (Pb), Cadmium (Cd), and Arsenic (As) are the most dangerous HMs. Furthermore, the deposition of HMs in plants and soil might result from sludge discharge in agricultural fields. In the last five decades, 22,000 tonnes of Cd, 939,000 tonnes of Copper (Cu), 783,000 tonnes of Pb and 1,350,000 tonnes of Zinc (Zn) have been released globally into the soil, atmosphere and water bodies (Cao et al. 2024). If heavy metals are present in aquatic ecosystems, they are tough to eliminate. They disturb aquatic ecosystems and human health (Lee et al. 2024). The main heavy metal and metalloid forms found in different ecosystems are cadmium (Cd), mercury (Hg), arsenic (As), lead (Pb), chromium (Cr), copper (Cu), zinc (Zn), nickel (Ni), and so forth, reported by Angon et al. (2023). Therefore, heavy metals are released into the environment due to various industrial operations and related activities. These channels include solid waste, wastewater, landfill leachate, and direct soil pollution. However, natural processes such as the weathering and erosion of rocks can release heavy metals (Upadhyay et al. 2024).

Elements with a density of more than 5.0 g/cm^3 are classified as heavy metals. Common sources of heavy metals include chromium, cadmium, arsenic, and so on (Modak et al. 2022). Because of inappropriate disposal, heavy metal contamination is a big concern for developed and developing countries (Roy and Bharadvaja 2020). Emenike et al. (2018) stated that commercial and industrial operations typically introduce heavy metals into wastewater. Higher concentrations of heavy metals can harm cell membranes, lower seed viability, reduce pollen grains, and negatively affect flora and fauna. Ahmad et al. (2022a) state that these are not biodegradable and are highly hazardous.

Adsorption, electrocoagulation, coagulation, ultrafiltration, reverse osmosis, and ion exchange are a few of the treatment technologies that have historically been employed to remove heavy metals from surface water and wastewater (Sarode et al. 2019).

Sources of Wastewater

The growing human population has led to a rise in the use of freshwater resources. Wastewater can be divided into three categories based on its source: municipal, agricultural, and industrial (Diniz et al. 2016). These wastewater types include nutrients like carbon, nitrogen, phosphorus, and other trace components suitable for microalgae growth. Numerous studies have demonstrated that using wastewater not only allows the reuse of wastewater itself but also transforms and produces a significant amount of biomass, which is particularly useful for producing microalgae biofuels and other applications with a wide range of potential uses (Hernandez-Garcia et al. 2019). Table 1 shows different wastewater sources, treatment methods used by different microalgae.

Municipal Wastewater

Municipal wastewater generated from domestic usage contains nutrients, organic matter, inorganic solids, and several potential pollutants. In general, there are four types of municipal wastewater: raw sewage before primary sedimentation, wastewater following primary sedimentation, wastewater treated with activated sludge (also known as secondary effluent), and centrate generated from sludge dehydration (Vajargah et al. 2023). According to Chen et al. (2023), *Chlorella sp.* grows more favourably in centrate than in the other three forms of wastewater and can adapt to the aforementioned four types of wastewaters. The rate at which *Chlorella* removed phosphorus, nitrogen, and chemical oxygen demand (COD) was also positively connected with the amount of nutrients in the effluent. In primary sewage, secondary sewage, and centrate, Al Momani et al. (2016) examined the growth and purification capacities of *Chlorella vulgaris*, *Neochloris obeoabundans* and mixed indigenous cultured algae *C. vulgaris (UTEX 265)* and *N. obeoabundans (UTEX 1185),* which has the removal efficiency 63.2-80.8% for inorganic nitrogen and 30.8-70% for phosphorus.

Wastewater from Agriculture

Animal manure wastewater and wastewater from the processing of agricultural products are among the wastewater that is released throughout the processes of growing crops, raising livestock, and producing agricultural products (Beltran-Flores et al. 2020, Khalid et al. 2019, Muhmood et al. 2019). Various findings have demonstrated that wastewater from the processing of agricultural products, such as potato, palm oil, and starch mills, can be utilized to produce microalgae (Tan et al. 2019, Li et al. 2019).

To accomplish the dual goals of lowering the cost of microalgae cultivation and resource utilisation of wastewater, it was discovered that animal manure-containing WW appears to be suitable as a medium for the development of microalgae when compared with the mineral composition of several generally used media for microalgae (Li et al. 2021). According to reports, *chlorella vulgaris* can be successfully adapted to remove contaminants from undiluted cow farm wastewater (WW). The removal rate can increase after two stages of *C. vulgaris*-based biological treatment (Lv

Table 1: Microalgae-based different WW with pollutant source and removal method

S. No.	Microalgae	Waste Water Treatment	Pollutant/nutrients source	Process of removal	Reference
1	*Spirulina platensis*	Dye-containing synthetic wastewater	Dye	Iron oxide nanoparticles using Microalgae	Shalaby 2021
2	Native microalgae	Biogas and concentrate waste water	NH_4^+ and PO^{3-}_4	Native microalgae consortium-snow system	Qiu et al. 2021
3	*Chlorella vulgaris* and *Scenedesmus rubescens*	Municipal WW	NA	Algal cell immobilization (twin-layer system)	Shi et al. 2007
4	Halochlorella rubescens	Municipal WW	Microalgae can reduce P and N concentrations up to 99%	Flocculation	Shi et al. 2014
5	*Chlorella sorokiniana*	Alcohol distillery WW	Effective in reducing nitrate and phosphate	Flocculation method	Solovchen et al. 2014
6	*Parachlorella kessleri*	Secondary effluent	Effective in reducing nitrate and phosphate	NA	Chen et al. 2020
7	*Chlorella vulgaris*	Waste water treatment effluent from tilapia (fish) culture pond	NA	Microalgae biomass production and fermentation	Bhuyar et al. 2021
8	*Nostoc sp.*, *Oscillatoria sp.*	Sewage	Both algae have the potential to remove Ca^{+2}, Mg^{+2}, K^+ and Na^+	NA	Azarpira et al. 2014
9	*Desmodesmus sp.*	Anaerobic digestion WW	High efficiency in the removal of N and P	NA	Ji et al. 2014

(Contd.)

(*Contd.*)

S. No.	Microalgae	Waste Water Treatment	Pollutant/nutrients source	Process of removal	Reference
10	*Thraustochytrid sp.*	Anaerobically digested swine lagoon wastewater	NH_4^+ and $PO^{3-}{}_4$	Fungal and microalgal co-cultivation	Wrede et al. 2014
11	*Chlorella sorokiniana* AK-1	Piggery wastewater	COD, TP and TN	NA	Chen et al. 2020
12	*Nannochloropsis gaditana*	Municipal wastewater	NA	Microalgae biomass production and fermentation	Onay et al. 2018
13	*Halochlorella rubescens*	Municipal WW	Microalgae can reduce phosphorus and nitrogen concentrations up to 99%	Twin-Layer photobioreactor	Shi et al. 2007
14	*Chlorella vulgaris Scenedesmus obliquus*,	Aquaculture effluents	Removal efficiencies in batch growth were 100% for total nitrogen and total phosphorus	Cultured in Erlenmeyer flasks sealed with hydrophobic cotton and agitated by an air flow	Viegas et al. 2021
15	*Dunaliella salina*	Municipal waste water	Nitrate ion and Ammonium ion	Photobioreactor	Liu and Yildiz 2018
16	*Synechococcus sp VDW*	Undiluted cattle farm wastewater	Nitrate ion and Ammonium ion	NA	Hariz et al. 2019
17	Mixed indigenous microalgae	Primary effluent	NA	Photobioreactor	Lv et al. 2018

et al. 2018). A sponge was utilized as a microalgae carrier to cleanse 50% intensity pig farm wastewater. Chen et al. (2020) revealed that *Chlorella. sorokiniana* AK-1 *Coelastrella sp.* (Lee et al. 2022), and *Tetradesmus obliquus* (Ferreira et al. 2022) exhibited strong resistance to pig farm wastewater.

Industrial Wastewater

Based on the various processing objectives, industrial wastewater is classified as metallurgical wastewater, papermaking WW, fertiliser-containing WW, tanning wastewater, textile or printing and dyeing wastewater, and pesticide WW. This kind of wastewater has a high organic content, is poorly biodegradable, and contains a range of contaminants, including grease, heavy metals, antibiotics, and certain other chemical poisons (Udaiyappan et al. 2017). Considering the aforementioned, numerous obstacles exist to overcome before using microalgae to treat this type of wastewater. Industrial wastewater is often regarded as unfavourable for microalgae production because of its features. Most research studies have shown that certain microalgae strains do exceptionally well in treating different industrial waste fluids (Mondal et al. 2022). For instance, it has been noted that *Chlorella vulgaris* is a promising microalgae species that can be used to treat textile wastewater. It can efficiently remove colour, COD, phosphate and nitrogen from wastewater (Javed et al. 2022).

In a different study, Pena et al. (2020) treated tannery wastewater with a consortium of microalgae primarily constituted of *Tetraselmis sp.* They discovered that, in continuous light, the removal rates of TN, TP, and COD could reach 71.74, 97.64, and 56.7%, respectively. Additionally, the microalgae consortium exhibits superior heavy metal biosorption performance in wastewater. According to Moreno-García et al. (2021), 99% of the Cr (III) in the tannery effluent can be absorbed by a local microalgae consortium that separated from a secondary settler.

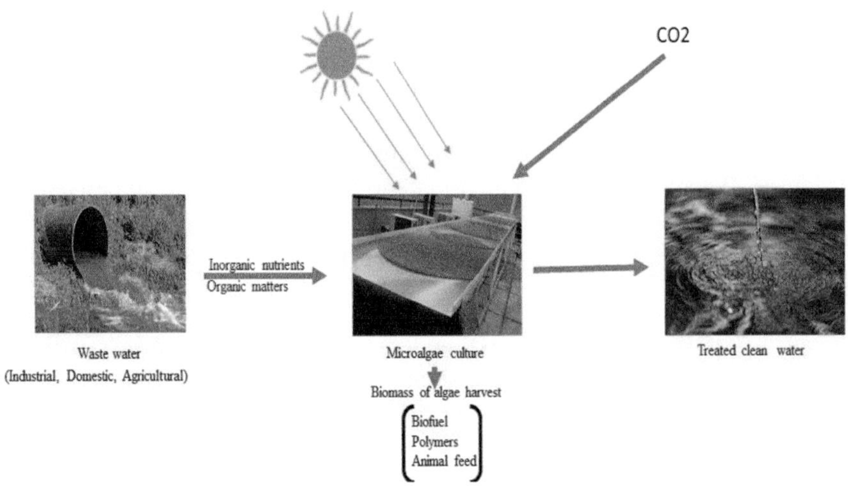

Figure 3: Showing the process of wastewater treatment by algae and production of bioproduct from algal biomass.

Mechanism for Pollutant Removal by Microalgae

The mechanism of wastewater treatment using microalgae involves different steps: biosorption, bioaccumulation, and biodegradation.

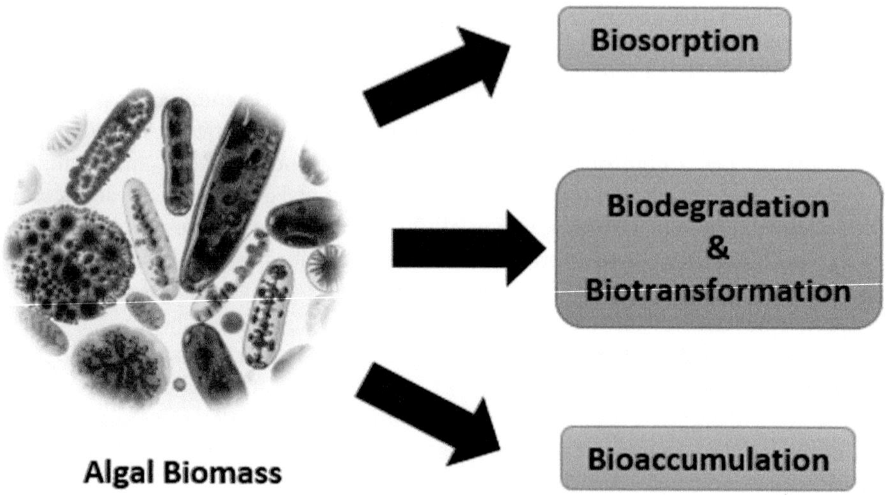

Figure 4: The mechanism of wastewater treatment by using microalgae involves different steps.

Biosorption

Biosorption is a passive technique in which a biological material that can bind and concentrate contaminants from water is used as the sorbent. In other words, biosorption is a process of mass transfer where a substance is pulled out of the liquid phase and adhered to the surface of a solid. According to Chia et al. (2020), biosorption comprises physical, chemical and metabolically independent processes and is facilitated by various mechanisms, including electrostatic contact, surface complexation, ion exchange, precipitation, absorption, and adsorption. A target sorbate that is dissolved or diffused in water and a biosorbent (solid-phase sorbent) are needed for the mechanisms behind this phenomenon. The biomaterial in question may consist of active or deceased microorganisms or even individual parts of them. The biosorbent's strong affinity for the target sorbate draws sorbate species, and the overall capacity determines how many sorbate molecules are present (Mustafa et al. 2021, El-Sheekh et al. 2020). Furthermore, the surfaces of microalgae have pores, and the surface charge promotes biosorption. Various chemical groups, including carboxyl (–COOH), hydroxyl (–OH), and sulphate (–SH), found in microalgae's cell wall (Figure 3), function as binding sites and efficient ion exchangers, aiding in the complexation of metal ions and the adsorption of organic materials from contaminated water (Ramírez et al. 2021).

The trace elements known as HMs, namely manganese (Mn), molybdenum (Mo), cobalt (Co), copper (Cu), zinc (Zn), iron (Fe), and boron (B), are crucial for enzyme functions and cell metabolism. Certain elements, like chromium (Cr), cadmium (Cd), mercury (Hg), lead (Pb), and arsenic (As), are generally harmful to microalgae; some of these are toxic in high quantities (Agnon et al. 2024). According to the hormesis phenomenon, HMs increase the development and metabolism of microalgae in the presence of trace amounts of harmful metal ions (Erofeeva et al. 2022). Furthermore, specific contaminants in water can form complexes with active binding sites on the cell surface, promoting flocculation and decreasing the overall amount of dissolved and suspended particles (Tohamy et al. 2022). The metal and the type of microalgae determine the ideal pH for biosorption. The pH ranges from 3.0 to 6.5 generally.

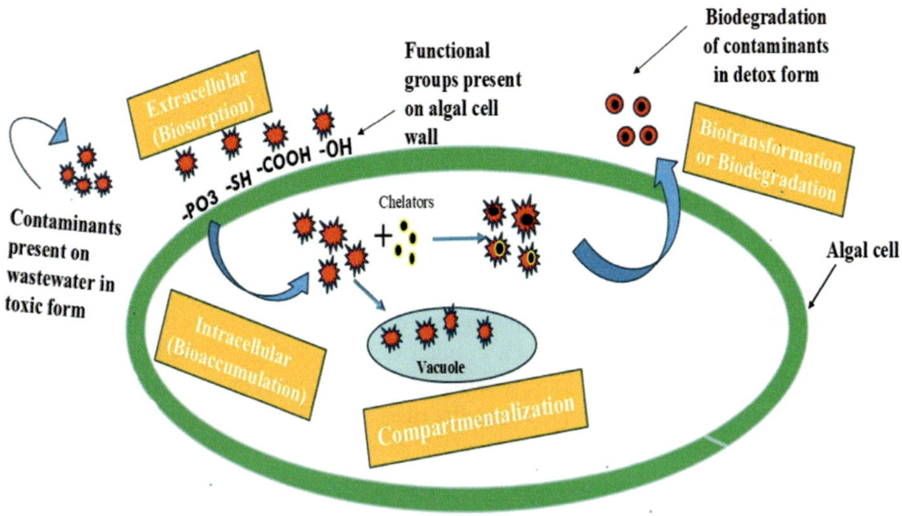

Figure 5: Mechanism of wastewater treatment by microalgae.

Antibiotics are captivated into the microalgal cell wall or attach themselves to organic materials, such as EPS (extracellular polymeric substances), which the algae excrete into their surroundings. This process is known as bioabsorption (Sutherland and Ralph 2019). The adsorption of 7-amino cephalosporanic acid was found to occur relatively quickly. The adsorption abilities of Chlorella, Chlamydomonas and Mychonastes species were 4.74, 3.09, and 2.95 mg/g, respectively, as reported by Guo et al. (2016). Furthermore, it was observed that Scenedesmus quadricauda and Tetraselmis suecica had maximal tetracycline adsorption capabilities, which are 295.34 mg/g and 56.25 mg/g, respectively, as reported by Daneshvar et al. (2018).

According to Norvill et al. (2016), the bioabsorption capabilities of microalgae were particular to their physical and chemical characteristics, such as surface chemistry and surface area. Because of the occurrence of major functional groups, including carboxyl (–COOH), hydroxyl (–OH), and phosphoryl (–PO_3), the cell walls of microalgae and EPS primarily carry negative charges (Xiong et al. 2018). Thus, electrostatic interactions can efficiently adsorb positive-charged antibiotics (Xiong et al. 2018). *Chlorella saccharophila* efficiently absorbed azithromycin (almost 100%), whereas trimethoprim's bio-absorption rate was less than 30% (Gojkovic et al. 2019). As the initial concentration of antibiotics increased, *Chlorella vulgaris's* efficacy in removing metronidazole was reduced (Hena et al. 2020).

While *Tetraselmis suecica* did not remove Tetracycline in a time-dependent manner, *Scenedesmus quadricauda's* bio-absorption of the drug increased significantly over 60 minutes (Daneshvar et al. 2018).

Bioaccumulation

Living cells cleanse wastes through bioaccumulation, taking in substances and digesting or accumulating them. According to Mustafa et al. (2019), this process is crucial for removing organic and inorganic pollutants such as pesticides, heavy metals, phosphates, nitrates and sulphates from the environment before they can enter into cells. Quantifying both absorbed and bioaccumulated contaminants is challenging due to the dynamic interplay between the two processes, even though bioaccumulation and biosorption are essentially distinct processes. Rempel et al. (2021) stated that microalgae collect several contaminants besides nutrients and microelements.

Moreover, microalgal species exhibit extraordinary resilience against various pollutants from different industries, agriculture, and household domains, augmenting their bioremediation capacity (Mojiri et al. 2020). Regardless of the process, bioaccumulation is significantly impacted by internal and external physicochemical parameters, such as temperature, pH, contact time, and concentration of pollutants (Xiong et al. 2020). For example, it was discovered that temperature has a significant impact on antibiotic bioaccumulation in the planktonic food web (Tang et al. 2020).

Biodegradation

The basic processes of biodegradation are grouped into two categories: (i) metabolic degradation, where toxicants act as a carbon source and donor/acceptor for microalgae, as well as an electron donor, and (ii) co-metabolism, where pollutants act as both a carbon source and an electron donor for non-living matter (Leng et al. 2020). Extracellular breakdown is followed by intracellular breakdown of the breakdown intermediates in microalgae-mediated biodegradation, which can occur extracellularly and intracellularly or in a combination of both locations (Tiwari et al. 2017). The primary source of exopolysaccharides (EPS) for extracellular degradation is microalgae. It stays confined to the outer walls of the cells, enabling microalgae to act as an external digestive system and mineralise contaminants when they are dissolved.

The two main types of biodegradation processes are (i) metabolic degradation, in which pollutants function as an electron donor, carbon source, and donor/acceptor for microalgae; and (ii) co-metabolism, in which pollutants function as an electron donor and carbon source for non-living matter (Leng et al. 2020). Microalgae-mediated biodegradation can occur extracellularly, intracellularly, or in a combination of both places. Extracellular breakdown is succeeded by intracellular breakdown of the breakdown intermediates (Tiwari et al. 2017). Microalgae are the main source of extracellular degradation products like exopolysaccharides (EPS). It remains restricted to the cell walls, allowing microalgae to function as an external digestive system and mineralise pollutants upon dissolution.

Furthermore, Xie et al. (2020) observed that *Chlamydomonas* sp. Tai-03 was 100% efficient in removing ciprofloxacin, with 65.05% of the removal coming from biodegradation. Some microalgal strains like *Haematococcus pluvialis*, *Selenastrum capricornutum*, *Scenedesmus quadricauda*, and *Chlorella vulgaris* were examined separately for their capacity to remove ten antibiotics including sulfamerazine (SMR), sulfamethoxazole (SMX), sulfamonomethoxine (SMM), trimethoprim (TMP), clarithromycin (CLA), azithromycin (AZI), roxithromycin (ROX), lomefloxacin (LOM), levofloxacin (LEV) and flumequine. Kiki et al. (2020) reported that biodegradation accounted for between 23% and 99% of the removal efficiency. Antibiotics are typically biodegraded by microalgae through a combination of extracellular and intracellular biodegradation, where the breakdown products are first metabolised extracellularly and then undergo additional intracellular metabolism (Tiwari et al. 2017, Xiong et al. 2018). Whereas external degradation depends on microalgal expulsion of EPS, intracellular degradation depends on the accumulated antibiotics in microalgal cells.

Bioproducts from Microalgae Biomass

Apart from biofuels, various valuable compounds such as diterpenes, sterols, exopolysaccharides (EPS), fucoxanthin, peptides, carotenoids, polyunsaturated fatty acids (PUFAs), and others can be recovered from microalgae, reported by Fawcett et al. (2022). As antiviral (Yim et al. 2004), immunomodulatory (Riccio et al. 2019), and antibacterial (Najdenski et al. 2013) agents, these compounds are incredibly potent. However, using biomass for human use is limited due to the possibility of unknown microbes growing and heavy metal accumulation in cultivations conducted on non-sterile media (Goswami et al. 2022).

Algal Biomass as Biofertilizer

Thus, the most feasible uses of microalgae biomass grown in wastewater effluent are for producing bioplastics, animal fodder, and fertilisers (Figure 6). Even in this situation, though, microalgae cultivation should be carried out in food-related wastewater, such as that from the beer or palm oil industries, because these wastewater media have fewer hazardous microorganisms than other wastewater media (Goswami et al. 2022). Because the microalgae-based biofertiliser is rich in micronutrients like nature-based trace elements and plant hormones, which are crucial for the development and growth of plants, it can be compared to or considered superior to inorganic fertilisers (Lorentz et al. 2020). Moreover, it supports aggregate stability and soil water retention (Khan et al. 2019). In contrast, biofertilisers with lower amounts of nitrogen can be developed to decrease nitrous oxide (N_2O) emissions from agriculture, where certain microalgae can be helpful as raw materials (E.P.R.I. 2013).

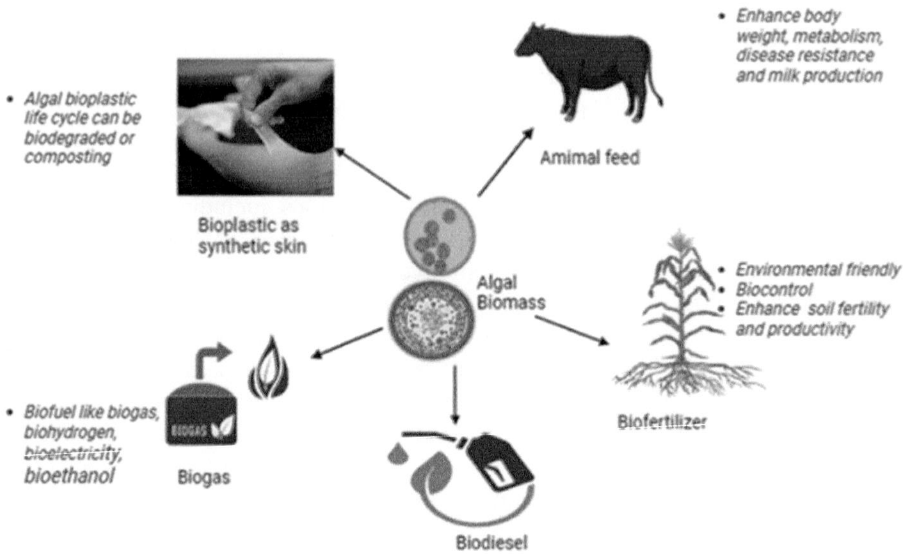

Figure 6: Algal bioproducts from microalgal biomass after wastewater treatment.

Algal Biomass as Biodiesel

The entire biomass from other processes, such as the leftover biomass from the manufacturing of biodiesel, which makes up about 65% of all applied biomass, can be utilised to produce biofertilisers, lowering process costs (Nayak et al. 2019). About 23% of arable land is used to produce oilseeds (6% for providing human food and 17% for feedstock). Oilseed flours comprise around 27% of animal feedstock, and their primary purpose is to supply protein to increase animal productivity (Fawcett et al. 2022). Thus, it has been demonstrated that microalgae contain more protein than oilseeds, such as soybeans. Along with their better fatty acid profile, vitamins, minerals, and amino acids, microalgae also include a high percentage of carbohydrates and have antibacterial qualities, which are advantageous to the intestinal system of the animal, hence reducing the need for costly synthetic supplements (Fawcett et al. 2022, Catone et al. 2021).

Algal Biomass as Feedstock

Before using biomass on a wider scale, it is crucial to assess its digestibility because microalgae are typically employed as a component of feed composition rather than as the sole source of nutrients. Animals differ in their rates of inclusion; generally speaking, birds (10% of microalgae inclusion) are more tolerant than pigs (5.51%) and cattle (1.18% of microalgae inclusion) (Fawcett et al. 2022). Adding 16% *Nannochloropsis oceanica* to commercial birds' diets caused the organs to enlarge and increased water intake; however, at 8%, the biomass could be utilised without causing any problems. It is important to emphasise the appropriate percentage of supplementation (Catone et al. 2021). Even though they are unconventional, biofertilisers, animal fodder, and biochar made from microalgae biomass can be utilised as NBS (nature-based solutions) by the purchaser of these goods to achieve carbon neutrality. Generally speaking, the buyer could produce carbon credits by comparing emissions managed by NBS with emissions that are not managed. In this way, by suppressing or inhibiting methanogenesis, animal feedstock can decrease enteric methane (CH_4) emissions from ruminant regimes (Sünkel et al. 2021).

Moreover, biochar is the most often used NBS technique derived from microalgae cultivation in WWTPs (wastewater treatment plants). In addition to being used in biodiesel manufacturing, biochar can be applied in soil management and forestry practices to raise soil carbon stocks and act as a sink over time due to carbon stability (Etter et al. 2021). Additionally, certain microalgae taxa, like *Chlorella* species and Cyanobacteria, can produce biodegradable polymers known as polyhydroxyalkanoates (PHA), whose synthesis is directly impacted by several variables, including temperature, pH, availability of nutrients, and cycles of light and dark. When the cells experience a shortage of certain nutrients, like nitrogen and phosphorus, they accumulate inside the microalgae as glycogen (Goswami et al. 2022, Ahmad et al. 2022b).

Algal Biomass as Bioplastic

Bioplastics derived from algal biomass are a sustainable alternative to traditional plastics because they are biodegradable and produced from renewable sources. Bioplastic is defined as bio-based polymers for plastic production and is biodegradable (Sudhakar et al. 2023, Vert et al. 2012). Microalgae can replace synthetic skins and absorbable sutures in sterile environments. PHA can also be used as a raw material for greenhouse gases (GHGs) including carbon dioxide and methane capture, and used in plastic ingredients to make valuable plastic because of its polymeric properties (Goswami et al. 2022, Ahmad et al. 2022b). As a result, bioplastics will reduce the manufacturing of traditional plastic and will eventually contribute to carbon immobilisation. The bioplastic manufacturing process must be less emission-intensive due to its biodegradable quality (Newlight Technologies 2019).

Scalability and Economic Viability

Large-scale implementation of microalgae systems faces several difficulties, particularly regarding energy expenses and cost savings. High energy consumption in industrial-scale production is mainly due to the need for constant light, mixing, and temperature control. Open ponds for growing algae require a large land area, making it challenging to maintain optimal growth conditions. In contrast, closed photobioreactors, although more effective, are very expensive to construct and maintain. Microalgae harvesting is another major problem. The conventional approaches include centrifugation and filtration, which use considerable energy and constitute a significant fraction of the cost. More efficient and affordable harvesting processes, including chemical flocculation and magnetic separation, must be adopted for the success of such systems. There has been some progress with algae that self-sediment or with the development of bioflocculants; however, these are not yet ready for commercial application.

Apart from energy and harvesting issues, nutrient management is also a problem. Wastewater can supply nutrients like phosphorus and nitrogen, which are essential for algae, TSS, and diatom concentrations. However, the correlation between growth yield and nutrient concentration is only linear to a point, after which the nutrients may limit growth—hence the need for additional resources. This raises operational expenses and makes the treatment of wastewater more difficult. In addition, what also complicates matters is that open systems require management of evaporation as well.

For microalgae to be cultured on a large scale and cost-effectively, photobioreactor design, highly energy-efficient biosolids recovery methods, and nutrient recycling technologies must be pursued. Incorporating these into current wastewater treatment and producing higher-value products such as biofuels, biopolymers, and biofertilisers will enhance economic performance, opening doors for circular economies. However, these constraints do not determine the effectiveness of microalgae systems for waste treatment and resource recovery, as the efficacy of ongoing technological and research development remains promising.

Environmental Regulations

There is no precise global regulation or framework for using microalgae in sludge for wastewater treatment, especially concerning pharmaceutical and microplastic concerns. In the United States, for instance, bodies such as the Environmental Protection Agency and other regional authorities, and the European Environment Agency, focus on using wastewater for irrigation purposes without concern about releasing pollutants into the environment through that medium. Pharmaceuticals such as antibiotics, hormones, personal care products, and microplastics pose bioaccumulation challenges in water bodies.

Pharmaceutical wastes, categorised as emerging pollutants in the past decades, are regulated to a limited extent in many regions. On the contrary, the Water Framework Directive (WFD) of the European Union attaches great importance to managing hazardous substances in surface waters, provided that the same may one day include the management of medicines. Also, in the example of the World Health Organisation, policies aimed at preventing and mitigating effects caused by microparticles of plastic pollution tend to emerge. Such interventions have been observed in countries such as Japan and Switzerland, where more stringent requirements for the type of effluents have been implemented.

The contribution of microalgae in pollutant removal through biosorption, bioaccumulation, and biodegradation also indicates how its regulatory requirements can be fulfilled. Nevertheless, laws are needed regarding the management of algal biomass to avoid its reuse after absorbing heavy metals and other pollutants.

There is also a need for global harmonisation to create recommendations for microalgae usage in WWTPs, allowing maximum pollutant removal while maintaining environmental integrity. More studies on the capacity of microalgae to capture specific pollutants could support the introduction of stricter regulations for these contaminants entering aquatic environments. Specifically, regulations for pharmaceuticals and microplastics could be established to help keep water bodies and surrounding ecosystems clean.

Conclusion

A wastewater treatment system must be environmentally benign, economically viable, and technically practicable to be developed sustainably. The microalgal-based wastewater treatment process is economically and technologically competitive compared to traditional treatment systems. Microalgae systems have negligible or no operating expenses, making them far more sustainable overall than traditional systems. Therefore, compared to traditional and biological wastewater treatment methods, employing microalgae to reduce nitrogenous, phosphorous, and carbonaceous material has the potential to function with a reduced footprint in terms of energy consumption and greenhouse gas (GHG) output. Algal biomass finds additional applications as feedstock, biofuel, and biofertiliser. However, the process's non-sterile surroundings remain a significant obstacle to the widespread use of microalgae in wastewater treatment. Therefore, standardised procedures for species-specific microalgae and optimal effluents for each culture are urgently needed. This would enable high biomass production capacity and promote the use of microalgae in large-scale WWTPs.

References

Abdelfattah, A., Ali, S.S., Ramadan, H., El-Aswar, E.I., Eltawab, R., Ho, S.-H. et al. 2023. Microalgae-based wastewater treatment: Mechanisms, challenges, recent advances, and future prospects. Environmental Science and Ecotechnology, 13(January): 100205.

Agoramoorthy, G. 2008. Can India meet the increasing food demand by 2020? Futures, 40(5): 503-506.

Ahemad, M. and Kibret, M. 2013. Recent trends in microbial biosorption of heavy metals: A review. Biochemistry & Molecular Biology, 1(1): 19.

Ahmad, A., Tariq, S., Zaman, J.U., Martin Perales, A.I., Mubashir, M. et al. 2022a. Recent trends and challenges with the synthesis of membranes: Industrial opportunities towards environmental remediation. Chemosphere, 306(November): 135634.

Ahmad, Ashfaq, Banat, F., Alsafar, H. and Hasan, S.W. 2022b. Algae biotechnology for industrial wastewater treatment, bioenergy production, and high-value bioproducts. Science of The Total Environment, 806(February): 150585.

Aksu, Z. 2001. Equilibrium and kinetic modelling of cadmium (II) biosorption by *C. Vulgaris* in a batch system: Effect of temperature. Separation and Purification Technology, 21(3) (January): 285-294.

Alcántara, C., Domínguez, J.M., García, D., Blanco, S., Pérez, R., García-Encina, P.A. et al. 2015. Evaluation of wastewater treatment in a novel anoxic–aerobic algal–bacterial photobioreactor with biomass recycling through carbon and nitrogen mass balances. Bioresource Technology, 191(September): 173-186.

Alengebawy, A., Abdelkhalek, S.T., Qureshi, S.R. and Wang, M.-Q. 2021. Heavy metals and pesticides toxicity in agricultural soil and plants: Ecological risks and human health implications. Toxics, 9(3)(February 25): 42.

Alix, A. and Capri, E. 2018. Modern agriculture in Europe and the role of pesticides. *In*: Advances in Chemical Pollution, Environmental Management and Protection, 1-22. Elsevier.

Al Momani, F.A. and Örmeci, B. 2016. Performance of *Chlorella Vulgaris, Neochloris Oleoabundans*, and mixed indigenous microalgae for treatment of primary effluent, secondary effluent and centrate. Ecological Engineering, 95(October): 280-289.

Almomani, F. (2020). Algal cells harvesting using cost-effective magnetic nano-particles. Science of The Total Environment, 720: 137621. https://doi.org/10.1016/j.scitotenv.2020.137621.

Al-Saleh, I., Shinwari, N. and Alsabbaheen, A. 2011. Phthalates residues in plastic bottled waters. The Journal of Toxicological Sciences, 36(4): 469-478.

Alshehri, M.M., Ouladsmane, M.A., Aouak, T.A., Alothman, Z.A. and Badjah Hadj Ahmed, A.Y. 2022. Determination of phthalates in bottled waters using solid-phase microextraction and gas chromatography tandem mass spectrometry. Chemosphere, 304(October): 135214.

Al-Tohamy, R., Ali, S.S., Li, F., Okasha, K.M., Mahmoud, Y.A.-G., Elsamahy, T. et al. 2022. A critical review on the treatment of dye-containing wastewater: Ecotoxicological and health concerns of textile dyes and possible remediation approaches for environmental safety. Ecotoxicology and Environmental Safety, 231(February): 113160.

Alvarez, A., Saez, J.M., Davila Costa, J.S., Colin, V.L., Fuentes, M.S., Cuozzo, S.A. et al. 2017. Actinobacteria: Current research and perspectives for bioremediation of pesticides and heavy metals. Chemosphere, 166(January): 41-62.

Angon, P.B., Anjum, N., Das, A., Kc, S., Poudel, A., Suchi, S.A. et al. 2023. Sources, effects and present perspectives of heavy metals contamination: Soil, plants and human food chain. California Digital Library (CDL).

Arun, S., Sinharoy, A., Pakshirajan, K. and Lens, P.N.L. 2020. Algae based microbial fuel cells for wastewater treatment and recovery of value-added products. Renewable and Sustainable Energy Reviews, 132(October): 110041.

Azarpira, H., Dhumal, K. and Pondhe, G. 2014. Application of phycoremediation technology in the treatment of sewage water to reduce pollution load. Adv. Environ. Biol, 5(7): 2419-2423.

Bajpai, A., Shukla, P., Dixit, B.S. and Banerji, R. 2007. Concentrations of organochlorine insecticides in edible oils from different regions of India. Chemosphere, 67(7)(April): 1403-1407.

Balakrishna, K., Rath, A., Praveenkumar Reddy, Y., Guruge, K.S. and Subedi, B. 2017. A review of the occurrence of pharmaceuticals and personal care products in Indian water bodies. Ecotoxicology and Environmental Safety, 137(March): 113-120.

Bang Truong, H., Nguyen, T.H.T., Ba Tran, Q., Son Lam, V., Thao Nguyen Nguyen, T., X. Cuong Nguyen et al. 2024. Algae-constructed wetland integrated system for wastewater treatment: A review. Bioresource Technology, 406(August): 131003.

Barra, R., Colombo, J.C., Eguren, G., Gamboa, N., Jardim, W.F. and Mendoza, G. 2006. Persistent organic pollutants (POPs) in eastern and western south American countries. Reviews of Environmental Contamination and Toxicology, 1-33.

Beltrán-Flores, E., Torán, J., Caminal, G., Blánquez, P. and Sarrà, M. 2020. The removal of diuron from agricultural wastewaters by trametes versicolor immobilized on pinewood in simple channel reactors. Science of The Total Environment, 728(August): 138414.

Bhuyar, P., Trejo, M., Dussadee, N., Unpaprom, Y., Ramaraj, R. and Whangchai, K. 2021. Microalgae cultivation in wastewater effluent from tilapia culture pond for enhanced bioethanol production. Water Science and Technology, 84(10-11)(May 20): 2686-2694.

Borowitzka, M.A. 2016. Algal physiology and large-scale outdoor cultures of microalgae. *In*: The Physiology of Microalgae, 601-652. Cham: Springer International Publishing.

Boucher, J., Faure, F., Pompini, O., Plummer, Z., Wieser, O. and Felippe de Alencastro, L. 2019. (Micro) Plastic fluxes and stocks in Lake Geneva Basin. TrAC Trends in Analytical Chemistry, 112(March): 66-74.

Cai, W., Zhao, Z., Li, D., Lei, Z., Zhang, Z. and Lee, D.-J. 2019. Algae granulation for nutrients uptake and algae harvesting during wastewater treatment. Chemosphere, 214: 55-59.

Cao, Y., Li, Y., Jia, L., Wang, Q., Niu, T., Yang, Q. et al. 2024. Long-term and combined heavy-metal contamination forms a unique microbiome and resistome: A case study in a Yellow River Tributary sediments. Environmental Research, 252(July): 118861.

Cartraud, A.E., Le Corre, M., Turquet, J. and Tourmetz, J. 2019. Plastic ingestion in seabirds of the western Indian Ocean. Marine Pollution Bulletin, 140(March): 308-314.

Catone, C.M., Ripa, M., Geremia, E. and Ulgiati, S. 2021. Bio-products from algae-based biorefinery on wastewater: A review. Journal of Environmental Management, 293(September): 112792.

Chawla, P., Malik, A., Sreekrishnan, T.R., Dalvi, V. and Gola, D. 2020. Selection of optimum combination via comprehensive comparison of multiple algal cultures for treatment of diverse wastewaters. Environmental Technology & Innovation, 18(May): 100758.

Chen, C.-Y., Kuo, E.-W., Nagarajan, D., Ho, S.-H., Dong, C.-D., Lee, D.-J. et al. 2020. Cultivating Chlorella *Sorokiniana* AK-1 with swine wastewater for simultaneous wastewater treatment and algal biomass production. Bioresource Technology, 302(April): 122814.

Chen, W., Liu, J., Chu, G., Wang, Q., Zhang, Y., Gao, C. et al. 2023. Comparative evaluation of four *Chlorella* species treating mariculture wastewater under different photoperiods: Nitrogen removal performance, enzyme activity, and antioxidant response. Bioresource Technology, 386(October): 129511.

Chia, W.Y., Ying Tang, D.Y., Khoo, K.S., Kay Lup, A.N. and Chew, K.W. 2020. Nature's fight against plastic pollution: Algae for plastic biodegradation and bioplastics production. Environmental Science and Ecotechnology, 4(October): 100065.

Chinnasamy, S., Bhatnagar, A., Hunt, R.W. and Das, K.C. 2010. Microalgae cultivation in a wastewater dominated by carpet mill effluents for biofuel applications. Bioresource Technology, 101(9)(May): 3097-3105.

Coelho, L.M., Rezende, H.C., Coelho, L.M., de Sousa, P.A.R., Melo, D.F.O. and Coelho, N.M.M. 2015. Bioremediation of polluted waters using microorganisms. *In*: Advances in Bioremediation of Wastewater and Polluted Soil. pp. 1-22.

Concha-Graña, E., Turnes-Carou, M.I., Muniategui-Lorenzo, S., López-Mahía, P., Prada-Rodríguez, D. and Fernández-Fernández, E. 2006. Evaluation of HCH isomers and metabolites in soils, leachates, river water and sediments of a highly contaminated area. Chemosphere, 64(4)(July): 588-595.

CPCB, 2018. Annual Report for the Year 2017–18 on Implementation of Plastic Waste Management Rules. Ministry of Environment, Forest and Climate Change, Govt. of India.

Daneshvar, E., Zarrinmehr, M.J., Hashtjin, A.M., Farhadian, O. and Bhatnagar, A. 2018. Versatile applications of freshwater and marine water microalgae in dairy wastewater treatment, lipid extraction and tetracycline biosorption. Bioresource Technology, 268(November): 523-530.

de-Bashan, L.E. and Bashan, Y. 2010. Immobilized microalgae for removing pollutants: Review of practical aspects. Bioresource Technology, 101(6)(March): 1611-1627.

Diniz, G.S., Silva, A.F., Araújo, O.Q.F. and Chaloub, R.M. 2016. The potential of microalgal biomass production for biotechnological purposes using wastewater resources. Journal of Applied Phycology, 29(2)(October 28): 821-832.

EEA. 2019. Industrial waste water treatment - Pressures on Europe's environment. EEA report. European Environment Agency.

E.P.R.I. Michigan State University, in: Quantifying N_2O Emissions Reductions in Agricultural Crops through Nitrogen Fertilizer Rate Reduction Approved VCS Methodology VM0022 Sectoral Scope 14, VERRA VCS, 2013, pp. 1-55.

El-Kassas, H.Y. and Mohamed, L.A. 2014. Bioremediation of the textile waste effluent by *Chlorella Vulgaris*. Egyptian Journal of Aquatic Research, 40(3): 301-308.

El-Sheekh, M.M., Metwally, M.A., Allam, N. and Hemdan, H.E. 2020. Simulation treatment of industrial wastewater using microbiological cell immobilization technique. Iranian Journal of Science and Technology, Transactions A: Science 44(3)(April 19): 595-604.

Emenike, C.U., Jayanthi, B., Agamuthu, P. and Fauziah, S.H. 2018. Biotransformation and removal of heavy metals: A review of phytoremediation and microbial remediation assessment on contaminated soil. Environmental Reviews, 26(2)(June): 156-168.

Erofeeva, E.A. 2022. Environmental hormesis: From cell to ecosystem. Current Opinion in Environmental Science & Health, 29(October): 100378.

Etter, H., Vera, A., Aggarwal, C., Delaney, M. and Manley, S. 2021. Methodology for biochar utilization in soil and non-soil applications 1. VERRA VCS, 1-55.

Fagerstone, K.D., Quinn, J.C., Bradley, T.H., De Long, S.K. and Marchese, A.J. 2011. Quantitative measurement of direct nitrous oxide emissions from microalgae cultivation. Environmental Science & Technology, 45(21)(October 6): 9449-9456.

Falkowski, P.G. and Raven, J.A. 2007. Aquatic Photosynthesis. Princeton University Press.

FAO/UNEP/OECD/SIB, 2001. Baseline Study on the Problem of Obsolete Pesticides Stocks. Rome.

Fawcett, C.A., Senhorinho, G.N.A., Laamanen, C.A. and Scott, J.A. 2022. Microalgae as an alternative to oil crops for edible oils and animal feed. Algal Research, 64(May): 102663.

Ferreira, A., Figueiredo, D., Cardeiras, R., Nabais, R., Ferreira, F., Ribeiro, B. et al. 2022. Exploring different pretreatment methodologies for allowing microalgae growth in undiluted piggery wastewater. Agronomy, 12(3)(February 26): 580.

FICCI, 2017. Knowledge Papaer on Plastic Industry for Infrastructure, in 3rd National Conference on Sustainable Infrastructure with Plastics. Federation of Indian Chambers of Commerce and Industry.

Fuentes, M.S., Benimeli, C.S., Cuozzo, S.A. and Amoroso, M.J. 2010. Isolation of pesticide-degrading actinomycetes from a contaminated site: Bacterial growth, removal and dechlorination of organochlorine pesticides. International Biodeterioration & Biodegradation, 64(6)(September): 434-441.

Gałuszka, A., Migaszewski, Z.M. and Manecki, P. 2011. Pesticide burial grounds in Poland: A review. Environment International, 37(7)(October): 1265-1272.

Ganeshkumar, V., Subashchandrabose, S.R., Dharmarajan, R., Venkateswarlu, K., Naidu, R. and Megharaj, M. 2018. Use of mixed wastewaters from piggery and winery for nutrient removal and lipid production by *Chlorella* sp. MM3. Bioresource Technology, 256(May): 254-258.

Garrison, A.W, Pope, J.D. and Allen, F.R. 1976. GC/MS analysis of organic compounds in domestic wastewaters. pp. 517-556. *In*: Keith, L.H. [ed.]. Identification and Analysis of Organic Pollutants in Water. Ann Arbor, Ann Arbor Science Pubishers.

Giacovelli, C., Zamparo, A., Wehrli, A. and Alverson, K. 2018. Single-use Plastics: A Roadmap for Sustainability. United Nations Environ. Program. Nairobi, p. 90.

Gimeno, P., Thomas, S., Bousquet, C., Maggio, A.-F., Civade, C., Brenier, C. and Bonnet, P.-A. 2014. Identification and quantification of 14 phthalates and 5 non-phthalate plasticizers in PVC medical devices by GC-MS. Journal of Chromatography B, 949-950(February): 99-108.

Gojkovic, Z., Lindberg, R.H., Tysklind, M. and Funk, C. 2019. Northern green algae have the capacity to remove active pharmaceutical ingredients. Ecotoxicology and Environmental Safety, 170(April): 644-656.

Gondi, R., Kavitha, S., Yukesh Kannah, R., Kumar, G. and Rajesh Banu, J. 2022. Wastewater based microalgae valorization for biofuel and value-added products recovery. Sustainable Energy Technologies and Assessments, 53(October): 102443.

Goswami, R.K., Mehariya, S., Karthikeyan, O.P., Gupta, V.K. and Verma, P. 2022. Multifaceted application of microalgal biomass integrated with carbon dioxide reduction and wastewater remediation: A flexible concept for sustainable environment. Journal of Cleaner Production, 339(March): 130654.

Gouveia, L., Graça, S., Sousa, C., Ambrosano, L., Ribeiro, B., Botrel, E.P. et al. 2016. Microalgae biomass production using wastewater: Treatment and costs. Algal Research, 16(June): 167-176.

Guieysse, B., Plouviez, M., Coilhac, M. and Cazali, L. 2013. Nitrous oxide (N_2O) production in axenic *Chlorella vulgaris* microalgae cultures: Evidence, putative pathways, and potential environmental impacts. Biogeosciences, 10(10)(October 25): 6737-6746.

Guo, W.-Q., Zheng, H.-S., Li, S., Du, J.-S., Feng, X.-C., Yin, R.-L. et al. 2016. Removal of cephalosporin antibiotics 7-ACA from wastewater during the cultivation of lipid-accumulating microalgae. Bioresource Technology, 221(December): 284-290.

Gutierrez, J., Kwan, T.A., Zimmerman, J.B. and Peccia, J. 2016. Ammonia inhibition in oleaginous microalgae. Algal Research, 19(November): 123-127.

Hena, S., Gutierrez, L. and Croué, J.-P. 2020. Removal of metronidazole from aqueous media by *C. vulgaris*. Journal of Hazardous Materials, 384(February): 121400.

Hernández, D., Riaño, B., Coca, M. and García-González, M.C. 2013. Treatment of agro-industrial wastewater using microalgae–bacteria consortium combined with anaerobic digestion of the produced biomass. Bioresource Technology, 135(May): 598-603.

Hignite, C. and Azarnoff, D.L. 1977. Drugs and drug metabolites as environmental contaminants: Chlorophenoxyisobutyrate and salicylic acid in sewage water effluent. Life Sciences, 20(2)(January): 337-341.

Huo, J., Yue, Q.L., Li, Z., Zhang, L.S. and Chen, J.Y. 2017. Analysis of 16 phthalates compounds in disposable plastic tableware sold on Chengdu market. Chinese J. Food Hyg, 29(4): 484-487.

Javed, F., Rehman, F., Khan, A.U., Fazal, T., Hafeez, A. and Rashid, N. 2022. Real textile industrial wastewater treatment and biodiesel production using microalgae. Biomass and Bioenergy, 165(October): 106559.

Ji, F., Liu, Y., Hao, R., Li, G., Zhou, Y. and Dong, R. 2014. Biomass production and nutrients removal by a new microalgae strain Desmodesmus sp. in anaerobic digestion wastewater. Bioresource Technology, 161(June): 200-207.

Ji, M.-K., Abou-Shanab, R.A.I., Kim, S.-H., Salama, E.-S., Lee, S.-H., Kabra, A.N. et al. 2013. Cultivation of microalgae species in tertiary municipal wastewater supplemented with CO2 for nutrient removal and biomass production. Ecological Engineering, 58(September): 142-148.

Kankanige, D. and Babel, S. 2020. Smaller-sized micro-plastics (MPs) contamination in single-use PET-bottled water in Thailand. Science of The Total Environment, 717(May): 137232.

Katam, K. and Bhattacharyya, D. 2018. Comparative study on treatment of kitchen wastewater using a mixed microalgal culture and an aerobic bacterial culture: Kinetic evaluation and FAME analysis. Environmental Science and Pollution Research, 25(21)(May 12): 20732-20742.

Khalid, A.A.H., Yaakob, Z., Abdullah, S.R.S. and Takriff, M.S. 2019. Assessing the feasibility of microalgae cultivation in agricultural wastewater: The nutrient characteristics. Environmental Technology & Innovation, 15(August): 100402.

Khan, M.N., Mobin, M., Abbas, Z.K. and Alamri, S.A. 2018. Fertilizers and their contaminants in soils, surface and groundwater. Encyclopedia of the Anthropocene, 225-240.

Khan, S.A., Sharma, G.K., Malla, F.A., Kumar, A., Rashmi and Gupta, N. 2019. Microalgae based biofertilizers: A biorefinery approach to phycoremediate wastewater and harvest biodiesel and manure. Journal of Cleaner Production, 211(February): 1412-1419.

Kiki, C., Rashid, A., Wang, Y., Li, Y., Zeng, Q., Yu, C.-P. et al. 2020. Dissipation of antibiotics by microalgae: Kinetics, identification of transformation products and pathways. Journal of Hazardous Materials, 387(April): 121985.

Kolpin, D.W., Furlong, E.T., Meyer, M.T., Thurman, E.M., Zaugg, S.D., Barber, L.B. et al. 2002. Pharmaceuticals, hormones, and other organic wastewater contaminants in U.S. streams, 1999–2000: A national reconnaissance. Environmental Science & Technology, 36(6)(March 1): 1202-1211.

Kumar, R., Sarmah, A.K. and Padhye, L.P. 2019. Fate of pharmaceuticals and personal care products in a wastewater treatment plant with parallel secondary wastewater treatment train. Journal of Environmental Management, 233(March): 649-659.

Kumari, B., Madan, V.K. and Kathpal, T.S. 2007. Status of insecticide contamination of soil and water in Haryana, India. Environmental Monitoring and Assessment, 136(1-3) (April 4): 239-244.

Lammel, G., Ghim, Y.-S., Grados, A., Gao, H., Hühnerfuss, H. and Lohmann, R. 2007. Levels of persistent organic pollutants in air in China and over the Yellow Sea. Atmospheric Environment, 41(3)(January): 452-464.

Lavrinovičs, A., Mežule, L. and Juhna, T. 2020. Microalgae starvation for enhanced phosphorus uptake from municipal wastewater. Algal Research, 52(December): 102090.

Lee, D.H., Lee, S.-I. and Kang, J.-H. 2024. Machine learning approaches to identify spatial factors and their influential distances for heavy metal contamination in downstream sediment. Science of The Total Environment, 948(October): 174755.

Lee, S.-A., Lee, N., Oh, H.-M. and Ahn, C.-Y. 2021. Stepwise treatment of undiluted raw piggery wastewater, using three microalgal species adapted to high ammonia. Chemosphere, 263(January): 127934.

Leng, L., Wei, L., Xiong, Q., Xu, S., Li, W., Lv, S. et al. 2020. Use of microalgae based technology for the removal of antibiotics from wastewater: A review. Chemosphere, 238(January): 124680.

Li, G., Zhang, J., Li, H., Hu, R., Yao, X. Liu, Y. et al. 2021. Towards high-quality biodiesel production from microalgae using original and anaerobically-digested livestock wastewater. Chemosphere, 273(June): 128578.

Li, S., Zhao, S., Yan, S., Qiu, Y., Song, C., Li, Y. et al. 2019. Food processing wastewater purification by microalgae cultivation associated with high value-added compounds production – A review. Chinese Journal of Chemical Engineering, 27(12)(December): 2845-2856.

Lima, S., Villanova, V., Grisafi, F., Caputo, G., Brucato, A. and Scargiali, F. 2020. Autochthonous microalgae grown in municipal wastewaters as a tool for effectively removing nitrogen and phosphorous. Journal of Water Process Engineering, 38(December): 101647.

Lorentz, J.F., Calijuri, M.L., Assemany, P.P., Alves, W.S. and Pereira, O.G. 2020. Microalgal biomass as a biofertilizer for pasture cultivation: Plant productivity and chemical composition. Journal of Cleaner Production, 276(December): 124130.

Lv, J., Liu, Y., Feng, J., Liu, Q., Nan, F. and Xie, S. 2018. Nutrients removal from undiluted cattle farm wastewater by the two-stage process of microalgae-based wastewater treatment. Bioresource Technology, 264(September): 311-318.

Magalhães, I.B., de P. Pereira, A.S.A., Silva, T.A., Ferreira, J., Braga, M.Q., Couto, E.A. et al. 2024. Advancements in high-rate algal pond technology for enhanced wastewater treatment and biomass production: A review. Journal of Water Process Engineering, 66(September): 105929.

Malik, A., Singh, K.P. and Ojha, P. 2007. Residues of organochlorine pesticides in fish from the Gomti River, India. Bulletin of Environmental Contamination and Toxicology, 78(5) (July 7): 335-340.

Miyazaki, A., Tung, R., Taing, B., Matsui, M., Iwamoto, A. and Cox, S.E. 2020. Frequent unregulated use of antibiotics in rural Cambodian infants. Transactions of The Royal Society of Tropical Medicine and Hygiene, 114(6)(April 20): 401-407.

Modak, S., Mokarizadeh, H., Karbassiyazdi, E., Hosseinzadeh, A. and Esfahani, M.R. 2022. The AI-assisted removal and sensor-based detection of contaminants in the aquatic environment. *In*: Artificial Intelligence and Data Science in Environmental Sensing, 211-244.

Mohd Udaiyappan, A.F., Abu Hasan, H., Takriff, M.S. and Sheikh Abdullah, S.R. 2017. A review of the potentials, challenges and current status of microalgae biomass applications in industrial wastewater treatment. Journal of Water Process Engineering, 20(December): 8-21.

Mohsenpour, S.F., Hennige, S., Willoughby, N., Adeloye, A. and Gutierrez, T. 2021. Integrating micro-algae into wastewater treatment: A review. Science of The Total Environment, 752(January): 142168.

Mojiri, A., Baharlooeian, M., Kazeroon, R.A., Farraji, H. and Lou, Z. 2020. Removal of pharmaceutical micropollutants with integrated biochar and marine microalgae. Microorganisms, 9(1)(December 22): 4.

Mondal, S., Bera, S., Mishra, R. and Roy, S. 2022. Redefining the role of microalgae in industrial wastewater remediation. Energy Nexus, 6(June): 100088.

Moreno-García, A.F., Neri-Torres, E.E., Mena-Cervantes, V.Y., Altamirano, R.H., Pineda-Flores, G., Luna-Sánchez, R. et al. 2021. Sustainable biorefinery associated with wastewater treatment of Cr (III) using a native microalgae consortium. Fuel, 290(April): 119040.

Muhmood, A., Lu, J., Dong, R. and Wu, S. 2019. Formation of struvite from agricultural wastewaters and its reuse on farmlands: Status and hindrances to closing the nutrient loop. Journal of Environmental Management, 230(January): 1-13.

Mustafa, S., Bhatti, H.N., Maqbool, M. and Iqbal, M. 2021. Microalgae biosorption, bioaccumulation and biodegradation efficiency for the remediation of wastewater and carbon dioxide mitigation: Prospects, challenges and opportunities. Journal of Water Process Engineering, 41(June): 102009.

Mutiyar, P.K., Gupta, S.K. and Mittal, A.K. 2018. Fate of pharmaceutical active compounds (PhACs) from river Yamuna, India: An ecotoxicological risk assessment approach. Ecotoxicology and Environmental Safety, 150(April): 297-304.

Najdenski, H.M., Gigova, L.G., Iliev, I.I., Pilarski, P.S., Lukavský, J., Tsvetkova, I.V. et al. 2013. Antibacterial and antifungal activities of selected microalgae and cyanobacteria. International Journal of Food Science & Technology, 48(7)(March 25): 1533-1540.

Nayak, M., Swain, D.K. and Sen, R. 2019. Strategic valorization of de-oiled microalgal biomass waste as biofertilizer for sustainable and improved agriculture of rice (*Oryza sativa* L.) crop. Science of The Total Environment, 682(September): 475-484.

Newlight Technologies. 2019. *In*: VCS Methodology VM0040 Methodology for Greenhouse Gas Capture and Utilization in Plastic Materials 1, VERRA VCS, pp. 1-38.

Ng, I., Tan, S., Kao, P., Chang, Y. and Chang, J. 2017. Recent developments on genetic engineering of microalgae for biofuels and bio-based chemicals. Biotechnology Journal, 12(10)(September 18).

Norvill, Z.N., Shilton, A. and Guieysse, B. 2016. Emerging contaminant degradation and removal in algal wastewater treatment ponds: Identifying the research gaps. Journal of Hazardous Materials, 313(August): 291-309.

Oswald, W.J. 2003. My sixty years in applied algology. Journal of Applied Phycology, 15(2/3)(March): 99-106.

Padgelwar, S., Nandan, A. and Mishra, A.K. 2019. Plastic waste management and current scenario in India: A review. International Journal of Environmental Analytical Chemistry, 101(13)(November 21): 1894-1906.

Park, C.J., Barakat, R., Ulanov, A., Li, Z., Lin, P.-C., Chiu, K. et al. 2019. Sanitary pads and diapers contain higher phthalate contents than those in common commercial plastic products. Reproductive Toxicology, 84(March): 114-121.

Pena, A.C.C., Agustini, C.B., Trierweiler, L.F. and Gutterres, M. 2020. Influence of period light on cultivation of microalgae consortium for the treatment of tannery wastewaters from leather finishing stage. Journal of Cleaner Production, 263(August): 121618.

Qiu, S., Yu, Z., Hu, Y., Chen, Z., Guo, J., Xia, W. and Ge, S. 2021. An evolved native microalgal consortium-snow system for the bioremediation of biogas and centrate wastewater: Start-up, optimization and stabilization. Water Research, 196: 117038.

Rajkumar, R. and Sobri Takriff, M. 2016. Prospects of algae and their environmental applications in Malaysia: A case study. Journal of Bioremediation & Biodegradation, 07(01).

Ram Namdev, G., Bajpai, A. and Malik, S. 2006. Effect of chemical fertilizers on water quality of irrigation reservoir (Kaliasote Reservoir) of Bhopal (M.P.). Current World Environment, 1(1)(June 25): 169-172.

Rees, G.M., Reyher, K.K., Barrett, D.C. and Buller, H. 2021. "It's Cheaper than a Dead Cow": Understanding veterinary medicine use on dairy farms. Journal of Rural Studies, 86(August): 587-598.

Rempel, A., Gutkoski, J.P., Nazari, M.T., Biolchi, G.N., Cavanhi, V.A.F., Treichel, H. et al. 2021. Current advances in microalgae-based bioremediation and other technologies for emerging contaminants treatment. Science of The Total Environment, 772(June): 144918.

Riccio, G. and Lauritano, C. 2019. Microalgae with immunomodulatory activities. Marine Drugs, 18(1)(December 18): 2.

Ridolfi, A.S., Álvarez, G.B. and Rodríguez Girault, M.E. 2014. Organochlorinated contaminants in general population of Argentina and Other Latin American Countries. *In*: Bioremediation in Latin America, 17-40. Cham: Springer International Publishing.

Rillig, M.C. 2012. Microplastic in terrestrial ecosystems and the soil? Environmental Science & Technology, 46(12)(May 31): 6453-6454.

Ross, M.E., Davis, K., McColl, R., Stanley, M.S., Day, J.G. and Semião, A.J.C. 2018. Nitrogen uptake by the macro-algae Cladophora Coelothrix and Cladophora Parriaudii: Influence on growth, nitrogen preference and biochemical composition. Algal Research, 30(March): 1-10.

Roy, A. and Bharadvaja, N. 2020. Removal of toxic pollutants using microbial fuel cells. *In*: Removal of Toxic Pollutants Through Microbiological and Tertiary Treatment, pp. 153-177.

Sangion, A. and Gramatica, P. 2016. Ecotoxicity interspecies QAAR models from daphnia toxicity of pharmaceuticals and personal care products. SAR and QSAR in Environmental Research, 27(10)(September 22): 781-798.

Sarode, S., Upadhyay, P., Khosa, M.A., Mak, T., Shakir, A., Song, S. et al. 2019. Overview of wastewater treatment methods with special focus on biopolymer chitin-chitosan. International Journal of Biological Macromolecules, 121: 1086-1100.

Schopel, B. and Stamminger, R. 2019. A comprehensive literature study on microfibres from washing machines. Tenside Surfactants Detergents, 56(2)(March 15): 94-104.

Shalaby, S.M., Madkour, F.F., El-Kassas, H.Y., Mohamed, A.A. and Elgarahy, A.M. 2021. Green synthesis of recyclable iron oxide nanoparticles using spirulina platensis microalgae for adsorptive removal of cationic and anionic dyes. Environmental Science and Pollution Research, 28(46)(July 28): 65549-65572.

Sharma, B.M., Bečanová, J., Scheringer, M., Sharma, A., Bharat, G.K., Whitehead, P.G. et al. 2019. Health and ecological risk assessment of emerging contaminants (pharmaceuticals, personal care products, and artificial sweeteners) in surface and groundwater (drinking water) in the Ganges River Basin, India. Science of The Total Environment, 646(January): 1459-1467.

Shi, J., Podola, B. and Melkonian, M. 2007. Removal of nitrogen and phosphorus from wastewater using microalgae immobilized on twin layers: An experimental study. Journal of Applied Phycology, 19(5)(February 17): 417-423.

Shi, J., Podola, B. and Melkonian, M. 2014. Application of a prototype-scale twin-layer photobioreactor for effective N and P removal from different process stages of municipal wastewater by immobilized microalgae. Bioresource Technology, 154(February): 260-266.

Simonne, E.H., Gazula, A., Ozores-Hampton, M., DeValerio, J. and Hochmuth, R.C. 2017. Localized application of fertilizers in vegetable crop production. *In*: Advances in Olericulture, 149-181. Cham: Springer International Publishing.

Singh, B., Bauddh, K. and Bux, F. 2015. Algae and Environmental Sustainability. 1st ed. Springer India, New Delhi.

Singh, K.P., Malik, A. and Sinha, S. 2006. Persistent organochlorine pesticide residues in soil and surface water of northern Indo-Gangetic alluvial plains. Environmental Monitoring and Assessment, 125(1-3): 147-155.

Singh, V. and Suthar, S. 2021. Occurrence, seasonal variations, and ecological risk of pharmaceuticals and personal care products in River Ganges at two holy cities of India. Chemosphere, 268(April): 129331.

Solovchenko, A., Pogosyan, S., Chivkunova, O., Selyakh, I., Semenova, L., Voronova, E. et al. 2014. Phycoremediation of alcohol distillery wastewater with a novel Chlorella Sorokiniana strain cultivated in a photobioreactor monitored on-line via chlorophyll fluorescence. Algal Research, 6(October): 234-241.

Soto-Ramírez, R., Lobos, M.-G., Córdova, O., Poirrier, P. and Chamy, R. 2021. Effect of growth conditions on cell wall composition and cadmium adsorption in Chlorella Vulgaris: A new approach to biosorption research. Journal of Hazardous Materials, 411(June): 125059.

Srimongkol, P., Sangtanoo, P., Songserm, P., Watsuntorn, W. and Karnchanatat, A. 2022. Microalgae-based wastewater treatment for developing economic and environmental sustainability: Current status and future prospects. Frontiers in Bioengineering and Biotechnology, 10. https://doi.org/10.3389/fbioe.2022.904046.

Sturm, B.S.M. and Lamer, S.L. 2011. An energy evaluation of coupling nutrient removal from wastewater with algal biomass production. Applied Energy, 88(10)(October): 3499-3506.

Subedi, B., Balakrishna, K., Sinha, R.K., Yamashita, N., Balasubramanian, V.G. and Kannan, K. 2015. Mass loading and removal of pharmaceuticals and personal care products, including psychoactive and illicit drugs and artificial sweeteners, in five sewage treatment plants in India. Journal of Environmental Chemical Engineering, 3(4)(December): 2882-2891.

Sunkel, M., Kand, D., Vrancken, H., Miller, M. and Riede, O. 2021. VCS Methodology for the reduction of enteric methane emissions from ruminants through the use of feed ingredients 1. VERRA VCS, 1-56.

Sutherland, D.L. and Ralph, P.J. 2019. Microalgal bioremediation of emerging contaminants opportunities and challenges. Water Research, 164: 114921.

Tan, X.-B., Zhao, X.-C. and Yang, L.-B. 2019. Strategies for enhanced biomass and lipid production by Chlorella Pyrenoidosa culture in starch processing wastewater. Journal of Cleaner Production, 236(November): 117671.

Tang, J., Wang, S., Tai, Y., Tam, N.F., Su, L., Shi, Y. et al. 2020. Evaluation of factors influencing annual occurrence, bioaccumulation, and biomagnification of antibiotics in planktonic food webs of a large subtropical river in South China. Water Research, 170(March): 115302.

Tiwari, B., Sellamuthu, B., Ouarda, Y., Drogui, P., Tyagi, R.D. and Buelna, G. 2017. Review on fate and mechanism of removal of pharmaceutical pollutants from wastewater using biological approach. Bioresource Technology, 224(January): 1-12.

Upadhyay, V., Kumari, A. and Kumar, S. 2024. From soil to health hazards: Heavy metals contamination in northern India and health risk assessment. Chemosphere, 354(April): 141697.

Ustun, I., Sungur, S., Okur, R., Sumbul, A.T., Oktar, S., Yilmaz, N. et al. 2014. Determination of phthalates migrating from plastic containers into beverages. Food Analytical Methods, 8(1)(May 18): 222-228.

Vajargah, M.F., Ramzanipour, M.M. and Mousavi, S.P. 2023. An overview of municipal wastewater and sludge treatment process. Ecology & Conservation Science: Open Access, 2(5).

Verma, A., Roy, A. and Bharadvaja, N. 2021. Remediation of heavy metals using nanophytoremediation. *In*: Advanced Oxidation Processes for Effluent Treatment Plants, 273-296. Elsevier. https://doi.org/10.1016/b978-0-12-821011-6.00013-x.

Vo, H.N.P., Ngo, H.H., Guo, W., Nguyen, K.H., Chang, S.W., Nguyen, D.D. et al. 2020. Micropollutants cometabolism of microalgae for wastewater remediation: Effect of carbon sources to cometabolism and degradation products. Water Research, 183(September): 115974.

Wang, L., Li, Y., Chen, P., Min, M., Chen, Y., Zhu, J. et al. 2010. Anaerobic digested dairy manure as a nutrient supplement for cultivation of oil-rich green microalgae Chlorella sp. Bioresource Technology, 101(8): 2623-2628.

World Health Organization. 2019. https://www.who.int/news-room/fact-sheets/detail/drinking-water #:~:text=Globally%2C%20at%20lea st%202%20billion,water %20 source%20contaminated%20with%20faeces.&text=Contaminated%20drinking %20 water%20is%20estimated,livin

Wrede, D., Taha, M., Miranda, A.F., Kadali, K., Stevenson, T., Ball, A.S. et al. 2014. Co-cultivation of fungal and microalgal cells as an efficient system for harvesting microalgal cells, lipid production and wastewater treatment. PLoS ONE, 9(11)(November 24): e113497.

Wu, Y., Li, T. and Yang, L. 2012. Mechanisms of removing pollutants from aqueous solutions by microorganisms and their aggregates: A review. Bioresource Technology, 107(March): 10-18.

Xie, P., Chen, C., Zhang, C., Su, G., Ren, N. and Ho, S.-H. 2020. Revealing the role of adsorption in ciprofloxacin and sulfadiazine elimination routes in microalgae. Water Research, 172(April): 115475.

Xiong, J.-Q., Kurade, M.B. and Jeon, B.-H. 2018. Can microalgae remove pharmaceutical contaminants from water? Trends in Biotechnology, 36(1)(January): 30-44.

Xiong, Q., Liu, Y.-S., Hu, L.-X., Shi, Z.-Q., Cai, W.-W., He, L.-Y. et al. 2020. Co-metabolism of sulfamethoxazole by a freshwater microalga *Chlorella Pyrenoidosa*. Water Research, 175(May): 115656.

Xiong, Q., Hu, L.-X., Liu, Y.-S., Zhao, J.-L. He, L.-Y. and Ying, G.-G. 2021. Microalgae-based technology for antibiotics removal: From mechanisms to application of innovational hybrid systems. Environment International, 155(October): 106594.

Yadav, I.C., Devi, N.L., Syed, J.H., Cheng, Z., Li, J., Zhang, G. et al. 2015. Current status of persistent organic pesticides residues in air, water, and soil, and their possible effect on neighboring countries: A comprehensive review of India. Sci. of The Total Environ, 511(April): 123-137.

Yan, C.R., Liu, E.K., Shu, F., Liu, Q., Liu, S. and He, W.Q. 2014. Review of agricultural plastic mulching and its residual PImution and prevention measures in China. J. Agr. Resour. Environ, 31(2): 95-102.

Yim, J.H., Kim, S.J., Ahn, S.H., Lee, C.K., Rhie, K.T. and Lee, H.K. 2004. Antiviral effects of sulfated exopolysaccharide from the marine microalga *Gyrodinium Impudicum* strain KG03. Marine Biotechnology, 6(1)(February 1): 17-25.

Yuan, N.J. 2009. Plastic. Chapter of Encyclopedia of China (2nd edition). Encyclopedia of China Publishing House, ISBN 9787500079583.

Zhao, G., Xu, Y., LI, W., Han, G. and Ling, B. 2007. PCBs and OCPs in human milk and selected foods from Luqiao and Pingqiao in Zhejiang, China. Science of The Total Environment, 3783(7): 281-292.

Zhou, W., Chen, P., Min, M., Ma, X., Wang, J., Griffith, R. et al. 2014. Environment-enhancing algal biofuel production using wastewaters. Renewable and Sustainable Energy Reviews, 36(8): 256-269.

Zhu, Y., Liu, H., Xi, Z., Cheng, H. and Xu, X. 2005. Organochlorine pesticides (DDTs and HCHs) in soils from the outskirts of Beijing, China. Chemosphere, 60(6): 770-778.

CHAPTER

7

Microalgae and Bacteria Consortia for the Removal of Emerging Contaminants from Wastewater and Biomass Production

Elakkiya Venugopal[1] and Sharangouda J. Patil[2]

[1] PG and Research Department of Biotechnology, Kongunadu Arts and Science College, Coimbatore, Tamil Nadu, India

[2] Department of Zoology, NMKRV College for Women (Autonomous), Bengaluru, Karnataka, India

Introduction

Water is the most basic and crucial element for all living things. Nearly 70% of the earth's surface area is covered with water, of which freshwater accounts for about 3%. Due to the overuse of natural water resources, water shortage has recently become a serious global problem. Industrialization and urbanization are the main causes of the ongoing pollution of the natural water environment. According to recent studies, people are becoming progressively more unhealthy with different water-borne illnesses due to drinking water polluted by wastewater treatment plants and industrial waste. Water bodies are becoming more saturated and contaminated owing to the addition of nutrients, harmful metals, colorants, pharmaceutical products, pesticides, antibiotic residues, and inorganic compounds from various sources, including municipal, industrial, and agricultural. Wastewater treatment systems are facing new challenges due to rapid changes in the quantity and quality of wastewater caused by population growth, shifting socioeconomic situations, and fluctuating consumption patterns. The United Nations World Water Development Report 2017 states that around 80% of all wastewater produced globally is disposed of without receiving the necessary treatment. There have been several traditional treatment techniques used, but none of them can be referred to as a universal technique because of their high cost, low effectiveness, and environmentally unfriendly characteristics. Therefore,

*Corresponding author: shajapatil@gmail.com

DOI: 10.1201/9781003527350-7

creating efficient, economical, and environmentally friendly wastewater treatment techniques is essential. Physical, chemical, and biological processes are used in primary, secondary, and tertiary stages of wastewater treatment. However, the primary disadvantages of these techniques are their high cost, high energy demand, and less environmentally friendly nature. Furthermore, these processes frequently waste resources and produce sludge and harmful byproducts, which lead to secondary contamination. After initial pre-treatment, biological treatment is considered a more sustainable and energy-efficient method of treating wastewater. Additionally, it helps remove the wastewater's phosphate and nitrate. Co-bioremediation systems mediated by bacteria and algae can lower greenhouse gas emissions and help convert CO_2 into bio-based chemical products like biofuels and bioalcohol. Algae and bacteria have a symbiotic relationship that is thought to form the structural foundation of the ecosystem, and their consortiums are helpful in treating wastewater. However, because wastewater has such a wide range of composition, its pH, temperature, and light intensity can all affect microalgae growth. Both flocculation and electrochemical precipitation techniques can be used to recover the algal biomass produced in wastewater treatment facilities.

This chapter outlines the most current developments in microalgal biotechnology to create a biorefinery strategy for employing microalgae to clean wastewater. Its primary focus is on treating wastewater streams through the use of bacterial and microalgal systems. The usage of microalgal consortia, the integration of microalgal biorefineries with wastewater treatment facilities, and the potential benefits and drawbacks of phycoremediation are also covered. In light of upcoming concerns about the cost-benefit ratios of algal biorefineries, the influence of wastewater-derived nutritional stress and its exploitation to modify the algal metabolite composition are also underlined.

Characteristics of Wastewater

Solids, nutrients, dissolved and particulate materials, microbes, heavy metals, and micropollutants are all present in wastewater. The origin of the wastewater has a significant impact on the concentration of these components. It can originate from the commercial, industrial, residential, or agricultural domains and may comprise various organic materials, including pesticides, washing waste, human excreta, and biodegradable waste with nutrients. Non-biodegradable inorganic waste makes up most of the wastewater's 0.1% suspended and dissolved solids. Proteins, carbohydrates, and lipids from food sector waste make up the volatile solids in sewage. A wide variety of microorganisms, including viruses, protozoa, antibiotic-resistant bacteria, and helminths, are found in sewage discharge. These microorganisms are primarily infectious and pose a risk to both human health and the environment. Wastewater can be treated by other environmentally benign microbes such as algae, Pseudomonas, and Zoogloeal. The right conditions for algae growth require nutrient-rich wastewater streams with high CO_2 levels, aiding lipid formation and improving nutrient recovery. Agricultural wastewater is produced by a range of farm activities and is inappropriate for algae growth due to its high ammonia content, high nitrogen load, suspended particles, and chroma content (Zhu et al. 2013). Similarly, the chemical oxygen demand values of starch processing effluent and dairy wastewater range from 10,000 to about 350,000 mg/L and 1,000 to approximately 70,000 mg/L, respectively. There have been reports of high chemical and biochemical oxygen demands in wastewater from various sources linked to high levels of ammonia and total phosphorus. These conditions can provide an excellent supply of nutrients for the expanding bacterial and algal biomass (Connor et al. 2022).

Wastewater Treatment

Wastewater Treatment by Algae

A wide range of eukaryotes, from basic blue-green algae to intricate seaweeds and kelps, are classified as algae. Eukaryotic microorganisms with one cell that live in fresh and saltwater environments are known as microalgae. The number of microalgal species discovered has averaged around 350,000. According to general estimates, they may include up to 70% fats, 60% carbs, 65% proteins, and 65% essential amino acids. As photosynthetic microorganisms, microalgae use light, carbon from CO_2, and inorganic nutrients to grow, create oxygen, and produce biomass. The production of biofuels, chemical extraction, and other bioproducts has recently made substantial use of microalgal biomass.

The microalgae absorb nutrients through cellular absorption, and the biomass they create can be utilized to recover resources. The best resource for producing algal biomass is wastewater for a number of reasons, including (i) less expensive growth media; (ii) supporting the production of bulk biomass and biofuels; (iii) having the capacity to supply an abundance of nutrients; and (iv) providing the opportunity to integrate algal cultivation with the wastewater treatment infrastructure already in place. In recent decades, several studies on the phycoremediation of wastewater using microalgae have been carried out, and techniques based on biorefineries have been suggested. Microalgae have been investigated for the treatment of both industrial and domestic wastewater, including wastewater from breweries, homes, textile industries, pharmaceutical waste streams, slaughterhouses, wastewater containing heavy metals, wastewater from palm oil mills, wastewater from starch-containing textile industries, and wastewater from agro-industrial sources. While there are many advantages to microalgae-based phycoremediation, a number of issues must be resolved.

Wastewater Treatment by Bacteria

Wastewater biological treatment is becoming more and more important, and it's seen as an economical and environmentally beneficial way to reduce pollution. Numerous mechanisms, including biosorption, biodegradation, biomineralization, bioaugmentation, and bioreduction of pollutants into less hazardous and innocuous compounds, can be used by bacteria to bioremediate wastewater. Generally speaking, biological wastewater treatment is linked to intricate metabolic biochemical processes that mostly result from interactions between various inorganic and organic contaminants and microorganisms. Wastewater treatment often involves the employment of a consortium of microorganisms that have a high biodegradation efficiency and are able to exploit a variety of substrates found in wastewater (Ramadoss et al. 2015, Morin et al. 2019).

Numerous enzymes, including chromate reductase and arsenate oxidase, are reported to be produced by various bacteria. These enzymes can change harmful arsenic As(III) and Cr(VI) into less hazardous Cr(II) and As(V). By altering the oxidation states of radionuclides, bacteria can also destroy them. Consequently, the substance may dissolve and be carried, precipitate, or become immobile. However, pH, temperature, and other environmental variables have a major impact on the effectiveness of bioremediation. The capacity of bacteria such as *Pseudomonas* species and *Lysinibacillus* species to eliminate potentially harmful components and serve as a carrier matrix for naturally occurring non-polymeric electrospun cyclodextrin fibers has been the subject of much research. The primary role of these microbial consortia and these cyclodextrin fibers is as a biosorbent for textile dyes and heavy metals. Triarylmethane dyes have been decolorized by isolates such as *Aeromonas hydrophila*. However, the effectiveness of this process was mostly influenced by temperature, pH, and oxygenation level (Imran et al. 2015, Safdari et al. 2018).

Binary Cultures for Wastewater Treatment

Polycultures enable us to build reliable ecosystems for wastewater treatment because they can be blended with different metabolic processes that allow themselves to thrive under environmental stress conditions. Further, integrated consortia such as algae-algae or algae-bacteria can absorb nutrients at greater levels because one strain can specifically remove heavy metals, and another can remove nitrogen. Maintaining only pure cultures according to field conditions is challenging, and it becomes even more difficult when applied to ponds receiving wastewater. This is why researchers are currently concentrating on the production of mixed cultures (Renuka et al. 2013). These consortia have a number of benefits, such as (i) resistance to pollution and predators owing to the secretion of allelochemicals; (ii) increased nutrient consumption, which ensures a sufficient supply of nutrients throughout the process; (iii) development of a settleable system for flocculation, which removes harvesting limitations; and (iv) boosted viability of phycoremediation, as the loss of one species is offset by the presence of other species. Allelochemicals are produced, which leads to contamination and predator resistance. It does, however, have certain limitations because it is challenging to form strong consortia due to the many possible combinations. Furthermore, it has been difficult to maintain consortia for extended periods, particularly in open-pond systems (Goncalves et al. 2017).

Algae-algae Consortia

Growing a variety of algal species that complement one another to increase productivity is a promising way to reduce problems and challenges associated with microalgal cultivation. This is because diverse communities may produce more biomass-specific lipids than corresponding monocultures. Furthermore, diversity and biomass stability are correlated, which is a crucial prerequisite for mass production. Additionally, greater diversity is correlated with a remarkable range of features exhibited by primary producer communities (Stockenreiter et al. 2016). Using a polyculture of 15 native algal strains, more than 96% of the nutrient removal efficiency, 6.82% lipid content, and 9.2–17.8 tons year^{-1} ha^{-1} of biomass productivity was obtained. When grown under photoautotrophic conditions, mixed microalgal cultures, primarily composed of *Chlorella* with small levels of *Scenedesmus*, were able to extract 481 mg L^{-1} of nitrogen and 31 mg L^{-1} of phosphorus from textile wastewater. *Chlorella* sp. polycultures showed that they could remove all nitrates in swine wastewater and that they could collect large amounts of protein and lipids, 34% and 59%, respectively. However, just 3% of the overall biomass content comprised lipids (Huy et al. 2018, Michelon et al. 2016).

Myco-algal Consortia

The synthetic aquaculture wastewater's phosphate, total ammonia, and nitrogen levels were nearly undetectable due to the combined propagation of *C. vulgaris* with *Mucor indicus*. Additionally, this method allowed 860 mg DW of myco-algal biomass to flocculate. Algal-fungal co-cultivation was used to treat household sewage wastewater and upgrade biogas at the same time, removing 81% of the total phosphorus (TP) and COD. Similarly, from swine wastewater treatment, co-culturing of *C. vulgaris* and *Ganoderma lucidum* decreased TP, TN, and COD contents by 84%, 74%, and 79%, respectively. Furthermore, *P. subcapitata* and *G. lucidum* worked together to effectively remove CO_2 for biogas upgradation, with an efficiency of 84%.

These investigations have unequivocally shown that fungal pellets can facilitate the simpler collection of algal biomass and that myco-algal relationships have the ability to improve wastewater treatment capacity. However, more thorough research is needed to assess the effect of myco-algal relationships on the microalgae's metabolite content. Since both extracellular and intracellular algal compounds can be significantly altered by the presence of fungi in the growth conditions, the quantity of any desired molecule can be increased or decreased. Furthermore, the subsequent processing of biomass from algae to obtain any required organic compounds may be hampered by the fungal pellets.

Algae-bacteria Consortia

Nutrient-rich wastewater, supporting microalgal growth and a variety of bacteria, can work in concert to recycle the nutrients in the wastewater and lower the amount of oxygen required for both biological and chemical processes. They also help in the wastewater's phosphate and nitrate removal processes. Algal- and bacterial-mediated co-bioremediation techniques can accelerate the transition of CO_2 to bio-based biochemical products, like biofuels and bioalcohols, as well as reduce greenhouse gases. The tests conducted in a recent study using a consortium of bacteria and algae to encourage the elimination of organic contaminants via bacteria and the reduction of nutrients with algae in wastewater are depicted in the figure. Five distinct communities were produced in this work by selecting various biomass proportions of bacteria and algae to determine which was best.

Wastewater can be efficiently treated by consortia of bacteria and algae, whose symbiotic connection is thought to represent the ecosystem's structural cornerstone. However, because wastewater has such a wide range of compositions, its pH, temperature, and light intensity can all affect the growth of microalgae. According to a number of studies, algae species including *Chlorella zofingiensis* and *Scenedesmus* spp. can be used to biologically treat various wastewaters, effectively lowering the levels of total nitrogen, total phosphorus, and chemical oxygen demand.

Both flocculation and electrochemical deposition techniques can be used to recover the algal biomass produced in wastewater treatment facilities. Because of this, the biomass is an extremely rich source of proteins, lipids, and carbohydrates that can be utilized as a precursor for microbial fermentation and to produce a variety of commercial goods, including carotenoids and polyhydroxyalkanoates. There are few studies on the benefits of using bacterial consortia and microalgae in the treatment of wastewater streams, despite some evaluations on the possible use of microalgae in the treatment of wastewater streams, such as agricultural and municipal effluents, having been published.

Algae to Bacteria Cell Adhesion

The formation of a consortium and the co-evolution of bacteria and algae are made possible by cell-to-cell adhesion. Even though the exact process by which toxic pollutants disrupt cell-to-cell adhesion is unclear, they disturb the equilibrium by interfering with the physical interaction between cells. Extracellular exudates from algae play a critical role in facilitating bacterial colonization and enhancing the host of coexisting microorganisms. Extracellular products that algae emit, including nutrients, chemoattractants, and antibacterial agents, are primarily responsible for controlling the bacteria-algae relationship. Toxic pollutants can potentially disrupt the discharge of extracellular polymeric compounds by bacteria and algae, disrupting the inter-organism connection (You et al. 2021).

Stronger bacterial colonization occurs in algae exposed to severe environmental conditions. The primary reason for this is a change in the release of extracellular substances and cell lysis, which releases intracellular components that bacteria use as nutrients. Furthermore, robust algae release antimicrobial compounds that prevent the growth of pathogenic bacteria and encourage the formation of productive consortia. Toxic pollutants may compromise this defense mechanism. Under stressful circumstances, such as when exposed to heavy metals like chromium, cadmium, mercury, and nanoparticles, a decrease in cell motility and chemotaxis is also seen (Cheng et al. 2019).

In comparison to isolated algae or bacteria, cell-to-cell adhesion in an algae-bacteria consortium may lessen the toxicity of contaminants, such as heavy metals and nanoparticles. It has been found that bacteria safeguard cell lysis and structure while also reducing the amount of hazardous substances that algae absorb. Environmental elements that can affect these defense mechanisms in a favorable or negative way include pH and the availability of nutrients. For instance, high concentrations of PO_4^{3-} enhance the toxicity of arsenic on the consortium between the bacterium *Alteromonas macleodii* and alga *Dunaliella salina*, and copper toxicity in *Chlorella* spp. development was reduced at acidic pH. The consortium equilibrium may be upset by these impacts, which could be caused by modifications in algal growth or bacterial motility and chemotaxis (Wang et al. 2016). Therefore, it may be concluded that the development of an algal-bacterial consortium depends on appropriate cell-to-cell adhesion in the right environmental conditions. Furthermore, this cell-to-cell attachment may shield the consortium from harmful pathogens and lengthen its shelf life under stressful circumstances.

Physio-chemical Factors Affecting Algal-bacterial System

The interactions between the bacteria and algae are influenced by a wide range of factors, including species-specific interactions, operating parameters like the hydraulic retention time and solids retention time, and environmental factors like temperature, pH, light intensity, photoperiod (the cycle of light and dark) and mixing conditions.

pH and Temperature

The pH and temperature of the system are fundamental requirements for all biological systems. The optimal pH range for the majority of microalgal species is 8.2 to 8.7, while it can be balanced between 7 and 9. Because of the chemical balance between H_2CO_3, HCO_3, and CO_3 species, the pH of the entire system is dependent on the amount of CO_2 provided. It is frequently seen that an increase in pH results from the algal biomass's consumption of CO_2. Additionally, an overabundance of CO_2 can occasionally cause the pH to drop to an acidic range, which will lyse algae cells. In order to promote algal development, the pH of the entire system should be maintained at its ideal range.

Light Intensity

As phototrophic organisms, microalgae synthesize organic matter by absorbing nutrients and dissolving inorganic carbon when exposed to light. Numerous investigations have demonstrated that the primary determinants of microalgae productivity are light intensity and the length of the light-dark cycle. The impact of photoperiod conditions (12:12, 36:12, and 60:12 h dark-light cycles) on nitrogen removal and biomass productivity was examined by Lee et al. (2015). The findings indicate that the amount of carbon removed had a positive relationship with the duration of the dark phase, but the relationships for nitrogen and phosphorus removal were inverse. Additionally, the intensity of the light influences the production of settleable algae-bacterial granules. Consequently, optimization of the light intensity and light/dark cycle is necessary for the practical implementation of algae-bacterial systems for wastewater treatment (Vergara et al. 2016).

Nutrient Ratio

The characteristic stoichiometric formula for microalgae biomass is C 106 H181 O 45N 16 P. Based on weight, the algal biomass's C/N/P ratio is roughly 50/8/1 (Molazadeh et al. 2019). The majority of wastewaters lack the same ratio needed for microalgae to develop to their best potential. Carbon may be a limiting factor for bacteria and microalgae that need organic carbon in mixotrophic or heterotrophic environments. Due to carbon constraints, most domestic wastewaters have a low C/N/P ratio compared to the composition of algal biomass, which restricts the removal of nutrients. Phosphorus limitation occurs when the N:P molar ratio is greater than 30:1, whereas nitrogen limitation occurs when the N:P molar ratio is approximately 5:1 (Larsdotter 2006).

Biomass Harvesting

Depending on the species diversity, culture conditions, and cell density, algae harvesting alone accounts for 20–30% of production expenses (Mennaa et al. 2015). Before releasing the treated water, the algal biomass that has formed in the wastewater must be separated.

Resource recovery can make use of the biomass that has been gathered. Mechanical, chemical, electrical, or biological approaches can all be used for harvesting. The effectiveness of using natural flocculants, such as Ecotan and Tanfloc, to collect microalgae from an HRAP treating urban wastewater was assessed by Gutierrez et al. (2015). The outcomes demonstrate that in just 10–20 minutes, over 90% recovery was attained.

After 27 minutes of interaction, a harvesting efficiency of over 95% was attained. The simplest and least expensive method of harvesting is sedimentation. Leite and Daniel (2020) investigated in a recent study how to induce a high pH in order to optimize the sedimentation characteristics for *Chlorella sorokiniana* harvesting. When the pH was 12, the velocity gradient was 250 s^{-1}, and the mixing period was 10 s, more than 98% efficiency was attained. Utilizing bio-flocculation for microalgae harvesting offers several benefits, including reduced energy usage, increased efficiency, and minimal toxicity. The interactions between algae and bacteria, algae and fungi, and algae and algae can result in bio-flocculation (Alam et al. 2016).

Future Prospective of the Algal-bacterial System for Wastewater Treatment

Algae-bacterial systems are not yet widely deployed as mainstream wastewater treatment systems, despite their many benefits, including nutrient recovery, aeration reduction, and lower operating costs. Numerous variables must be optimized for algae-bacterial wastewater treatment systems to be competitive with traditional biological treatment systems. The main component of algal-bacterial systems is the availability of enough light intensity to support photosynthetic activity. By lowering the waste component of a wastewater treatment system, wastewater treatment coupled with the concept of an algal biorefinery enables more effective utilization of the algal biomass grown on wastewater, contributing to the system's long-term economic viability (Bodenes et al. 2019). Significant developments in microalgal biology, strain production, downstream processing, and biofuels and bioproducts obtained from microalgae are required in order to achieve commercially viable results. Mass culture-based conventional algae research takes a long time and requires a lot of manpower. Recently, there has been a lot of interest in the application of microfluidics to algal biofuel research and microalgal production. Even though microalgae-bacterial systems have been the subject of numerous studies, more investigation is still required to optimize the parameters for large-scale units. The stability of the consortia is the primary obstacle to employing algae-bacterial systems for real-scale systems. Creating a sustainable wastewater treatment system with algal-bacterial consortia requires a thorough grasp of the relationships between the culture conditions, biochemical composition, operational parameters, harvesting, and biomass conversion to fuel.

Conclusion

The relationship between the bacteria and algae is complicated by their competition for resources. There are few studies on the relationship between bacteria and algae in wastewater treatment. Algal-bacterial systems are challenging to handle since their dynamics are mainly unknown. Therefore, it is necessary to create reliable models that can faithfully replicate the bioprocesses in algal-bacterial systems. The majority of modeling research uses restricted processes related to algae and is limited to laboratory experiments. To incorporate into modeling, a deeper comprehension of intricate algal-bacterial interactions is required, along with consideration of the systems' hydrodynamic elements. To make reactor operation simpler, more research is required to integrate the data from modeling studies with intelligent automated control technology.

References

Alam, M.A., Vandamme, D., Chun, W., Zhao, X., Foubert, I., Wang, Z. et al. 2016. Bioflocculation as an innovative harvesting strategy for microalgae. Reviews in Environmental Science and Biotechnology, 15: 573-583.

Bodenes, P., Wang, H.Y., Lee, T.H., Chen, H.Y. and Wang, C.Y. 2019. Microfluidic techniques for enhancing biofuel and biorefinery industry based on microalgae. Biotechnology for Biofuels, 12: 1-25.

Cheng, H., Guan, Q., Villalobos, L.F., Peinemann, K.V., Pain, A. and Hong, P.Y. 2019. Understanding the antifouling mechanisms related to copper oxide and zinc oxide nanoparticles in anaerobic membrane bioreactors. Environmental Science: Nano, 6(11): 3467-3479.

Connor, J., Bolan, N.S., Kumar, M., Nitai, A.S., Ahmed, M.B., Bolan, S.S. et al. 2022. Distribution, transformation and remediation of poly- and per-fluoroalkyl substances (PFAS) in wastewater sources. Process Safety and Environmental Protection, 164: 91-108.

Gutierrez, R., Passos, F., Ferrer, I., Uggetti, E. and García, J. 2015. Harvesting microalgae from wastewater treatment systems with natural flocculants: Effect on biomass settling and biogas production. Algal Research, 9: 204-211.

Huy, M., Kumar, G., Kim, H.W. and Kim, S.H. 2018. Photoautotrophic cultivation of mixed microalgae consortia using various organic waste streams towards remediation and resource recovery. Bioresource Technology, 247: 576-581.

Imran, M., Crowley, D.E., Khalid, A., Hussain, S., Mumtaz, M.W. and Arshad, M. 2015. Microbial biotechnology for decolorization of textile wastewaters. Reviews in Environmental Science and Bio/Technology, 14: 73-92.

Larsdotter, K. 2006. Wastewater treatment with microalgae – A literature review. Vatten, 62(1): 31.

Lee, C.S., Lee, S.A., Ko, S.R., Oh, H.M. and Ahn, C.Y. 2015. Effects of photoperiod on nutrient removal, biomass production, and algal-bacterial population dynamics in lab-scale photobioreactors treating municipal wastewater. Water Research, 68: 680-691.

Leite, L.D.S. and Daniel, L.A. 2020. Optimization of microalgae harvesting by sedimentation induced by high pH. Water Science and Technology, 82(6): 1227-1236.

Mennaa, F.Z., Arbib, Z. and Perales, J.A. 2015. Urban wastewater treatment by seven species of microalgae and an algal bloom: Biomass production, N and P removal kinetics and harvestability. Water Research, 83: 42-51.

Michelon, W., Da Silva, M.L.B., Mezzari, M.P., Pirolli, M., Prandini, J.M. and Soares, H.M. 2016. Effects of nitrogen and phosphorus on biochemical composition of microalgae polyculture harvested from phycoremediation of piggery wastewater digestate. Applied Biochemistry and Biotechnology, 178: 1407-1419.

Molazadeh, M., Ahmadzadeh, H., Pourianfar, H.R., Lyon, S. and Rampelotto, P.H. 2019. The use of microalgae for coupling wastewater treatment with CO_2 biofixation. Frontiers in Bioengineering and Biotechnology, 7: 42.

Morin-Crini, N., Lichtfouse, E., Torri, G. and Crini, G. 2019. Applications of chitosan in food, pharmaceuticals, medicine, cosmetics, agriculture, textiles, pulp and paper, biotechnology, and environmental chemistry. Environmental Chemistry Letters, 17(4): 1667-1692.

Ramadass, K., Megharaj, M., Venkateswarlu, K. and Naidu, R. 2015. Ecological implications of motor oil pollution: Earthworm survival and soil health. Soil Biology and Biochemistry, 85: 72-81.

Renuka, N., Sood, A., Ratha, S.K., Prasanna, R. and Ahluwalia, A.S. 2013. Evaluation of microalgal consortia for treatment of primary treated sewage effluent and biomass production. Journal of Applied Phycology, 25: 1529-1537.

Safdari, M.S., Kariminia, H.R., Rahmati, M., Fazlollahi, F., Polasko, A., Mahendra, S. et al. 2018. Development of bioreactors for comparative study of natural attenuation, biostimulation, and bioaugmentation of petroleum-hydrocarbon contaminated soil. Journal of Hazardous Materials, 342: 270-278.

Stockenreiter, M., Haupt, F., Seppälä, J., Tamminen, T. and Spilling, K. 2016. Nutrient uptake and lipid yield in diverse microalgal communities grown in wastewater. Algal Research, 15: 77-82.

Vergara, C., Muñoz, R., Campos, J.L., Seeger, M. and Jeison, D. 2016. Influence of light intensity on bacterial nitrifying activity in algal-bacterial photobioreactors and its implications for microalgae-based wastewater treatment. International Biodeterioration & Biodegradation, 114: 116-121.

Wang, Y., Zhang, C.H., Lin, M.M. and Ge, Y. 2016. A symbiotic bacterium differentially influences arsenate absorption and transformation in Dunaliella salina under different phosphate regimes. Journal of Hazardous Materials, 318: 443-451.

You, K., Ge, F., Wu, X., Song, K., Yang, Z., Zhang, Q. et al. 2021. Nutrients recovery from piggery wastewater and starch wastewater via microalgae-bacteria consortia. Algal Research, 60: 102551.

Zhu, L., Wang, Z., Shu, Q., Takala, J., Hiltunen, E., Feng, P. and Yuan, Z. 2013. Nutrient removal and biodiesel production by integration of freshwater algae cultivation with piggery wastewater treatment. Water Research, 47(13): 4294-4302.

CHAPTER

8

Role of Bio-filtration Techniques in the Removal of Pollutants from Wastewater

Bhawana Yadav[1#], Akanksha Dhoundiyal[1#], Priyanshu Kumar[1], Ananta Gupta[2], Shivangi Chamoli[1*] and Payal Gupta[1*]

[1] Department of Biotechnology, Graphic Era Deemed to be University, Dehradun - 248001, Uttarakhand, India

[2] Department of Microbiology, Graphic Era Deemed to be University, Dehradun - 248001, Uttarakhand, India

Introduction

Water is indispensable for life, yet with the increasing demands of a growing global population, combined with industrialisation and urbanisation, this vital finite resource is being used extensively, leading to enormous amounts of wastewater generation. Wastewater is a term used to describe water contaminated by human activities which cannot be reused without treatment (Sonune and Ghate 2004). As a result, wastewater treatment has become a critical process in ensuring the availability of clean and safe water. The importance of wastewater treatment cannot be overstated. One of the vital challenges in wastewater treatment is the diversity of pollutants present in it. Wastewater can be categorised into domestic, industrial and agricultural (Ali et al. 2021) (Figure 1A). Domestic wastewater, also known as sewage, is primarily generated from households and residential areas, containing organic matter, nutrients, pathogens and household chemicals. Industrial wastewater, generated by various industrial processes, can contain harmful contaminants like heavy metals and toxic chemicals (Anderson et al. 2022).

On the other hand, agricultural wastewater is primarily generated from irrigation runoff and livestock operations (Yadav et al. 2021). When released directly into water bodies, this polluted water can have devastating effects on ecosystems, human health and the economy. For instance, untreated sewage can lead to the spread of waterborne diseases such as cholera and dysentery, while industrial effluents can

*Corresponding author: payalgupta.bt@geu.ac.in; shivangichamoli.ls@geu.ac.in

# Authors contributed equally

DOI: 10.1201/9781003527350-8

cause long-term damage to aquatic life through the accumulation of toxic substances (Chahal et al. 2016, Yadav et al. 2021). Furthermore, the release of nutrient-rich wastewater into water bodies can result in eutrophication, a process that leads to excessive growth of algae and other plants, ultimately depleting oxygen levels and harming aquatic life (Akinnawo 2023).

Conventional wastewater treatment methods employ a combination of physical, chemical and biological processes to remove contaminants and purify the water for safe discharge or reuse. These methods include filtration, sedimentation, activated sludge, photocatalysis, coagulation and flocculation (Ali et al. 2021). The aforesaid methods are effective in removing contaminants but have limitations that affect their efficiency, cost-effectiveness and environmental sustainability. Filters can become clogged over time, leading to amplified maintenance costs and limited effectiveness (Morris and Knowles 2011). Sedimentation, on the other hand, has limited removal efficiency and requires proper disposal of vast volumes of sludge (Demirbas et al. 2017). Activated sludge requires continuous aeration and energy consumption, posing environmental challenges. Meanwhile, photocatalysis is costly and necessitates specialised catalysts and light sources (Iervolino et al. 2020). Coagulation and flocculation involve chemical usage and residues, demanding vigilant disposal and fluctuating effectiveness based on water quality(Dotto et al. 2019).

Conventional treatment methods can have undesirable environmental impacts, such as greenhouse gas emissions and the generation of by-products. Therefore, the growing demand for clean water, coupled with the increasing scarcity of freshwater resources, has led to the emergence of innovative approaches to wastewater treatment. Addressing these challenges, biofiltration provides an effective alternative that leverages biological processes to purify wastewater, thus offering a natural and sustainable approach to removing contaminants from wastewater. The need for biofiltration in wastewater treatment arises from the growing demand for eco-friendly, cost-effective and efficient methods to handle the complex and diverse pollutants in wastewater. Biofiltration, additionally, relies on natural processes that are less resource-intensive and produce fewer by-products (Ergas and Cardenas-Gonzalez 2004). This aligns with the growing emphasis on sustainability and environmental stewardship in wastewater management practices. Biofiltration is particularly effective at removing many contaminants, including organic matter, nutrients (such as nitrogen and phosphorus) and certain inorganic compounds (Sheoran et al. 2022).

Furthermore, biofiltration can be integrated with other treatment processes to enhance efficiency. For example, biofilters can be used as a refining stage after conventional treatment to diminish nutrient levels further or treat specific contaminants resistant to other methods. Biofiltration systems typically require lower operational and maintenance costs than traditional treatment methods (Pachaiappan et al. 2022, Shah et al. 2021). Once established, the biofilm in a biofilter can function for extended periods with minimal intervention, reducing the need for constant monitoring and chemical inputs (Bar-Yosef 2008). The relatively simple design and operation of biofilters also make them accessible to communities with limited resources or technical expertise, providing an affordable wastewater treatment solution in developed and developing regions.

The present chapter discusses the process of biofiltration, focusing on its mechanisms for treating wastewater, the various types of biofiltration systems and the key factors that influence their efficiency. Additionally, it examines the conditions that impact the performance of biofilters, offering a comprehensive understanding of how to optimise this sustainable treatment method for wastewater management.

History of Biofiltration

Biofiltration technology is one of the oldest forms of biotechnologically used methods for removing undesired gaseous components from air, unwanted particles, and contaminants from water (Dwivedi and Dwivedi 2022). The principles of biofiltration have been used as part of ancient practices, even if not fully understood. Early biofilters were made by digging trenches, laying an air distribution system and refilling the trenches with permeable soil, wood chips or compost (Bajpai and Bajpai 2014). In the early 1800s, slow sand filtration was developed in Europe. These systems used layers of sand and gravel through which water passed slowly, allowing biological activity to break down contaminants. A significant advancement in the biological filter design to enhance performance reliability was achieved in the U.S. at the Lawrence Experimental Station (Gill et al. 2018). It marked the inception of a pioneering wastewater treatment facility, showcasing successful trials that exhibited a near-total elimination of organic compounds and ammonium oxidation. Sporadic soil filtration research was conducted during the final years of the 1800s. Other nations, like England in 1893, pioneered the bio-trickling filter method to treat sewage water (Pachaiappan et al. 2022). The 20th century saw significant advancements in biofiltration technology. The history of usage of biofilters dates back to the 1920s when biofilters were first used to remove the odour of gas emissions from wastewater or animal farming (Morral et al. 2021). Later on, the concept of treatment of odorous hydrogen sulphide gas and contaminated water from sewage treatment plant using a first-ever biofilter was found in literature as early as 1923 by German scientist Bach (Leson and Winer 1991, Pachaiappan et al. 2022, Sheoran et al. 2022). Then, in 1953, the first patented biofilter soil beds were installed in Long Beach, California, USA. The patent was given to Richard Pomeroy for engaging microorganisms in soil beds to combat sewage odour (Pachaiappan et al. 2022, Sheoran et al. 2022). Later, in 1959, soil beds were installed at a municipal sewage treatment plant in Nuremberg, West Germany. The first systematic research on the biofiltration of hydrogen sulphide gas was conducted in the early 1960s by Carlson and Leiser in the U.S. (Parmar et al. 2022). They successfully installed several soil filters at a wastewater treatment plant in Seattle and demonstrated biodegradation rather than sorption for odour removals. During the following two decades, several researchers in the U.S. have further studied the biofiltration concept and showed its usefulness in several full-scale applications. The knowledge and experience of this soil bed concept is owed by Hinrich Bohn, who has investigated the theory and potential applications of soil beds for more than 15 years (Leson and Winer 1991). After that, biofiltration technology rapidly increased, and the primary focus was on the industrial sector's toxic volatile organic compounds (VOCs) from Europe and the U.S. After this, a vast industrial success with good operating efficiency of biofiltration was seen in Europe and Japan (Pachaiappan et al. 2022). Since the 1960s, European countries like West Germany and the Netherlands developed biofiltration technology into a widely used air pollution control (APC) technology, which is now considered the best available control technology in a variety of VOCs and odour control applications (Showqi et al. 2016). Many approaches and designs were investigated, such as the air distribution system and several filter materials with higher biological activities and lower flow resistance than soil. Compost derived from municipal solid waste was used as a filter material in early 1966. Developing or designing a biofilter system was based entirely on experiments and practice, not on any theories. Jager and von Jager (1978) discussed the economic rewards of biofilters compared to other APC technologies in treating odorous off-gases from composting facilities for municipal wastes.

Biofiltration Versus Conventional Methods

Urbanisation, along with human activities on a large scale worldwide, has resulted in the enormous discharge of wastewater into rivers and coastal zones, dumping significant amounts of heavy metals, dyes, microplastics, persistent organic pollutants (POCs), pharmaceutical compounds and VOCs into the waterbodies (Wang et al. 2020). Several conventional (non-biological) wastewater treatment processes have targeted the removal of these elements and compounds. Various physical and chemical processes such as membrane separation, coagulation, sedimentation, adsorption and photocatalytic degradation methods have been used to mitigate the increasing levels of numerous pollutants in wastewater (Qasem et al. 2021). Conventional removal frameworks have increased operational costs, substantial energy utilisation, and production of secondary wastes, rendering them inefficient for treatment plants (Katheresan et al. 2018). For instance, the membrane separation method, which is a pressure-driven filtration technique, suffers from limitations such as fouling, clogging and tearing by organic pollutants, impeding its overall efficiency (Shen et al. 2023). Moreover, wastewater contains a variety of contaminants in a broader range of particle sizes. Therefore, the process is not sustainable, as averaging the membrane's pore size means that fouling or blockage is inevitable (Amin et al. 2014). The clogging or fouling of the separation membrane can be prevented by a pretreatment step involving the coagulation-flocculation process, eradicating larger particles in the water. Yet, this process comes with other downsides, such as the colossal sludge volume produced, which could be hazardous by environmental standards and would require costly disposal (Ejimofor et al. 2021). Also, the coagulation of pollutants is effective only within a specific pH range and coagulant dosage. The limitation associated with the process of catalytic oxidation is catalytic poisoning induced by the presence of chlorinated organic compounds and sulphides (González-Velasco et al. 2014). Activated carbon-based adsorption of emerging pollutants from wastewaters is inarguably a popular approach; however, accumulating pollutants on carbon subsequently forms a secondary waste requiring additional methods for mitigation, alongside high regeneration costs, which is challenging (El-Baz et al. 2020). In the majority of industrial emissions, the concentration of pollutants is relatively trace in wastewater compared to other constituents. Therefore, low adsorption capacity and the suboptimal efficiency-to-cost ratio of adsorbents are other limiting factors, rendering the process impractical for routine wastewater treatment (Ngo et al. 2015).

Some of the aforementioned downsides accompanying conventional wastewater treatment methods can potentially be addressed by employing biological water treatment techniques. Biological methods encompass various approaches such as biofiltration, bioreactors, biosorption, activated sludge, aerobic and anaerobic treatment, trickling filters, biofilms, phytoremediation, phytoaccumulation, phytovolatilisation and vermifiltration, which have proven highly effective in treating wastewater (De Sanctis et al. 2022, Pachaiappan et al. 2022, Samer 2015). Out of these methods, biofilters have arisen as a chiefly apt and sustainable nature-based solution (NBS) for removing most kinds of contaminants found in water and air, achieving removal efficiencies exceeding 90%. The biofiltration process is perhaps the most prominent and practical example of an underutilised green technology-based attractive alternative for wastewater pollutant mitigation. Biofiltration leverages biological organisms like bacteria, algae, plants, protozoa, yeast, viruses and mixed microbes as catalysts (Hemalatha et al. 2024, Thanigaivel et al. 2022). These organisms grow on diverse natural and synthetic media, including sand, crushed rock, ceramic materials, river gravel, plastic, beads and rings. The polluted water or air is directed through the filter media, where the microorganisms break down the contaminants into more straightforward and less harmful compounds, which they utilise as energy and nutrients for their growth and development (Sinha and Mukherji 2022). The biofiltration method differs from the conventional filtration technique, as it actively fosters sustaining biological activity within and/or on filter media to eliminate suspended particles, pathogens and dissolved organics from the water phase (Wang et al. 2021). This mitigation method capitalises on natural microbial processes, typically not necessitating additional energy inputs, and mainly produces minimal waste compared to alternative treatment methods aimed at similar purposes. Acceptance of biofiltration as a conventional treatment technology in some regions has been hindered due to concerns regarding the health risks attributable to microbially mediated treatment, difficulties in operation and unlikely regulatory approvals (Brown et al. 2015).

Mechanism of Biofiltration

The primary function of the biofilter is to remove dissolved organic compounds, while the elimination of suspended solids is achieved through a conventional filter before the wastewater is applied to the biofiltration process (Kumar et al. 2013). Conventional filters in wastewater treatment plants typically consist of various granular materials such as sand, anthracite, expanded ceramics and granular activated carbon (GAC). However, biofiltration is more suitable when the media bed is composed of GAC (Basu et al. 2016). GAC is widely utilised in water treatment due to its ability to adsorb specific contaminants from solution and its suitability to support the growth of advantageous microorganisms. GAC also provides a more robust and conducive surface for microbial attachment and development compared to anthracite or sand, attributed to factors like adsorption capacity, surface charge, surface area, porosity, surface roughness and fewer adsorption variations with changes in water temperature (Basu et al. 2016). When the adsorption capacity of the GAC filter is depleted, the media filter is replaced with a new one. However, in the case of biofilters, the transition from GAC to biologically activated carbon (BAC) filters transpires, with increased biomass accumulation occurring predominantly at the upper regions of the filter rather than at the lower strata (Ni et al. 2022, Van der Aa et al. 2006, Zhang et al. 2017). The topmost layer, also known as the schmutzdecke, is the most biologically active layer in terms of decontamination of pollutants, attributed to the fact that the layer is supplemented with active microbial growth due to better accessibility of oxygen in the top layer compared to the bottom layers (Maurya et al. 2020). The two principal mechanisms of biofiltration are sorption and biodegradation (Adler 2001, Schroeder 2002). Wastewater treatment plants where filter media is not replaced naturally transition into biologically active filters (biofilters), gradually depleting the adsorption filter's adsorption capacity. Simultaneously, continuous nutrient influx into the filter sustains microbial growth, leading to a marked escalation in the proportion of biological removal within the overall removal efficacy (Bai et al. 2022). Microorganisms exhibit strong adhesion to activated carbon through the utilisation of EPS, which concurrently acts as a medium for molecular diffusion and the extracellular transformation of dissolved organic materials into smaller molecules conducive to microbial growth (Du et al. 2020). During the colonisation phase of microbes on GAC media, significant removal of dissolved organic carbon (DOC) can be achieved, as the treatment process is initially dominated by adsorption until the transition to biofiltration occurs upon reaching the saturation point of adsorption capacity.

Biological removal consistently progresses alongside physical–chemical removal in biofiltration processes, albeit one mechanism may surpass the other (Wu et al. 2014). Concurrently, a layer rich in nutrients forms gradually on the filter surface due to suspended particles, instigating colonisation of the surface by microorganisms. Over time, biofilms emerge on the surface, marking the advent of biological removal primarily by biodegradation as the primary mechanism (Brown et al. 2015, Delhoménie and Heitz 2005). Biofilms represent complex and dense congregations of microbial cells affixed to a material substrate through a secreted polymeric matrix, referred to as an extracellular polymeric substance (EPS) (Rabbani and Bharti 2020). Biofilm-forming microorganisms (including aerobic and anaerobic bacteria, algae, fungi, protozoa, and facultative organisms) on the substrate's surface are suspended within the water matrix of the medium's constituent components (Tripathi and Hussain 2022). The transport of microorganisms within the aqueous medium to the surface of the substrate is governed by mechanisms such as active motility of the microbial cells, diffusion, sedimentation, and convection (Sun et al. 2020). Contaminants serve as carbon sources to microorganisms. Subsequently, through effective metabolic processes, these microorganisms proliferate, establishing colonies that facilitate the degradation of pollutants (Adler 2001). The bacterial community is widely recognised as the primary consumer of carbon within the overall microbial consortium (Jones et al. 2009). Collectively, the bioactive attributes of biofilters have been associated with iron-oxidising bacteria, nitrifiers, and manganese-oxidising bacteria. The infusion of substantial nutrients into the filter bed fosters the swift accumulation of the microbial community, comprising nitrifying bacteria, iron-oxidizing bacteria, methane-oxidizing bacteria and heterotrophic bacteria, crucial in the decomposition and elimination of methane, ammonium, bentazone and various other organic micropollutants through co-metabolism (Song et al. 2020, Zhao et al. 2020). The efficacy of biological elimination pivots on the biological activity of the accumulated biotic community in filters (Lukwambe et al. 2019). In biofilters, the advantages conferred by more mature biofilms in treatment applications include enhanced biomass, elevated metabolic activity and improved resistance to specific toxic contaminants. Patterns and configurations of cell attachment, the organization of microbial cells, cell density, and species diversity are pivotal characteristics that define the structure of microbial communities within biofilms. Diverse operational factors, like filter media, influent water nutrients, rate of water flow, temperature and maintenance practices (e.g. filter media replacement and backwashing), can regulate the biotic community, yielding distinct patterns of elimination mechanisms. *Acidobacteria, Proteobacteria*, *Nitrospirae, Planctomycetes* and *Bacteroidetes* constitute the fundamental community observed at all drinking water treatment plants with a mean relative abundance exceeding 1%. *Proteobacteria,* in particular, dominate the bacterial landscape in most biofilters, possibly due to the adaptability of certain *Proteobacteria* strains to environments with low concentrations of dissolved organic carbon (Guarin and Pagilla 2021, Zanacic et al. 2017).

The biofilm formation process can be delineated into several distinct stages whereby microorganisms colonise and acclimatise to the surface milieu of the filter media. At the cellular scale, the development of biofilms can be systematically classified into four distinct phases (Alotaibi and Bukhari 2021, Sauer et al. 2022, Thuptimdang et al. 2021, Zhu et al. 2022) (Figure 1B). (1) In the initial phase, planktonic cells utilise their appendages, such as flagella, to navigate and adhere to the substrate; this initial phase is termed reversible attachment due to the transient nature of adhesion at this juncture. (2) Subsequently, the microbial cells employ additional organelles, including fimbriae or pili, to anchor themselves more securely to the surface, culminating in what is defined as irreversible attachment. (3) Once the cells have adequately adapted to the surface environment, there is an enhancement in both population proliferation and EPS synthesis. This progression leads to the maturation phase of biofilms. (4) The concluding phase is characterised by the dispersion or destabilisation of the biofilms, which occurs under unfavourable conditions, thereby liberating the microbial cells from the biofilm matrix and reverting them to a planktonic state. The liberated microbial cells subsequently re-engage in surface attachment. Nevertheless, in a more extensive bioreactor apparatus that provides optimal growth conditions, these recurrent cycles yield increased biomass in the maturation phase over time, resulting in a more significant accumulation of biomass on the surface compared to a more diminutive system (Iorhemen et al. 2016, Srivastava and Majumder 2008, Thuptimdang et al. 2021). The immobilisation of microorganisms can be accomplished through two distinct methods: natural attachment or artificial immobilisation of the biofilter bed materials (Baltrėnas et al. 2020). The glycocalyx, which constitutes the outer layer of the cell membrane of microorganisms, is primarily composed of polysaccharides. This layer

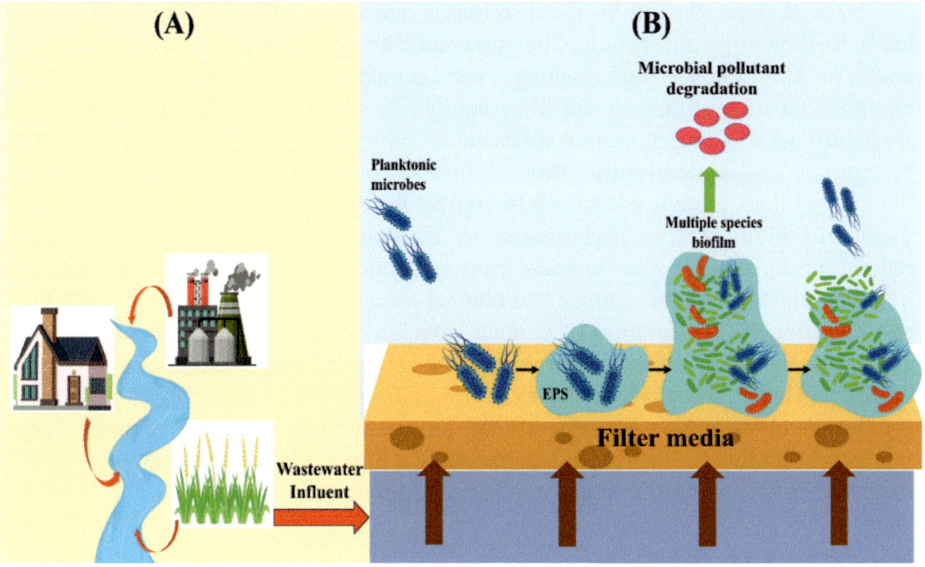

Figure 1: (A) Domestic sewage, industrial effluents and agricultural wastewater polluting waterbodies; (B) Diagrammatic illustration of biofiltration mechanism; Planktonic microbes reversibly attach to filter media, formation of micro-colonies by EPS production, multiple species biofilm maturation which is responsible for biodegradation of pollutants in wastewater, dispersion of microbes to planktonic stage.

plays a critical role in maintaining the structural integrity of microorganisms and facilitating their adhesion to surfaces (Soccol et al. 2003). Recent research indicates that rhamnolipid secreted by microorganisms demonstrates significant efficacy in promoting biofilm formation through microbial adhesion in wastewater treatment processes (Peng et al. 2020). A confluence of forces contributes to the self-attachment of microorganisms, including covalent bond formation, electrostatic interactions, and hydrophobic interactions (Malhautier et al. 2005, Soccol et al. 2003). Microbial cells are comprised of various reactive functional groups that readily interact with ligands present within the biofilter bed material via covalent bonding interactions (Malhautier et al. 2005). Microbial cells can also be encapsulated within these three-dimensional polymeric matrices composed of materials such as polyurethane, polyester, cellulose, polystyrene, resin, and agar. The dimensions of the pores within the polymer matrices are observed to be smaller than those of the microbial cells, thereby facilitating their entrapment (Cohen 2001). The process of entrapment immobilisation of cells presents several advantages, including plasmid stability, heightened metabolic activity of the entrapped cells, mild experimental conditions, economic feasibility, physical segregation and immobilisation of diverse microbial species, as well as restriction to toxic compounds, along with being biodegradable and environmentally sustainable.

Nevertheless, the build-up of biomass and metal precipitates in the filter leads to clogging, diminishing the filtration rate progressively over time (Rowe and VanGulck 2004). Backwashing exerts a notable influence on the structural composition of biofilms on the filter media. To sustain the efficacy of biofilters, frequent backwashing at certain intervals is required to eradicate the accumulated precipitates and concurrently allow microbial communities to thrive undisturbed within the filters, thereby enhancing biological removal. Effective backwashing helps maintain biological filter performance by facilitating biomass detachment, media redistribution, mitigation of impacts from oxidants in backwash water, and removal of accumulated particles. Empty bed contact time (EBCT) is expressed as the flow rate determining the duration of contact between water and carbon (Fundneider et al. 2021). EBCT is a crucial parameter for predicting organic matter removal in biofilters. However, discrepancies exist regarding the timeframe, suggesting a need for further investigation into operational parameters that influence organic matter removal in biofilters. Earlier studies indicated a direct relationship between organic matter removal and EBCT (Halle et al. 2015, Moona et al. 2021).

Biofiltration Methods

Biofiltration methods are used widely in biological wastewater treatment. They are effective, environmentally friendly and have good removal efficiency. Some of the filters are discussed below.

Trickling Filter

The trickling filter technique is an improvement of the ancient Egyptian method, where they filtered water by digging along the Nile River, using river sand as a filter (Maurya et al. 2020). Trickling filters are a biological wastewater treatment system in which wastewater is sprayed over a bed of media in which anaerobic bacteria, fungi, protozoa, and insects are used to make biofilm over pebbles or stones that act as a filter (Hamza et al. 2016) (Figure 2). Microbes can quickly grow and multiply on wet surfaces and have a metabolic capacity to remove pollutants (Wang et al. 2015). Wastewater is sprinkled over the medium, and microbes present act on organic pollutants, heavy metals, and nutrients, degrading them to carbon dioxide and water and maintaining a high growth rate of biofilm bacteria, fungi, and protozoa. Nitrifying bacteria present at the end of the film carry out nitrification; protozoa feed on the biofilm; and fungi can reduce pollutants, but only at low pH. Due to the deposition of biomass, sloughing occurs, but it is removed shortly after the incoming flow of waste and a new biofilm is created (Ali et al. 2017, Metcalf et al. 1991). Trickling filters are cost-friendly, reliable, and have the ability to retain large amounts of biomass with low operating and maintenance requirements (Ali et al. 2017). Despite its benefits, it has its own design and operational limitations, and problems like clogging of solids in media, foul odour caused by anaerobic conditions, biofilm sloughing, the development of snails (which reduces nitrifying bacteria), maintaining a proper hydraulic and organic loading rate, etc., affect its efficiency. Therefore, addressing these issues can enhance its performance and reliability (Ali et al. 2017).

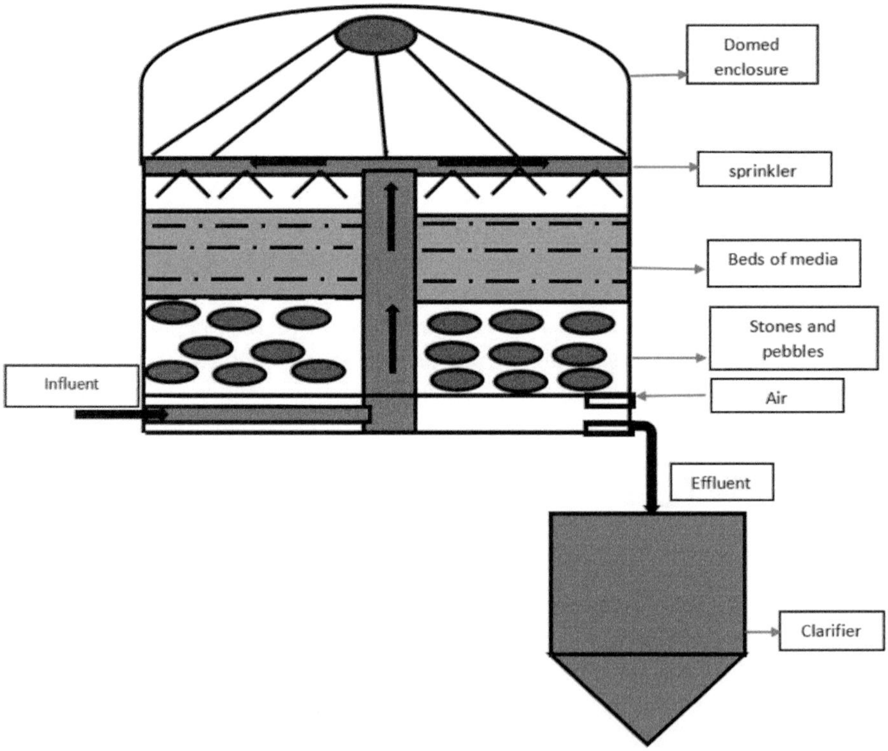

Figure 2: A systematic diagram of trickling filter units for wastewater treatment plants showing wastewater treatment through different filter media layers.

Construction Wetland

A wetland is an area wholly saturated with water throughout the year or for a specific season. Constructed wetlands are artificially constructed wetlands that use natural processes like biodegradation, phytoremediation, sorption, natural attenuation, etc., for the breakdown of pollutants in wastewater (Hassan et al. 2021). This includes physical processes like filtration and sedimentation, biological processes like biodegradation, and chemical processes like adsorption and precipitation (Barik 2018). It can be used to treat municipal wastewater, industrial wastewater, textile wastewater, etc. (Vymazal and Březinová 2016). Effluent from various sources is distributed through pipes and passed to roots and rhizomes (Padalkar and Kumar 2018). These roots and rhizomes create aerobic and anaerobic conditions that are suitable for the growth of various microbes (Schmidt et al. 2002). These microbes carry out the transformation and degradation of contaminants (Brovelli et al. 2011). Heterotrophic bacteria degrade organic contaminants like petroleum hydrocarbons, while autotrophic bacteria expand their cells by using inorganic pollutants. In this way, both organic and inorganic pollutants can be utilised (Stefanakis 2020). High-density vascular plants are used here to improve sorption and produce attachment sites for microbes (Thorslund et al. 2017), while phytoplankton creates a balance between living and non-living components (Pandey et al. 2004). In constructed wetlands, a broad spectrum of pollutants can be treated by the contribution of numerous processes, which include physical processes like filtration, precipitation, and sedimentation; biological processes carried out by plants like phytoremediation, biosorption, and predation; and chemical processes like nitrification, denitrification, and chemical precipitation (Barik 2018, Spieles 2022). Germany was the first nation to adapt this technique for wastewater treatment (Barik 2018), and now countries like Austria, China, South Africa, Tanzania, Kenya, and various European nations are using constructed wetlands (Rezania et al. 2019). Constructed wetlands are becoming popular worldwide due to their advantages, like low construction costs, lesser environmental impact, sustainability, aesthetics, and ecological values (Brovelli et al. 2011). Despite its significant advantages, there are challenges like large area requirements, climate sensitivity, an increase in methane emissions, clogging of the soil, etc., that need to be tackled to improve its efficiency (Brovelli et al. 2011, Vymazal and Březinová 2016).

Rotating Biological Contactor

A rotating biological contactor is an aerobic and anaerobic fixed-film (Irfan et al. 2022a) wastewater treatment system that operates on the attached growth principle (Mizyed 2021), where a series of cylindrical discs are partially immersed in wastewater and partially exposed to the atmosphere (Waqas and Bilad 2019) (Figure 3). The discs are subjected to steady and continuous rotation that ensures the supply of oxygen and nutrients, leading to the development of microorganisms and biofilm formation. These microbes carry out the biodegradation of organic matter and the nitrification and denitrification of nitrogen content (Waqas et al. 2021, Waskar et al. 2012). RBCs are widely used to treat municipal and industrial wastewater, where wastewater is pumped into the channels, and biofilm removes the contaminants (Mizyed 2021). The biofilm contains both anaerobic and aerobic microbes, including bacteria, fungi, and protozoa (Kesaano and Sims 2014). The outer layer of the biofilm is exposed to oxygen and supports the growth of aerobic bacteria. These bacteria break down organic material into CO_2 and H_2O (Martin and Nerenberg 2012). Inner layers of the biofilm are not exposed to oxygen, so under anoxic conditions, anaerobic microbes break down complex organic compounds (Najafpour et al. 2006). The presence of these layers efficiently treats the wastewater (Waqas et al. 2022), and after treatment, the effluent tank receives contamination-free water (Vasiliadou et al. 2014). DRS is a parameter that affects the amount of oxygen being delivered to the biofilm (Vairavel and Murty 2020), the proper thickness of the biofilm, and the shear force that acts on the biofilm (Rana et al. 2018), so it must be carefully chosen. HRT and SRT refer to the amount of time wastewater and microbes spend in a bioreactor, respectively, and influence the efficiency of the RBC (Hassard et al. 2020, Sirianuntapiboon and Chuamkaew 2007). RBCs are popular because of their minimal operational cost, energy consumption, and maintenance needs. They are easy to handle with a basic design and are resistant to hazardous loads and shocks (Mizyed 2021). RBCs can handle contaminants like xenobiotics, hydrocarbons, textile dyes, heavy metals, etc. They have low energy demands and can operate even on wind and solar energy. Despite their advantages, they come with limitations, such as the high cost of large-scale installation (Waqas et al. 2021) and, over time, the accumulation of solids and biofilm on discs, leading to clogging and maintenance issues (Waqas and Bilad 2019).

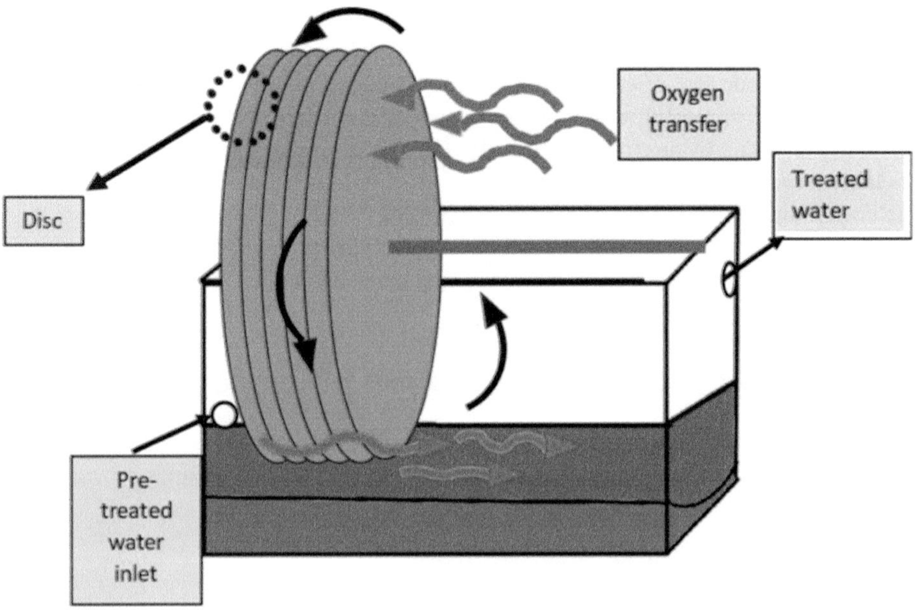

Figure 3: A schematic diagram of a rotating biological contactor for wastewater treatment.

Membrane Bioreactors

A membrane bioreactor is a wastewater treatment technology that integrates biological processes (aerobic and anaerobic) with membrane filtration systems (Al-Asheh et al. 2021). In this process, the bioreactor is used for the degradation of biomass (Marrot et al. 2004), and microfiltration or ultrafiltration is used for the separation of treated wastewater from biological treatments (Eddy et al. 2014). It is a mature technology that can remove soluble and suspended compounds from municipal and industrial wastewater and produces high-quality effluent (Brown 2005, Marrot et al. 2004). Before entering the MBR, wastewater is pretreated using ultrasound radiation to prevent the blocking of pores and the production of excessive sludge, as high fabrication cost and brittleness make polymeric membranes more suitable (Arefi-Oskoui et al. 2019). Bunches of membranes are joined together to form a membrane module. This membrane module is then immersed in a bioreactor, and it is provided with high vacuum pressure, causing water to enter the hollow container while the remaining solids are deposited on the outer part of the membrane (Asante-Sackey et al. 2022, Ben Aim and Semmens 2003). Ceramic and polymeric materials like alumina and silicon carbide are used in the construction of membranes.

Ceramics have impressive fouling resistance with mechanical and chemical stability, but their fabrication cost and brittle nature make polymeric substances a more popular choice, although their hydrophobic nature makes them prone to fouling (Rahman et al. 2021, Tetteh et al. 2019). Both physical and chemical methods are required to clean membranes in MBR. Physical cleaning is done either by back flushing or by membrane relaxation. Chemical cleaning for organic matter uses essential solutions like sodium hypochlorite and sodium hydroxide, whereas inorganic substances use acidic solutions (Poojamnong et al. 2020). The main problem with MBR is fouling. Fouling depends on operating conditions, membrane cleaning, and membrane properties. To control fouling pretreatments, substrate modifications and proper cleaning are done (Maaz et al. 2019). MBRs have several benefits over conventional treatment methods; their SRT is higher and HRT is lower than that of the traditional activated sludge process (Al-Asheh et al. 2021). They have good organic and inorganic separation capabilities, high resistance to loading, and high effluent quality. However, this technology still faces challenges in achieving high fouling control, cost-effectiveness, and low energy consumption (Al-Asheh et al. 2021).

Techniques to Study Biofiltration Efficiency

To analyze various parameters of wastewater before and after its treatment, it is essential to know the efficiency of the system. Various techniques are used for its assessment.

Analytical Techniques and Methods

Various analytical tools, techniques and methods are used to determine the physical, chemical, and biological parameters of the treatment system, and they allow us to evaluate the efficiency of the biofiltration system (Y. Dong et al. 2020, Li et al. 2020). Physical parameters like turbidity are measured using a turbidimeter, which evaluates the amount of suspended solid present in water (Amran et al. 2021). Removal efficiency of turbidity varied in different biofiltration systems. It was observed that in UASB reactors, the removal efficiency was best at 38°C with a sludge age of 180 days and HRT of 12 hours (Rizvi et al. 2015), whereas in construction wasteland system, turbidity removal efficiency varied between 50 and 99% depending upon the treatment used (Parde et al. 2021). TSS in biofilm and effluent is measured according to the standard system, and it indicates the number of suspended solids in the waste water (Sinha and Mukherji 2022). Chemical parameters like BOD, COD, TN, TP, etc., are also analyzed according to standard methods (He et al. 2007) or in some cases, they can be directly analyzed using multi-parameter water quality testers (Y. Dong et al. 2020). Whereas, for ammonium nitrogen Nessler's reagent spectrophotometry and phosphorous calorimetry is used for analysis. pH and DO levels are checked by pH meter and DO meter, respectively (Y. Dong et al. 2020). Advanced chromatography and spectrometry techniques are used to assess specific pollutants like heavy metals, pesticides, etc. However, optical biosensors have been used to provide quantitative or semi-quantitative analytical data (Razzak et al. 2022). Software like CAD and CFD are used for simulation and analysis of data.

Microbial and Biofilm Analysis

Advanced molecular biology tools and techniques are being used to analyse the growth of microorganisms within the biofilm and study the efficiency of the biofiltration system (Loh et al. 2022). For microbe growth analysis, firstly, DNA of the microbes present within the biofilm is extracted using DNA extraction kits like the Omega DNA Kit, E.Z.N.A® Soil DNA Kit, etc., and then its purity, concentration, and quality are checked by spectrophotometry (Chi et al. 2018). The amplification process is carried out using PCR techniques by adding suitable primers, and these amplicons are sequenced on sequencing platforms like MiSeq and then sent for taxonomic profile analysis (He et al. 2018). However, during this analysis, dead and inactive microorganisms, which are not involved in the biodegradation of organic matter, are also detected. Therefore, for more accurate efficiency analysis, advanced genome analysis techniques like metagenomics and advanced PCR are used (Singer et al. 2017). Metagenomics is an advanced molecular biology technique that analyses the genome directly from environmental samples without cultivation. It is now being used for the identification and quantification of DNA along with advanced amplification techniques such as RNA-based techniques (e.g. RT-PCR), qPCR, fluorescent quantitative PCR, etc. (Inkinen et al. 2019, Ni et al. 2022, Xin et al. 2021). To identify the inter-relationships between various bacterial species at the environmental and genus levels, Principal Coordinate Analysis (PCoA) is used (Dong et al. 2020).

Advanced Mathematical Simulation and Software

Mathematical simulation and software tools are used to study biofiltration efficiency, providing accurate results and predictions, as well as detailed analysis, modelling, and optimisation of operational parameters. They are used for dynamic study of wastewater systems in a shorter time and at a lower cost (Hendy et al. 2020, Moragaspitiya et al. 2017a). Many commercial software programs are being used in biofiltration systems for modelling and optimisation, helping to assess and improve removal efficiency. These simulation software programs are handy; example COSMOL Multiphysics software is widely used by scientists and engineers due to its flexibility, efficiency, and optimized modelling. It helps in maximizing efficiency by modelling physical, chemical, and biological processes and design parameters (Demirer et al. 2023, Samsó and Garcia 2013). COMSOL Multiphysics software was used by Samsó and Garcia (2013) for building the BIO_PORE model to increase construction wetland development and efficiency, and by Hamisi et al. (2024) for observing the dynamic flow of water and nutrient adsorption in sand filters. HYDRUS is another software used to simulate and analyse heat, solute, and water through a biofiltration system (Šimůnek et al. 2006), like HYDRUS 1D, which was used by Rahman et al. (2021) to study the simulation of water and solute through the system and to develop a variable saturated model of biochar amended biofiltration, while Hasan et al. (2022) used HYDRUS 2D to optimize onsite wastewater treatment. BioWin is also a simulation software used to design, upgrade, and optimize all types of wastewater treatment facilities and can be used to optimize treatment processes and predict effluent quality (Moragaspitiya et al. 2017). BioWin software was used to study two different integrated aerobic and anaerobic systems, and maximize their efficiency (Hendy et al. 2020).

Factors Affecting Biofiltration Efficiency

The efficiency of the biofiltration system is central because it directly influences the ability to decrease harmful pollutants. Better removal efficiencies can be achieved by improving biofiltration processes, making biofiltration a sustainable and cost-effective solution for long-term environmental management. The efficiency of biofiltration is intricately influenced by several key factors that impact its capacity to remove contaminants from wastewater (Figure 4).

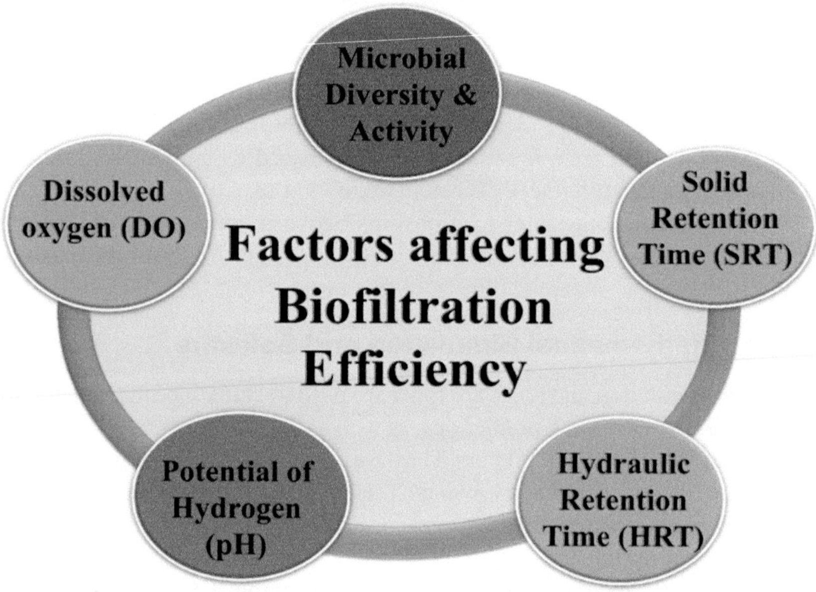

Figure 4: Critical factors affecting efficiency of biofiltration.

Microbial Diversity and Activity

The performance and the efficiency of the biofilters largely depend on microbe diversity and its activity. It is essential that the filter media has a diverse number of microbes, as its diversity improves the effluent quality, biological treatment capacity, stability of the system and adaptation towards environmental stress like pH changes, temperature shifts, nutrient scarcity, etc. It also ensures that the performance is consistent even when some microbe species are removed or inhibited, thus increasing its overall efficiency (Irfan et al. 2022a). Microbes are essential for biochemical degradation, and a large community of microbes, including bacteria, fungi, protozoa and algae, ensures that the complex mixture of various pollutants can be treated (Samal et al. 2017). Microbial activity directly influences the rate of biomass degradation; the higher the microbial activity, the faster the degradation rate (Chen et al. 2017).

Dissolved Oxygen (DO)

The amount of dissolved oxygen plays a vital role in affecting the efficiency and effectiveness of the biofiltration system (Loh et al. 2021). An adequate amount of oxygen must be present in the filter media. DO levels are responsible for microorganisms' healthy growth and activity (Zhang et al. 2019). At high DO levels, aerobic bacteria's growth and activity are enhanced, leading to better metabolization of organic matter and an increase in the degradation of pollutants, which increases overall efficiency (Hasan and Muhammad 2020). Low DO levels reduce the biological activity of aerobic bacteria and can impede the activity of bacteria such as AOB (Ammonia-oxidising bacteria) as well as the proliferation of NOB (Nitrite-oxidising bacteria). DO levels also influence biological processes such as nitrification, and it has been discovered that the activity of nitrifying bacteria improves at 2–3 mg/L (Hasan and Muhammad 2020). Nitrification drops at low DO levels, often less than 0.9 mg/L, resulting in high nitrogen content in water (Kalkan et al. 2011). As a result, maintaining adequate DO is critical for improving the efficiency and performance of the biofiltration system.

Temperature

Maintaining optimal temperature is very important for the biofiltration system's performance and efficiency. Temperature directly affects the performance of biofilters (Cheng et al. 2020), microbe structure and biological activity (Halle et al. 2015), nitrification and denitrification processes (Zhang et al. 2017), and the diffusion of pollutants within biofilters. Microbial growth and degradation are temperature sensitive (Selbes et al. 2016). If a biofiltration system suffers from long-term low temperatures, the microbial activity, diversity, metabolism and degradation of the biofiltration system are affected, leading to a decline in performance and efficiency (Shah et al. 2021). Very high temperatures can lower the dissolved oxygen levels and oxidation-reduction potential of the system, affecting microorganism and biofilm efficiency (Christofilopoulos et al. 2019, Nguyen et al. 2019). It was observed that when the temperature was above 18°C, the nitrification and denitrification performance was stable, and on decreasing temperature from 18°C to 13°C, nitrogen removal efficiency declined from 98.0% to 78.1% (He et al. 2007). So, maintaining optimal temperature in a biofiltration system is a very crucial factor for good performance and efficiency.

Potential of Hydrogen (pH)

pH directly influences the efficiency and performance of the biofiltration system as it affects microbial activity, nutrient intake, biofilm stability, and pollutant removal efficiency (Barzgar et al. 2017, Rybarczyk 2022). Although microbes present in the biofiltration system can function over a wide range of pH values (Van der Aa et al. 2006), heterogeneous bacteria are most efficient at neutral pH, whereas fungi thrive at slightly acidic pH. For aerobic systems, the optimal pH ranges from neutral to slightly alkaline, whereas for anaerobic systems, the optimal pH ranges between 6.6 and 7.5 (Chen et al. 2017). In biofiltration systems like constructed wetlands, soil pH impacts the solubility and availability of nutrients and trace elements to microorganisms, affecting contaminant degradation efficiency (Locatelli et al. 2013). In the case of membrane bioreactors, pH impacts membrane fouling; it was observed that membrane fouling increased when pH was reduced, decreasing the efficiency of the system (Burman and Sinha 2018, Iorhemen et al. 2016). pH is also important for the adsorption of pollutants and the attachment of microorganisms to the biofilm; a drastic pH shift can affect the holding capacity for pollutants and disrupt the biofilm (Tee et al. 2022). Therefore, maintaining optimal pH is crucial for the biofiltration system's effective performance and high efficiency.

Operational Parameters

Operational parameters are a significant factor influencing the efficiency and effectiveness of the biofiltration system. Hydraulic Retention Time (HRT) and Solid Retention Time (SRT) are operational parameters that affect the performance of the biofiltration system.

HRT is an operational parameter that defines the average time that wastewater spends in a bioreactor (Mohameda et al. 2022). HRT is the ratio of reactor volume and influent flow rate, which is directly linked to the food/microbe ratio, also called F/M ratio (Mohameda et al. 2022). Higher HRT improves the pollutant removal efficiency as it increases the time of contact between the microbe and the wastewater; 24 to 36 hours of HRT can improve the biofilm thickness ((Ebrahimi et al. 2018). But if we increase the HRT adversely, it can accumulate net materials and unwanted microbe growth, which reduces the bioreactor efficiency (Waqas et al. 2022). However, lower HRT decreases efficiency as it decreases the contact time between microbe and wastewater but it improves the performance by reducing the growth of less efficient bacteria and the accumulation of inert material (Daudpoto et al. 2021). So, the HRT should be carefully chosen to avoid these shortcomings. HRT also impacts COD, turbidity, and total nitrogen removal. With an increase in HRT, the COD, turbidity removal efficiency, and total nitrogen removal efficiency also increase (Irfan et al. 2022).

SRT is also an operational parameter that defines the average time a microorganism spends in a bioreactor (Hassard et al. 2020). It affects the growth and proliferation of microbes present in the system, ultimately affecting biomass production and biomass activity (Gopi Kiran et al. 2021, Limpiyakorn et al. 2005). If SRT is short, there is less time for interaction between microorganisms and biomass, which promotes the growth of faster-growing microorganisms like heterotrophic bacteria responsible for the degradation of organic matter and prevents the accumulation of less efficient bacteria. If SRT is long, slow-growing microorganism growth is promoted, like nitrifying and denitrifying bacteria, promoting more stable and consistent microbial community (Greenham et al. 2019, Najafpour et al. 2006).

Conclusion

The biofiltration technique is an innovative solution for the treatment of wastewater. However, many key challenges must be addressed for the long-term sustainable implementation, adoption, and success of biofiltration-based wastewater treatment plans. The primary challenge associated with wastewater treatment is the variability in the composition and characteristics of pollutants present in wastewater. Therefore, a biofilm unit should be developed that effectively removes a wide range of pollutants and responds efficiently to changes in the environmental conditions of sewage. Another major challenge with the scalability of biofiltration techniques is its high cost and energy demands. Therefore, biofiltration units should be designed to handle large volumes of wastewater with the lowest energy inputs. Cost-effectiveness and energy efficiency are two essential features that govern the adoption of a technique at a large industrial scale. Moreover, proper maintenance and monitoring of the biofiltration unit are required to avoid microbial contamination and maximize the growth of microbial communities to achieve maximum performance output.

The scalability and success of biofiltration techniques can be increased by developing new innovative support media for biofilm production. Incorporation of advanced treatment technologies can improve the efficiency of contaminant removal. Moreover, the shift to a circular economy emphasizes obtaining valuable products from wastewater and treating it.

References

A Mohameda, M., A Fouad, H. and M ElHefny, R. 2022. Reviewing rotating biological contactor's different aspects for wastewater treatment with experiment. Engineering Research Journal – Faculty of Engineering (Shoubra), 51: 180-187.

Adler, S.F. 2001. Biofiltration – A Primer. Chem Eng Prog, 97: 33-41.

Akinnawo, S.O. 2023. Eutrophication: Causes, consequences, physical, chemical and biological techniques for mitigation strategies. Environmental Challenges, 12: 100733.

Al-Asheh, S., Bagheri, M. and Aidan, A. 2021. Membrane bioreactor for wastewater treatment: A review. Case Studies in Chemical and Environmental Engineering, 4: 100109.

Ali, I., Khan, Z., Peng, C., Naz, I., Sultan, M., Ali, M. et al. 2017. Identification and elucidation of the designing and operational issues of trickling filter systems for wastewater treatment. Pol J Environ Stud, 26.

Ali, I., Naz, I., Peng, C., Abd-Elsalam, K.A., Khan, Z.M., Islam, T. et al. 2021. Sources, classifications, constituents, and available treatment technologies for various types of wastewater: An overview. Aquananotechnology, 11-46.

Alotaibi, G.F. and Bukhari, M.A. 2021. Factors influencing bacterial biofilm formation and development. Am. J. Biomed. Sci. Res, 12: 617-626.

Amin, M.T., Alazba, A.A. and Manzoor, U. 2014. A review of removal of pollutants from water/wastewater using different types of nanomaterials. Adv. Mater. Sci. Eng, 2014, 825910.

Amran, A.H., Zaidi, N.S., Syafiuddin, A., Zhan, L.Z., Bahrodin, M.B., Mehmood, M.A. et al. 2021. Potential of Carica papaya seed-derived bio-coagulant to remove turbidity from polluted water assessed through experimental and modeling-based study. App. Sci, 11: 5715.

Anderson, A., Anbarasu, A., Pasupuleti, R.R., Manigandan, S., Praveenkumar, T.R. and Kumar, J.A. 2022. Treatment of heavy metals containing wastewater using biodegradable adsorbents: A review of mechanism and future trends. Chemosphere, 295: 133724.

Arefi-Oskoui, S., Khataee, A., Safarpour, M., Orooji, Y. and Vatanpour, V. 2019. A review on the applications of ultrasonic technology in membrane bioreactors. Ultrason Sonochem, 58: 104633.

Asante-Sackey, D., Rathilal, S., Tetteh, E.K. and Armah, E.K. 2022. Membrane bioreactors for produced water treatment: A mini-review. Membranes (Basel), 12: 275.

Bai, X., Dinkla, I.J.T. and Muyzer, G. 2022. Microbial ecology of biofiltration used for producing safe drinking water. Appl Microbiol Biotechnol, 106: 4813-4829.

Bajpai, P. and Bajpai, P. 2014. Biological methods for the elimination of odourous compounds. Biological Odour Treatment, 17-44.

Baltrėnas, P., Baltrėnaitė, E., Baltrėnas, P. and Baltrėnaitė, E. 2020. Natural and inoculated microorganisms as important component for sustainability of biofiltration system. Sustainable Environmental Protection Technologies: Contaminant Biofiltration, Adsorption and Stabilization, 433-482.

Barik, D. 2018. Energy from Toxic Organic Waste for Heat and Power Generation. Woodhead Publishing.

Bar-Yosef, B. 2008. Fertigation management and crops response to solution recycling in semi-closed greenhouses. Soilless Culture: Theory and Practice, 343-424.

Barzgar, S., Hettiaratchi, J.P., Pearse, L. and Kumar, S. 2017. Inhibitory effects of acidic pH and confounding effects of moisture content on methane biofiltration. Bioresour Technol, 245: 633-640.

Basu, O.D., Dhawan, S. and Black, K. 2016. Applications of biofiltration in drinking water treatment – A review. Journal of Chemical Technology & Biotechnology, 91: 585-595.

Ben Aim, R.M. and Semmens, M.J. 2003. Membrane bioreactors for wastewater treatment and reuse: A success story. Water Science and Technology, 47: 1-5.

Brovelli, A., Carranza-Diaz, O., Rossi, L. and Barry, D.A. 2011. Design methodology accounting for the effects of porous medium heterogeneity on hydraulic residence time and biodegradation in horizontal subsurface flow constructed wetlands. Ecol Eng, 37: 758-770.

Brown, J., Summers, R.S., LeChevallier, M., Collins, H., Roberson, J.A., Hubbs, S. et al. 2015. Biological drinking water treatment? Naturally. Journal – American Water Works Association, 107, 20-30.

Brown, J.A. 2005. The early history of wastewater treatment and disinfection. *In*: Impacts of Global Climate Change, 1-7.

Burman, I. and Sinha, A. 2018. A review on membrane fouling in membrane bioreactors: Control and mitigation. *In*: Environmental Contaminants: Measurement, Modelling and Control, 281-315.

Chahal, C., Van Den Akker, B., Young, F., Franco, C., Blackbeard, J. and Monis, P. 2016. Pathogen and particle associations in wastewater: Significance and implications for treatment and disinfection processes. Adv Appl Microbiol, 97: 63-119.

Chen, Y., Lan, S., Wang, L., Dong, S., Zhou, H., Tan, Z. et al. 2017. A review: Driving factors and regulation strategies of microbial community structure and dynamics in wastewater treatment systems. Chemosphere, 174: 173-182.

Cheng, Q., Liu, Z., Huang, Y., Li, F., Nengzi, L. and Zhang, J. 2020. Influence of temperature on CODMn and Mn^{2+} removal and microbial community structure in pilot-scale biofilter. Bioresour Technol, 316: 123968.

Chi, X., Li, A., Li, M., Ma, L., Tang, Y., Hu, B. et al. 2018. Influent characteristics affect biodiesel production from waste sludge in biological wastewater treatment systems. Int Biodeterior Biodegradation, 132: 226-235.

Christofilopoulos, S., Kaliakatsos, A., Triantafyllou, K., Gounaki, I., Venieri, D. and Kalogerakis, N. 2019. Evaluation of a constructed wetland for wastewater treatment: Addressing emerging organic contaminants and antibiotic resistant bacteria. N Biotechnol, 52: 94-103.

Cohen, Y. 2001. Biofiltration – The treatment of fluids by microorganisms immobilized into the filter bedding material: A review. Bioresour Technol, 77: 257-274.

Daudpoto, M.R., Talpur, M.G.H., Shah, F. and Khooharo, A. 2021. A residual analysis for the removal of biological oxygen demand through rotating biological contactor. Mehran University Research Journal of Engineering & Technology, 40: 459-464.

De Sanctis, M., Murgolo, S., Altieri, V.G., De Gennaro, L., Amodio, M., Mascolo, G. et al. 2022. An innovative biofilter technology for reducing environmental spreading of emerging pollutants and odour emissions during municipal sewage treatment. Science of The Total Environment, 803: 149966.

Delhoménie, M.-C. and Heitz, M. 2005. Biofiltration of air: A review. Crit Rev Biotechnol, 25: 53-72.

Demirbas, A., Edris, G. and Alalayah, W.M. 2017. Sludge production from municipal wastewater treatment in sewage treatment plant. Energy Sources, Part A: Recovery, Utilization, and Environmental Effects, 39: 999-1006.

Demirer, E., Coene, E., Iraola, A., Nardi, A., Abarca, E., Idiart, A. et al. 2023. Improving the performance of reactive transport simulations using artificial neural networks. Trans Porous Media, 149: 271-297.

Dong, J., Wang, Y., Wang, L., Wang, S., Li, S. and Ding, Y. 2020. The performance of porous ceramsites in a biological aerated filter for organic wastewater treatment and simulation analysis. Journal of Water Process Engineering, 34: 101134.

Dong, Y., Lin, H. and Zhang, X. 2020. Simultaneous ammonia nitrogen and phosphorus removal from micro-polluted water by biological aerated filters with different media. Water Air Soil Pollut, 231: 1-15.

Dotto, J., Fagundes-Klen, M.R., Veit, M.T., Palácio, S.M. and Bergamasco, R. 2019. Performance of different coagulants in the coagulation/flocculation process of textile wastewater. J Clean Prod, 208: 656-665.

Du, Z., Jia, R., Li, C., Cui, P., Song, W. and Liu, J. 2020. Pilot-scale UV/H2O2-BAC process for drinking water treatment – Analysis and comparison of different activated carbon columns. Chemical Engineering Journal, 382: 123044.

Dwivedi, N. and Dwivedi, S. 2022. Biofiltration: A modern technology for wastewater treatment. *In*: An Innovative Role of Biofiltration in Wastewater Treatment Plants (WWTPs), 389-410. Elsevier.

Ebrahimi, A., NAjafpour, G.D., Anazadeh, M. and Ghavami, M. 2018. Optimization of whey treatment in rotating biological contactor: Application of Taguchi method. Iranica Journal of Energy & Environment, 9: 146-152.

Eddy, M., Abu-Orf, M., Bowden, G., Burton, F.L., Pfrang, W., Stensel, H.D. et al. 2014. Wastewater Engineering: Treatment and Resource Recovery. McGraw Hill Education.

Ejimofor, M.I., Ezemagu, I.G. and Menkiti, M.C. 2021. Physiochemical, instrumental and thermal characterization of the post coagulation sludge from paint industrial wastewater treatment. S Afr J Chem Eng, 37: 150-160.

El-Baz, A.A.A., Hendy, I.A., Dohdoh, A.M. and Srour, M.I. 2020. Adsorption technique for pollutants removal; current new trends and future challenges – A review. The Egyptian International Journal of Engineering Sciences and Technology, 32: 1-24.

Ergas, S.J. and Cardenas-Gonzalez, B. 2004. Biofiltration: Past, present and future directions. Biocycle, 48: 35-38.

Fundneider, T., Alonso, V.A., Abbt-Braun, G., Wick, A., Albrecht, D. and Lackner, S. 2021. Empty bed contact time: The key for micropollutant removal in activated carbon filters. Water Res, 191: 116765.

Gill, S., Al-Shankiti, A. and Azam, F.E. 2018. A comprehensive review on sewage collection and treatment: Historical perspective. Acta Scientific Agriculture, 2: 166-176.

González-Velasco, J.R., Aranzabal, A., Pereda-Ayo, B., González-Marcos, M.P. and González-Marcos, J.A. 2014. Catalytic oxidation of volatile organic compounds: Chlorinated hydrocarbons. *In*: Handbook of Advanced Methods and Processes in Oxidation Catalysis: From Laboratory to Industry. World Scientific, pp. 91-131.

Gopi Kiran, M., Das, R., Behera, S.K., Pakshirajan, K. and Das, G. 2021. Modelling a rotating biological contactor treating heavy metal contaminated wastewater using artificial neural network. Water Supply, 21: 1895-1912.

Greenham, R.T., Miller, K.Y. and Tong, A. 2019. Removal efficiencies of top-used pharmaceuticals at sewage treatment plants with various technologies. J Environ Chem Eng, 7: 103294.

Guarin, T.C. and Pagilla, K.R. 2021. Microbial community in biofilters for water reuse applications: A critical review. Science of The Total Environment, 773: 145655.

Halle, C., Huck, P.M. and Peldszus, S. 2015. Emerging contaminant removal by biofiltration: Temperature, concentration, and EBCT impacts. Journal – American Water Works Association, 107: E364-E379.

Hamisi, R., Renman, A., Renman, G., Wörman, A. and Thunvik, R. 2024. Treatment efficiency and recovery in sand filters for onsite wastewater treatment: Column studies and reactive modelling. J Clean Prod, 142696.

Hamza, R.A., Iorhemen, O.T. and Tay, J.H. 2016. Advances in biological systems for the treatment of high-strength wastewater. Journal of Water Process Engineering, 10: 128-142.

Hasan, H.A. and Muhammad, M.H. 2020. A review of biological drinking water treatment technologies for contaminants removal from polluted water resources. Journal of Water Process Engineering, 33: 101035.

Hasan, M.S., Trapp, J. and Geza, M. 2022. Improvement of onsite wastewater systems performance: Experimental and numerical investigation. Clean Technologies, 4: 824-840.

Hassan, I., Chowdhury, S.R., Prihartato, P.K. and Razzak, S.A. 2021. Wastewater treatment using constructed wetland: Current trends and future potential. Processes, 9: 1917.

Hassard, F., Biddle, J., Cartmell, E., Coulon, F. and Stephenson, T. 2020. Biosolids recycling impact on biofilm extracellular enzyme activity and performance of hybrid rotating biological reactors. Science of The Total Environment, 706: 135865.

He, Q., Chen, L., Zhang, S., Wang, L., Liang, J., Xia, W. et al. 2018. Simultaneous nitrification, denitrification and phosphorus removal in aerobic granular sequencing batch reactors with high aeration intensity: Impact of aeration time. Bioresour Technol, 263: 214-222.

He, S., Gall, D.L. and McMahon, K.D. 2007. "Candidatus Accumulibacter" population structure in enhanced biological phosphorus removal sludges as revealed by polyphosphate kinase genes. Appl Environ Microbiol, 73: 5865-5874.

Hemalatha, M., Kashyap, P., Sana, M.F., Mohanasrinivasan, V. and Devi, C.S. 2024. Biofiltration approach for emerging contaminants in organic wastes. *In:* Environmental Nexus Approach. CRC Press, pp. 200-223.

Hendy, I., Dohdoh, A.M. and Abdo, A. 2020. Utilization of BioWinTM simulator in simulating the integrated anaerobic fluidized bed-UASB/aerobic moving bed biofilm system. Desalination Water Treat, 183: 276-289.

Iervolino, G., Zammit, I., Vaiano, V. and Rizzo, L. 2020. Limitations and prospects for wastewater treatment by UV and visible-light-active heterogeneous photocatalysis: A critical review. Heterogeneous Photocatalysis: Recent Advances, 225-264.

Inkinen, J., Jayaprakash, B., Siponen, S., Hokajärvi, A.-M., Pursiainen, A., Ikonen, J. et al. 2019. Active eukaryotes in drinking water distribution systems of ground and surface waterworks. Microbiome, 7: 1-17.

Iorhemen, O.T., Hamza, R.A. and Tay, J.H. 2016. Membrane bioreactor (MBR) technology for wastewater treatment and reclamation: Membrane fouling. Membranes (Basel), 6: 33.

Irfan, M., Waqas, S., Khan, J.A., Rahman, S., Kruszelnicka, I., Ginter-Kramarczyk, D. et al. 2022a. Effect of operating parameters and energy expenditure on the biological performance of rotating biological contactor for wastewater treatment. Energies (Basel), 15: 3523.

Jager, B. and von Jager, J. 1978. Geruchsbekampfung in Kompostwerken am Beispiel Heidelberg. Mull und Abfall, 10: 48-54.

Jones, S.E., Newton, R.J. and McMahon, K.D. 2009. Evidence for structuring of bacterial community composition by organic carbon source in temperate lakes. Environ Microbiol, 11: 2463-2472.

Kalkan, Ç., Yapsakli, K., Mertoglu, B., Tufan, D. and Saatci, A. 2011. Evaluation of biological activated carbon (BAC) process in wastewater treatment secondary effluent for reclamation purposes. Desalination, 265: 266-273.

Katheresan, V., Kansedo, J. and Lau, S.Y. 2018. Efficiency of various recent wastewater dye removal methods: A review. J Environ Chem Eng, 6: 4676-4697.

Kesaano, M. and Sims, R.C. 2014. Algal biofilm based technology for wastewater treatment. Algal Res, 5: 231-240.

Kumar, K.V., Sridevi, V., Harsha, N., Lakshmi, M.C. and Rani, K. 2013. Biofiltration and its application in treatment of air and water pollutants – A review. International Journal of Application or Innovation in Engineering and Management, 2: 2319–4847.

Leson, G. and Winer, A.M. 1991. Biofiltration: An innovative air pollution control technology for VOC emissions. J Air Waste Manage Assoc, 41: 1045-1054.

Li, Y., Guo, J., Li, H., Song, Y., Chen, Z., Lu, C. et al. 2020. Effect of dissolved oxygen on simultaneous removal of ammonia, nitrate and phosphorus via biological aerated filter with sulfur and pyrite as composite fillers. Bioresour Technol, 296: 122340.

Limpiyakorn, T., Shinohara, Y., Kurisu, F. and Yagi, O. 2005. Communities of ammonia-oxidizing bacteria in activated sludge of various sewage treatment plants in Tokyo. FEMS Microbiol Ecol, 54: 205-217.

Locatelli, A., Spor, A., Jolivet, C., Piveteau, P. and Hartmann, A. 2013. Biotic and abiotic soil properties influence survival of Listeria monocytogenes in soil. PLoS One, 8: e75969.

Loh, Z.Z., Zaidi, N.S., Syafiuddin, A., Yong, E.L., Boopathy, R., Hong Kueh, A.B. et al. 2021. Shifting from conventional to organic filter media in wastewater biofiltration treatment: A review. Applied Sciences, 11: 8650.

Loh, Z.Z., Zaidi, N.S., Yong, E.L., Syafiuddin, A., Boopathy, R. and Kadier, A. 2022. Current status and future research trends of biofiltration in wastewater treatment: A bibliometric review. Curr Pollut Rep, 8: 234-248.

Lukwambe, B., Zhao, L., Nicholaus, R., Yang, W., Zhu, J. and Zheng, Z. 2019. Bacterioplankton community in response to biological filters (clam, biofilm, and macrophytes) in an integrated aquaculture wastewater bioremediation system. Environmental Pollution, 254: 113035.

Maaz, M., Yasin, M., Aslam, M., Kumar, G., Atabani, A.E., Idrees, M. et al. 2019. Anaerobic membrane bioreactors for wastewater treatment: Novel configurations, fouling control and energy considerations. Bioresour Technol, 283: 358-372.

Malhautier, L., Khammar, N., Bayle, S. and Fanlo, J.-L. 2005. Biofiltration of volatile organic compounds. Appl Microbiol Biotechnol, 68: 16-22.

Marrot, B., Barrios-Martinez, A., Moulin, P. and Roche, N. 2004. Industrial wastewater treatment in a membrane bioreactor: A review. Environmental Progress, 23: 59-68.

Martin, K.J. and Nerenberg, R. 2012. The membrane biofilm reactor (MBfR) for water and wastewater treatment: Principles, applications, and recent developments. Bioresour Technol, 122: 83-94.

Maurya, A., Singh, M.K. and Kumar, S. 2020. Biofiltration technique for removal of waterborne pathogens. *In*: Waterborne Pathogens. Elsevier, pp. 123-141.

Metcalf, L., Eddy, H.P. and Tchobanoglous, G. 1991. Wastewater Engineering: Treatment, Disposal, and Reuse. McGraw-Hill, New York.

Mizyed, A. 2021. Review on application of rotating biological contactor in removal of various pollutants from effluent. Technium BioChemMed, 2: 41-61.

Moona, N., Holmes, A., Wünsch, U.J., Pettersson, T.J.R. and Murphy, K.R. 2021. Full-scale manipulation of the empty bed contact time to optimize dissolved organic matter removal by drinking water biofilters. ACS ES&T Water, 1: 1117-1126.

Moragaspitiya, C., Rajapakse, J., Senadeera, W. and Ali, I. 2017a. Simulation of dynamic behaviour of a biological wastewater treatment plant in South East Queensland, Australia using bio-win software. Engineering Journal, 21: 1-22.

Morral, E., Gabriel, D., Dorado, A.D. and Gamisans, X. 2021. A review of biotechnologies for the abatement of ammonia emissions. Chemosphere, 273: 128606.

Morris, R.H. and Knowles, P. 2011. Measurement techniques for wastewater filtration systems. *In*: Waste Water-Treatment and Reutilization, 109-132.

Najafpour, G.D., Zinatizadeh, A.A.L. and Lee, L.K. 2006. Performance of a three-stage aerobic RBC reactor in food canning wastewater treatment. Biochem Eng J, 30: 297-302.

Ngo, H.H., Guo, W., Zhang, J., Liang, S., Ton-That, C. and Zhang, X. 2015. Typical low cost biosorbents for adsorptive removal of specific organic pollutants from water. Bioresour Technol, 182: 353-363.

Nguyen, P.M., Afzal, M., Ullah, I., Shahid, N., Baqar, M. and Arslan, M. 2019. Removal of pharmaceuticals and personal care products using constructed wetlands: Effective plant-bacteria synergism may enhance degradation efficiency. Environmental Science and Pollution Research, 26: 21109-21126.

Ni, J., Yang, H., Chen, L., Xu, J., Zheng, L., Xie, G. et al. 2022. Metagenomic analysis of microbial community structure and function in a improved biofilter with odorous gases. Sci Rep, 12: 1731.

Pachaiappan, R., Cornejo-Ponce, L., Rajendran, R., Manavalan, K., Femilaa Rajan, V. and Awad, F. 2022. A review on biofiltration techniques: Recent advancements in the removal of volatile organic compounds and heavy metals in the treatment of polluted water. Bioengineered, 13: 8432-8477.

Padalkar, A.V. and Kumar, R. 2018. Common effluent treatment plant (CETP): Reliability analysis and performance evaluation. Water Science and Engineering, 11: 205-213.

Pandey, B.N., Hussain, S., Ambasta, O.P. and Poddar, S.K. 2004. Phytoplankton and its correlation with certain physico-chemical parameters of Ramjan river of Kishanganj, Bihar.

Parde, D., Patwa, A., Shukla, A., Vijay, R., Killedar, D.J. and Kumar, R. 2021. A review of constructed wetland on type, treatment and technology of wastewater. Environ Technol Innov, 21: 101261.

Parmar, S., Shrivastav, A. and Bhattacharya, S. 2022. Insights on broad spectrum applications and pertinence of biofiltration in various fields. *In*: An Innovative Role of Biofiltration in Wastewater Treatment Plants (WWTPs). Elsevier, pp. 189-206.

Peng, C., Huang, H., Gao, Y., Fan, X., Peng, P., Zhang, X. et al. 2020. A novel start-up strategy for mixotrophic denitrification biofilters by rhamnolipid and its performance on denitrification of low C/N wastewater. Chemosphere, 239: 124726.

Poojamnong, K., Tungsudjawong, K., Khongnakorn, W. and Jutaporn, P. 2020. Characterization of reversible and irreversible foulants in membrane bioreactor (MBR) for eucalyptus pulp and paper mill wastewater treatment using fluorescence regional integration. J Environ Chem Eng, 8: 104231.

Qasem, N.A.A., Mohammed, R.H. and Lawal, D.U. 2021. Removal of heavy metal ions from wastewater: A comprehensive and critical review. NPJ Clean Water, 4: 1-15.

Rabbani, A. and Bharti, P. 2020. Microbial biofilms application in environmental monitoring, bioremediation and waste water treatment. Journal of Medicinal and Aromatic Plant Sciences, 42: 30-50.

Rahman, M.Y.A., Cooper, R., Truong, N., Ergas, S.J. and Nachabe, M.H. 2021. Water quality and hydraulic performance of biochar amended biofilters for management of agricultural runoff. Chemosphere, 283: 130978.

Rana, S., Gupta, N. and Rana, R.S. 2018. Removal of organic pollutant with the use of rotating biological contactor. Mater Today Proc, 5: 4218-4224.

Razzak, S.A., Faruque, M.O., Alsheikh, Z., Alsheikhmohamad, L., Alkuroud, D., Alfayez, A. et al. 2022. A comprehensive review on conventional and biological-driven heavy metals removal from industrial wastewater. Environmental Advances, 7: 100168.

Rezania, S., Park, J., Rupani, P.F., Darajeh, N., Xu, X. and Shahrokhishahraki, R. 2019. Phytoremediation potential and control of Phragmites australis as a green phytomass: An overview. Environmental Science and Pollution Research, 26: 7428-7441.

Rizvi, H., Ahmad, N., Abbas, F., Bukhari, I.H., Yasar, A., Ali, S. et al. 2015. Start-up of UASB reactors treating municipal wastewater and effect of temperature/sludge age and hydraulic retention time (HRT) on its performance. Arabian Journal of Chemistry, 8: 780-786.

Rowe, R.K. and VanGulck, J.F. 2004. Filtering and drainage of contaminated water. *In*: Proc. 4th Int. Conf. on GeoFilters, Stellenbosch. Citeseer, pp. 1-63.

Rybarczyk, P. 2022. Removal of volatile organic compounds (VOCs) from air: Focus on biotrickling filtration and process modeling. Processes, 10: 2531.

Samal, K., Dash, R.R. and Bhunia, P. 2017. Treatment of wastewater by vermifiltration integrated with macrophyte filter: A review. J Environ Chem Eng, 5: 2274-2289.

Samer, M. 2015. Biological and chemical wastewater treatment processes. Wastewater Treatment Engineering, 150: 61250.

Samsó, R. and Garcia, J. 2013. BIO_PORE, a mathematical model to simulate biofilm growth and water quality improvement in porous media: Application and calibration for constructed wetlands. Ecol Eng, 54: 116-127.

Sauer, K., Stoodley, P., Goeres, D.M., Hall-Stoodley, L., Burmølle, M., Stewart, P.S. et al. 2022. The biofilm life cycle: Expanding the conceptual model of biofilm formation. Nat Rev Microbiol, 20: 608-620.

Schmidt, I., Sliekers, O., Schmid, M., Cirpus, I., Strous, M., Bock, E. et al. 2002. Aerobic and anaerobic ammonia oxidizing bacteria – Competitors or natural partners? FEMS Microbiol Ecol, 39: 175-181.

Schroeder, E.D. 2002. Trends in application of gas-phase bioreactors. Rev Environ Sci Biotechnol, 1: 65-74.

Selbes, M., Amburgey, J., Peeler, C., Alansari, A. and Karanfil, T. 2016. Evaluation of seasonal performance of conventional and phosphate-amended biofilters. Journal – American Water Works Association, 108: E523-E532.

Shah, M.P., Rodriguez-Couto, S. and Biswas, J.K. 2021. An Innovative Role of Biofiltration in Wastewater Treatment Plants (WWTPs). Elsevier.

Shen, M., Zhao, Y., Liu, S., Hu, T., Zheng, K., Wang, Y. et al. 2023. Recent advances on micro/nanoplastic pollution and membrane fouling during water treatment: A review. Science of The Total Environment, 881: 163467.

Sheoran, K., Siwal, S.S., Kapoor, D., Singh, N., Saini, A.K., Alsanie, W.F. et al. 2022. Air pollutants removal using biofiltration technique: A challenge at the frontiers of sustainable environment. ACS Engineering Au, 2: 378-396.

Showqi, I., Lone, F.A., Ashraf, M., Mehmood, M.A. and Rashid, A. 2016. Biofilters in mitigation of odour pollution – A review. Nature Environment & Pollution Technology, 15.

Šimůnek, J., Van Genuchten, M.T. and Šejna, M. 2006. The HYDRUS software package for simulating two- and three-dimensional movement of water, heat, and multiple solutes in variably-saturated media. Technical Manual, version 1: 241.

Singer, E., Wagner, M. and Woyke, T. 2017. Capturing the genetic makeup of the active microbiome in situ. ISME J, 11: 1949-1963.

Sinha, P. and Mukherji, S. 2022. Biofiltration process for treatment of water and wastewater. Transactions of the Indian National Academy of Engineering, 7: 1069-1091.

Sirianuntapiboon, S. and Chuamkaew, C. 2007. Packed cage rotating biological contactor system for treatment of cyanide wastewater. Bioresour Technol, 98: 266-272.

Soccol, C.R., Woiciechowski, A.L., Vandenberghe, L.P., Soares, M., Neto, G.K. and Thomas-Soccol, V. 2003. Biofiltration: An emerging technology. Indian Journal of Biotechnology, 2(3): 396-410.

Song, W., Lee, L.Y., You, H., Shi, X. and Ng, H.Y. 2020. Microbial community succession and its correlation with reactor performance in a sponge membrane bioreactor coupled with fiber-bundle anoxic bio-filter for treating saline mariculture wastewater. Bioresour Technol, 295: 122284.

Sonune, A. and Ghate, R. 2004. Developments in wastewater treatment methods. Desalination, 167: 55-63.

Spieles, D.J. 2022. Wetland construction, restoration, and integration: A comparative review. Land (Basel), 11: 554.

Srivastava, N.K. and Majumder, C.B. 2008. Novel biofiltration methods for the treatment of heavy metals from industrial wastewater. J Hazard Mater, 151: 1-8.

Stefanakis, A.I. 2020. Constructed wetlands for sustainable wastewater treatment in hot and arid climates: Opportunities, challenges and case studies in the Middle East. Water (Basel), 12: 1665.

Sun, J., Ran, R., Muftu, S., Gu, A.Z. and Wan, K.-T. 2020. The mechanistic aspects of microbial transport in porous media. Colloids Surf A Physicochem Eng Asp, 603: 125169.

Tee, G.T., Gok, X.Y. and Yong, W.F. 2022. Adsorption of pollutants in wastewater via biosorbents, nanoparticles and magnetic biosorbents: A review. Environ Res, 212: 113248.

Tetteh, E.K., Rathilal, S., Chetty, M., Armah, E.K. and Asante-Sackey, D. 2019. Treatment of water and wastewater for reuse and energy generation-emerging technologies. Water and Wastewater Treatment, 53-80.

Thanigaivel, S., Priya, A.K., Gnanasekaran, L., Hoang, T.K.A., Rajendran, S. and Soto-Moscoso, M. 2022. Sustainable applicability and environmental impact of wastewater treatment by emerging nanobiotechnological approach: Future strategy for efficient removal of contaminants and water purification. Sustainable Energy Technologies and Assessments, 53: 102484.

Thorslund, J., Jarsjo, J., Jaramillo, F., Jawitz, J.W., Manzoni, S., Basu, N.B. et al. 2017. Wetlands as large-scale nature-based solutions: Status and challenges for research, engineering and management. Ecol Eng, 108: 489-497.

Thuptimdang, P., Siripattanakul-Ratpukdi, S., Ratpukdi, T., Youngwilai, A. and Khan, E. 2021. Biofiltration for treatment of recent emerging contaminants in water: Current and future perspectives. Water Environment Research, 93: 972-992.

Tripathi, S. and Hussain, T. 2022. Biofiltration treatment of wastewater through microbial ecology. *In*: An Innovative Role of Biofiltration in Wastewater Treatment Plants (WWTPs). Elsevier, pp. 19-44.

Vairavel, P. and Murty, V.R. 2020. Decolorization of Congo red dye in a continuously operated rotating biological contactor reactor. Desalin. Water Treat, 196: 299-314.

Van der Aa, L.T.J., Magic-Knezev, A., Rietveld, L.C. and Van Dijk, J.C. 2006. 35 Biomass development in biological activated carbon filters. Recent Progress in Slow Sand and Alternative Biofiltration Processes, 293.

Vasiliadou, I.A., Molina, R., Martínez, F. and Melero, J.A. 2014. Experimental and modeling study on removal of pharmaceutically active compounds in rotating biological contactors. J Hazard Mater, 274: 473-482.

Vymazal, J. and Březinová, T. 2016. Accumulation of heavy metals in aboveground biomass of Phragmites australis in horizontal flow constructed wetlands for wastewater treatment: A review. Chemical Engineering Journal, 290: 232-242.

Wang, H., Ji, G., Bai, X. and He, C. 2015. Assessing nitrogen transformation processes in a trickling filter under hydraulic loading rate constraints using nitrogen functional gene abundances. Bioresour Technol, 177: 217-223.

Wang, H., Wang, G. and Gu, W. 2020. Macroalgal blooms caused by marine nutrient changes resulting from human activities. Journal of Applied Ecology, 57: 766-776.

Wang, M., Zhu, J. and Mao, X. 2021. Removal of pathogens in onsite wastewater treatment systems: A review of design considerations and influencing factors. Water (Basel), 13: 1190.

Waqas, S. and Bilad, M.R. 2019. A review on rotating biological contactors. Indonesian Journal of Science and Technology, 4: 241-256.

Waqas, S., Bilad, M.R., Aqsha, A., Harun, N.Y., Ayoub, M., Wirzal, M.D.H. et al. 2021. Effect of membrane properties in a membrane rotating biological contactor for wastewater treatment. J Environ Chem Eng, 9: 104869.

Waqas, S., Harun, N.Y., Bilad, M.R., Samsuri, T., Nordin, N.A.H.M., Shamsuddin, N. et al. 2022. Response surface methodology for optimization of rotating biological contactor combined with external membrane filtration for wastewater treatment. Membranes (Basel), 12: 271.

Waskar, V.G., Kulkarni, G.S. and Kore, V.S. 2012. Review on process, application and performance of rotating biological contactor (RBC). International Journal of Science and Research Publications, 2: 1-6.

Wu, T., Fu, G.Y., Sabula, M. and Brown, T. 2014. Bacterial community in the biofilm of granular activated carbon (GAC) PreBiofilter in bench-scale pilot plants for surface water pretreatment. World J Microbiol Biotechnol, 30: 3251-3262.

Xin, X., Liu, S., Qin, J., Ye, Z., Liu, W., Fang, S. et al. 2021. Performances of simultaneous enhanced removal of nitrogen and phosphorus via biological aerated filter with biochar as fillers under low dissolved oxygen for digested swine wastewater treatment. Bioprocess Biosyst Eng, 44: 1741-1753.

Yadav, P., Gupta, R.K., Singh, R.P., Yadav, P.K., Patel, A.K. and Pandey, K.D. 2021. Role of cyanobacteria in green remediation. *In*: Sustainable Environmental Clean-Up. Elsevier, pp. 187-210.

Zanacic, E., McMartin, D.W. and Stavrinides, J. 2017. From source to filter: Changes in bacterial community composition during potable water treatment. Can J Microbiol, 63: 546-558.

Zhang, Q., Chen, X., Luo, W., Wu, H., Liu, X., Chen, W. et al. 2019. Effects of temperature on the characteristics of nitrogen removal and microbial community in post solid-phase denitrification biofilter process. Int J Environ Res Public Health, 16: 4466.

Zhang, S., Gitungo, S.W., Axe, L., Raczko, R.F. and Dyksen, J.E. 2017. Biologically active filters – An advanced water treatment process for contaminants of emerging concern. Water Res, 114: 31-41.

Zhao, Y., Wang, H., Dong, W., Chang, Y., Yan, G., Chu, Z. et al. 2020. Nitrogen removal and microbial community for the treatment of rural domestic sewage with low C/N ratio by A/O biofilter with Arundo donax as carbon source and filter media. Journal of Water Process Engineering, 37: 101509.

Zhu, T., Yang, C., Bao, X., Chen, F. and Guo, X. 2022. Strategies for controlling biofilm formation in food industry. Grain & Oil Science and Technology, 5: 179-186.

CHAPTER
9

Nature-based Solutions for the Mitigation of Emerging Pollutants in Landfill Leachate

Bharti Sadhiyan[1], Sangeeta Madan[1]*, Monika Kumari[1], Ritika Maurya[1] and Pradeep Kumar Sharma[2]

[1] Department of Environmental Science, Kanya Gurukula Campus, Gurukul Kangri Deemed to be University, Haridwar - 249404, Uttarakhand, India

[2] Department of Environmental Science, Graphic Era Deemed to be University, Clement Town, Dehradun - 248002, Uttarakhand, India

Introduction

Waste disposal is one of the important problems in the world. With the increasing amount of solid waste, landfills have become the most popular disposal method. When water percolates through the solid waste, leachate is produced (Sawaya et al. 2021). Leachate generation is a significant environmental issue. Numerous factors affect the process of generating leachate from landfills in varying amounts and quality. Factors such as the landfill's depth, initial moisture content, waste composition and density, annual precipitation, runoff, infiltration, evaporation, transpiration, and ambient temperature are among them (Hussein et al. 2024).

Landfill leachate contains a large number of recalcitrant and toxic organic matters, xenobiotics, heavy metals, pathogens, and high ammonium nitrogen, and has the characteristics of high pollution load, high biodegradability and slow biodegradation rate (Abdel-Shafy et al. 2024). Direct discharge of untreated leachate will cause eutrophication of water bodies, which threatens the health of humans and the ecological environment. Landfill leachate cannot be discharged without treatment due to its complex and harsh treatment conditions. However, the traditional *ex-situ* treatment method has high capital investment and operational costs and often ends with receiving water treatment (Laskar et al. 2022). An *in-situ* bioreactor allows direct injection of exogenous microorganisms through connected low pressure and has high potential in treating leachate under a closed environment with low cost,

*Corresponding author: sangeeta.madan@gkv.ac.in

DOI: 10.1201/9781003527350-9

great operational stability, and easy maintenance. The *in-situ* treatment effect is also closely related to the leachate quality of the landfills.

Background

Wastes may be categorised into a variety of classifications, and each class has its own landfills and methods of disposal. Prior to the 1950s, waste materials were disposed of in open dumpsites, and the effects of waste disposal on the environment and public health were not taken into consideration. It was commonly believed that the leachate from waste sites was remedied by groundwater and soil and did not affect the ecology around dumpsites in the short or long term. However, landfills have a negative impact on soil, air, water, and marine life despite laws and regulations designed to reduce the effect of waste disposal.

In open dumpsites, leachate penetration became one of the most challenging parts of waste disposal since there was no mechanism in place to collect leachate. In the early 1930s, Europe expressed concern about the need for improved waste disposal methods, but by 1959, little progress had been made in this area. The first widely recognised definition of a sanitary landfill was developed in 1959 by the American Society of Civil Engineers (ASCE). Wastes are often classified into two categories: hazardous and non-hazardous. According to this rule, non-hazardous waste disposal locations are referred to as sanitary landfills, and throughout the design and operation phases, engineering activities must adhere to strict criteria. The presence of rodents like rats, smells, air pollution, scavenger birds like crows, and insects at these dumpsites has a detrimental effect on the environment and public health. This chapter reviews the leachate characteristics, treatment techniques, and the environmental impacts (Li et al. 2023).

Landfill Leachate

In a landfill, water percolates into the waste, transmitting certain kinds of contaminants and creating leachate (Mojiri et al. 2017). The presence of several contaminants, including organic and inorganic chemicals, heavy metals, ammonium, and others, poses a contamination concern in landfill leachate. During rainfall, falling water infiltrates the waste, and waste biodegradation produces a percolate called leachate (Anqi et al. 2020). Antoñanzas et al. (2020) classified three factors that determine the quantity and quality of the leachate that is generated. The first factor is meteorological and includes the amount of infiltration water received and the action of leachate on the buried waste. Hydrological, hydraulic, thermal, and biological processes occur during this decomposition (Ma et al. 2022). The second factor is the temporal evolution of the landfill. During the operational phase, there are the greatest amounts of co-infiltration and disposal water. The quantity is less, but the quality is generally greater after the closure of the landfill. Finally, the last factor is the waste management strategy, which includes the lining, capping, recirculation and regeneration of the leachate, and the treatment of the leachate generated (Reshadi et al. 2021, Ergene et al. 2022, Ma et al. 2022).

Characteristics of Landfill Leachate

In addition to high concentrations of refractory compounds like humic substances and contaminants of emerging concern, heavy metals, organic contaminants, toxic materials, ammonia, and inorganic materials, landfill leachate is characterised by a high chemical and biological oxygen demand (Chávez et al. 2019). Depending on the age, hydrology, climate, and decomposition process of the dump, the characteristics of the leachate may change. Improper handling of landfill leachate is frequently linked to health problems and environmental damage (Mojiri et al. 2016). The landfill leachate's biggest environmental risk is contamination of surface and groundwater. The characteristics of landfill leachate are obviously a controversial topic because of the involved factors and the disposal practices. The design of a management system for this high-strength leachate is dependent on many other factors: solid wastes generally include household wastes, commercial wastes and demolition wastes from a variety of sources, where the determinants of the composition and quality of leachate are the biological, chemical, and physical nature (Luo et al. 2020).

Composition of Landfill Leachate

In a mixed municipal waste dump, the kind of waste, landfill size, and waste life cycle affect leachate composition and pollutant concentration (Abdel-Shafy et al. 2024). Leachates that come into contact with decomposed solid waste cause the precipitated percolating water to become extremely polluted and start leaking solid waste products. The term "leachate" is used. The type and makeup of the municipal solid waste that was deposited, the landfill's age and rate of degradation, the hydrogeology of the landfill area, the climate, and the waste degradation stage—also referred to as the aerobic phase, anaerobic acid phase, methanogenic phase, and stabilisation phase—all have an impact on the chemical composition of leachate (Mandal et al. 2017, Naveen et al. 2017). To better understand how the physical properties of different landfills, which have an average age of 20 to 23 years, may be utilised to assess the characteristics of the leachate, these characteristics are detailed (Gautam and Kumar 2021, Abdel-Shafy et al. 2024).

Landfill Leachate and Emerging Pollutants

Researchers discovered that 365 reports of 180 distinct emerging contaminants were made in landfill leachate; the most common were antibiotics, such as sulfadiazine (0.2–29.2 μg/L) and erythromycin (0.3–20.2 μg/L). However, a large portion of these studies has only examined the effects of pollutants released into water sources by landfill leachate; very few have examined the effects of these pollutants on the soil, crops, or biota surrounding the landfill locations, and as a result, there is a dearth of knowledge. Recently, there have been reports of other emerging pollutants found in landfill leachate, such as microplastics (MPs), which have the potential to adsorb trace organics to their surface and cause additional environmental impacts (Rodríguez-Narvaez et al. 2021, Narevski et al. 2021).

Landfill leachate contains a range of organic pollutants, such as chlorinated compounds including polychlorinated biphenyls (PCBs), organochlorinated pesticides (OCPs), and polychlorinated dibenzo-p-dioxins and polychlorinated dibenzofurans (PCDD/Fs), in addition to total phenols (TP), halogenated phenols (HP), linear, branched, cyclic alkanes, alkenes, aromatic hydrocarbons, and linear, cyclic, and carbocyclic organic compounds. Due to their chemical stability, durability, and longevity, these are considered to be compelling environmental contaminants, and they, along with pharmaceuticals and biopharmaceuticals, are classified as emerging pollutants.

Transportation of Pollutants from Leachate to Environment

Landfill leachate has a very complex composition due to the high heterogeneity of the waste that can be deposited in landfills. This fact, allied to the physicochemical changes that occur inside the landfills, generates complex leachates with highly variable compositions (Mojiri et al. 2021). Notwithstanding this variability, it is possible to characterise different types of landfill leachate produced in the various stages of the landfilling process. The biodegradability of these compounds in natural systems is low, and due to the solubilisation processes, they do not settle and become highly soluble (Abdel-Shafy et al. 2024). In the case of inorganic pollutants, when dissolved in water, they tend to be transported by leachates. These features contribute to the mobilisation and transport of these compounds from the landfill solid residue to the surrounding environment, polluting it. This fact is related to the mode of waste deposition and its dynamics. In the case of landfills located in developing countries, a higher accumulation of organic matter allows organic compounds to run in the leachate (Wijekoon et al. 2022, Iravanian and Ravari 2020).

Sources of Emerging Pollutants

Heavy metals are a class of compounds usually found in landfills since they result from a wide variety of processes such as synthetic, industrial, and mining activities. The continuous increase in the production and use of heavy metals capable of producing alterations in human beings and the environment raises awareness about the environmental risks they may carry (Thongyuan et al. 2021, Ahmad et al. 2023). On the other hand, organic pollutants, which are produced by the incomplete combustion of fossil fuels or are present in industrial products like pharmaceuticals, as well as petroleum products, plastic, rubber production, soil additives—pesticides, herbicides—or as incidental products in industrial processes, have spread around the world, mainly due to atmospheric deposition (Shumbula et al. 2021). It is difficult to establish the presence of an emerging pollutant in leachate because these compounds exist in concentrations that can vary greatly over time and also due to the presence of existing pollutants in leachate (e.g. heavy metals, organic compounds, etc.) that can mask their presence. Emerging pollutants generally reach the landfill through contamination events (Propp et al. 2021, Gong et al. 2023). That is to say, they are not pollutants routinely involved in solid waste disposal. These compounds reach the landfill due to utility and household practices, such as flushing solvents, medicines, etc., down the toilet. This is the reason why it is possible to find a variety of compounds reaching the landfill, ranging from heavy metals to drugs, hormones, and other endocrine-disrupting chemicals, antimicrobials, personal care products, pesticides, biocides, and surfactants.

Environmental Impacts of Landfill Leachate

Landfills are potential sources of environmental pollution that have the ability to contaminate soil, water, air, and other natural areas. Because the elements of the ecosystem are intricately linked to one another, pollution from any of the aforementioned sources has the potential to aggravate the impacts of pollution and contaminate other sources (Iravanian and Ravari 2020). Landfill leachate is an economic and environmental threat that could cause a substantial impact on all living organisms. It is often observed that landfill leachate can generate specific compounds characterised by low molecular weight, mainly volatile fatty acids (VFAs). As a result, dissolving this acid can affect the biochemical structure and activity of aquatic organisms, reduce the reproductive capacity of fish, and degrade the ecological structure of the recipient community, worsening the quality of living water and threatening both the ecological environment and human health (Boonnorat et al. 2021, Montvydienė et al. 2020). The environmental effects of MSW in landfills around Europe were recently measured by Sauve and Van Acker (2020), taking into consideration several variables such as waste type, meteorological conditions, and landfill management techniques. The study's findings point to comparatively large effects on freshwater resources as well as other ecosystems, and it attributes these effects to the landfilling of sludge produced by leachate treatment. Numerous previous research have shown that the resultant water quality metrics often surpass regional and global standards set by the World Health Organisation (WHO) for irrigation and drinking. In particular, these investigations have highlighted a common problem: while metals in leachate can contaminate soil, surface waters, and groundwater near landfills, groundwater may be more likely to become unfit for human use due to insufficient replenishment and flushing from new inflows (Alemayehu et al. 2019, Rezapour et al. 2018).

Groundwater and Surface Water Contamination

A large volume of mixed municipal, commercial, and industrial waste is placed in one location in the majority of landfills. Each of these causes significant leachate accumulations and groundwater contamination. Contamination of surface and groundwater is the primary environmental risk linked to landfill leachate. The leachate from landfills that contaminates groundwater has the most negative environmental effects. As seen in Fig. 1, rainfall is a major source of this pollution as it spreads the contaminated water laterally and to the ground. The substances present in landfill leachate could also cause risks to the environment, such as oxygen depletion, ammonia, phosphate and nitrate contamination, toxicity for aquatic organisms, contamination of soil, groundwater, and surface water, greenhouse gas emissions, and odour production in the vicinity of the landfill. If harmful pollutants are exposed to populated areas, both people and the environment are at risk (Iravanian and Ravari 2020, Parvin and Tareq 2021).

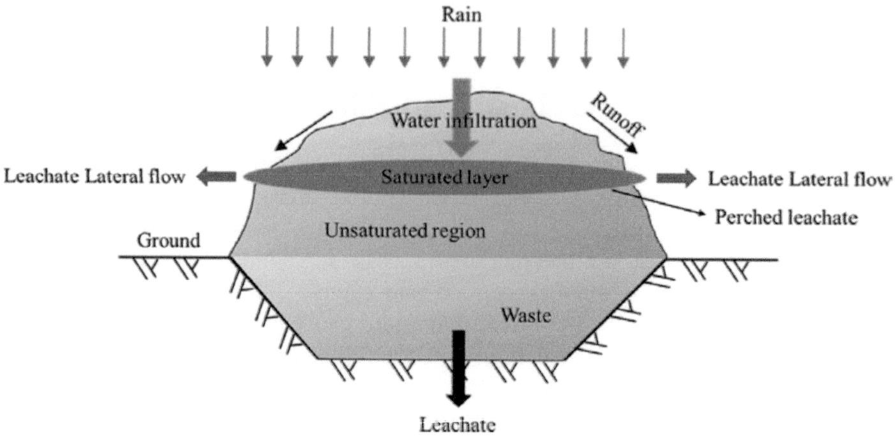

Figure 1: Effect of rainfall on landfill contamination (H.I. Abdel-Shafy, A.M. Ibrahim, A.M. Al-Sulaiman et al., Ain Shams Engineering Journal 15 (2024) 102293).

Surface water pollution is a key concern due to its increased impacts, particularly in locations close to landfill sites, as precipitation accelerates the discharge process of contaminants from the landfill site to the nearby canals, streams, and rivers through surface and overland flows (Ma et al. 2022). Leachate poses a threat to surface water bodies due to the presence of dissolved and suspended pollutants that were leached from solid waste. Surface water contamination can also occur through interaction with poorly constructed and maintained landfill facilities, which can contribute contaminants to the surface sources (Trihadiningrum et al. 2023). The discharge of leachate water into surface water bodies without any pre-treatment may not only cause fish, plankton, and algae kills but also generate excessive foam accumulation, turbidity, and colour effects that cause adverse aesthetic alteration in the receiving water, while also impacting local aquatic ecosystems—particularly in river downstreams and dam reservoirs—which worsens with impaction in the wet season. Additionally, leachate discharging directly without any treatment may also cause non-fatal damage to aquatic life, in addition to slim inaccuracy that pollutes river basin ecosystems (Britz 2020).

Nature-based Solutions for Pollution Mitigation

Numerous approaches, including chemical (Fenton process, chemical precipitation), biological (activated sludge and fluidised bed reactor processes), and physical-chemical (adsorption and membrane processes), can be used to treat landfill leachate. Nevertheless, a trade-off between the effectiveness of the treatment and the expense of these methods is usually necessary (Chunying Teng et al. 2021). Nature-based treatment systems have been highly successful for the removal of emerging pollutants such as pharmaceuticals, personal care products (PPCPs), artificial sweeteners, or illicit drugs from wastewater, showing degradation rates above 90%.

Biological Treatment Methods

The metabolic activities of microorganisms cause contaminants to be broken down biologically (Gotvajn and Pavko 2015). Biological methods are often used since they are inexpensive and effective in removing organic chemicals and nutrients (such as ammonia); nevertheless, they may not be able to effectively remove heavy metals and non-biodegradable organics (Miao et al. 2018). The two primary categories of biological approaches are as follows (Dabaghian et al. 2019):

(i) Aerobic biological procedures
(ii) Anaerobic biological procedures

Bioreactors

Bioreactors have been utilised for wastewater treatment for a number of years because they are very cost-effective, dependable, and user-friendly (Gotvajn and Pavko 2015). Leachate toxicity for microbial populations and temperature problems are the primary disadvantages of bioreactor treatments, according to Lippi et al. (2018).

Suspended-growth Biomass Processes via "Lagooning"

Aerated lagoons, like the one above, are considered a more cost-effective way to treat wastewater because they eliminate germs as well as organic and inorganic contaminants. Based on anaerobic-aerobic lagoons and artificial wetlands, biological treatment of diluted landfill leachate removed more than 70% of nitrogen, phosphorus, and iron (Luo et al. 2020). Phenolic and organic waste were treated using the lagooning approach, which resulted in a 55–64% reduction in COD and an 80–88% reduction in phenol. Nonetheless, due to its strong influence on microbial activity, temperature dependency restricts extensive lagooning (Bouaouda et al. 2023).

Activated Sludge Processes

This treatment is mostly used for co-treatment of residential wastewater, leachate, and sewage. However, it has been shown to be inadequate for treating landfill leachate. While useful in removing nutrients, organic carbon, and ammonia levels, there are a few other considerations: microbial activity is hampered by longer aeration times, challenging sludge settleability, high energy needs, and excessive sludge production at high ammonium-nitrogen strengths (Nath and Debenath 2022).

Attached-growth Biomass Systems

Biofilm was used to introduce attached-growth biomass aerobic systems, which ensure adequate biomass activity, in response to the shortcomings of conventional aerobic systems, which included sludge bulking and poor separability. However, nitrification is lower than in suspended-growth systems due to inhibition caused by the high nitrogen content and low temperatures (Song et al. 2023).

Moving Beds Biofilm Reactors (MBBR)

A biofilm of the active biomass may develop on the surfaces of the aeration tank thanks to the use of suspended carriers of porous polymers that move constantly within. The method's main benefits are reduced susceptibility to harmful substances, increased biomass levels, and sufficient sludge-settling times, in addition to high ammonia and organic rejections in a single cycle. Additionally, at high ammonia concentrations, no inhibition of nitrification was observed (Badmus et al. 2021). A stable equilibrium between the adsorption and biodegradation processes may be achieved by using granular activated carbon (GAC), which has been demonstrated to work well as a porous surface for the adsorption of organic matter and to offer favourable conditions for enhanced biodegradation. With an efficient biological AC fluidised bed, more than 70% of refractory organics can be removed. Additionally, it was demonstrated that 60–81% of COD and 85–90% of ammonia could be reduced.

Up-flow Anaerobic Sludge Blanket (UASB) Reactors

This method is preferred for volumetric organic loading ratios between 20 and 35°C because it ensures both a shorter hydraulic retention period and more effective treatment. In the 20–23°C temperature range, the average drop in COD was more than 70%, and at 35°C, it was more than 80% (Khurshid et al. 2021). The drop in COD by up to 92% was caused by low to moderate organic loading rates. The study's conclusion limited heat consumption before high-rate treatment since it showed no significant impact on COD rejection efficiency at temperatures between 15 and 35°C. The usual COD removal values in an anaerobic lab-scale suspended-growth digester were 80–90% at ambient temperature and around 55% at 35°C. That being said, the technique is not resistant to harmful materials (Safitri et al. 2022).

Attached-growth Biomass Processes

The following methods are used in this treatment process:

(i) Anaerobic filter
(ii) Hybrid bed filter
(iii) Fluidized bed reactor

Anaerobic Filter

Anaerobic filters are regarded as a high-rate system that lessens the disadvantages of other anaerobic systems while combining their advantages. As biofilms on a supporting medium, like plastic rings, biomass is stored in reserve via up-flow anaerobic filters (Nath and Debnath 2022). Anaerobic filters with loading rates between 1.26 and 1.45 have been shown to remove 90% COD per day for landfills of varying ages. The overall amount of biogas produced was 400–500 L kg^{-1}; methane production ranged from 75–85% (Reddy et al. 2022).

Hybrid Bed Filter

An anaerobic top filter and an up-flow sludge blanket (UASB) at the bottom make up the system. Increasing the concentration of biomass may improve the process performance. Depending on the degree of organic input, one study indicated a steady-state COD elimination of 81–97% during methanogenic digestion. The additional cost of the support medium is a disadvantage of the hybrid reactor (Simeonov et al. 2025).

Fluidized Bed Reactor

Fluidized beds with carbon assistance were examined. The removal of different organic compounds may be achieved effectively by combining the processes of adsorption and biodegradation. The leachate of previous landfills was shown to be significantly more effectively treated by the biological AC fluidized bed than by more conventional techniques such as fixed film and activated sludge processes.

Phytoremediation

Phytoremediation is probably one of the best-known nature-based solutions (NBS) as a green technology that uses the power of plants to remove, destroy, or isolate contaminants in environmental matrices. Phytoremediation techniques make use of plant–soil systems' capacity to break down and render potentially hazardous substances in leachate inactive (Song et al. 2018). The literature analysis reveals that many methods have been used to break down and detoxify landfill leachate by mixing phytoremediators (mostly macrophytes or halophytes) with mycorrhizal fungi (MF). While *Mentha australis*, *Phragmites australis*, and *Phalaris arundinacea* remove organic pollutants somewhat or significantly, *Typha latifolia* serves as an alternative instrument for degrading various organic contaminants, even to the point of total elimination for some. Phytoremediation has many advantages: (1) cheap energy and installation costs; and (2) removal of contaminants from landfill leachate (Madera-Parra 2016).

To remediate landfill leachate, Daud et al. (2018) employed *Lemna minor*. During a 15-day contact period, more than 70% of metals, 39% of COD, and 47% of BOD were eliminated. Leachate can be treated with a variety of aquatic plants, including those listed by Daud et al. (2018) and Song et al. (2018) as *Colocasia esculenta, Pistia stratiotes, Eichhornia crassipes, Phragmites australis, Azolla filiculoides, Typha domingensis, Hydrilla verticillata, Azolla caroliniana, Salvinia Cucullata, Heliconia psittacorum, Azolla pinnata, L. minor, Lemna gibba, Lemna aequinoctialis, Gynerum sagittatum,* and *Spirodela polyrhiza.* Hyperaccumulator plants are those that have an exceptional capacity to store metal. A key component of phytoremediation's effectiveness is hyperaccumulation (Alaboudi et al. 2018). According to Almaboudi et al. (2018), hyperaccumulation is essential to the effectiveness of phytoremediation. Translocation Factor (TF) and Bioconcentration Factor (BCF) are two indicators of hyperaccumulator plants. The plant's capacity to transfer metals from its root to its shoot is indicated by TF (Equation (2)) (Ndimele et al. 2014). Metal buildup in plant tissues is shown by BCF (Equation (3)). Mellam et al. (2009) state that plants with BCF values greater than two or TF values greater than one are considered hyperaccumulators. The concentration of metals in plant roots and shoots after metal removal using artificial wetlands or phytoremediation (Siyar et al. 2022).

TF = Concentration of metals in aerial parts/Concentration of metal in roots

BCF = Concentration of metal in plant tissues/ Concentration of metal in substrate (water)

Bioremediation

According to Moris et al. (2018), bioremediation is the process of biologically eliminating pollutants from the environment. Techniques that are both economical and ecologically beneficial are among its advantages. The bioremediation of landfill leachate using microalgae, algae, and different fungi and bacteria has been documented in the literature (Moris et al. 2018, Spina et al. 2018). According to Paskuliakova et al. (2018), although some complicated chemicals could go through some degree of biotransformation, algae are capable of eliminating inorganic and simple organic molecules. According to Paskuliakova et al. (2018), landfill leachate has been remedied using cyanobacteria and other phylogenetic microalgae, as well as members of the Scenedesmus, Chlamydomonas, and Chlorella genera. Furthermore, the primary bacteria used to remediate landfill leachate include Brevibacillus panacihumi strain ZB1, Firmicutes, Actinobacteria, Proteobacteria, and Pseudomonas putida (Moris et al. 2018, Michalska et al. 2020).

Physical and Chemical Treatment Methods

Chemical Precipitation

The chemical precipitation technique is a widely utilised pre-treatment for leachate due to its low equipment cost, simplicity, and capability. This process effectively eliminates heavy metals, non-biodegradable substances, and NH_3-N from landfill leachate (Dash and Singh 2023).

Adsorption

The adsorption technique is well recognised as a viable and efficacious fundamental approach in wastewater treatment processes. It can be applied separately or in combination with a biological process as part of a comprehensive chemical, physical, and biological approach to leachate treatment. Granular activated carbon, or PAC, is the most widely used adsorbent (Abbas et al. 2009).

With PAC as the adsorbent, the SBR process operated more effectively. The findings showed that the PAC-SBR yielded better sludge characteristics and increased removal efficiency of pollutants. The goal of treating wastewater leachate using modified activated carbon (MAC) was to create a solution of 30 mg/L KMnO4 by immersing 2.0 g of granular activated carbon in it for six hours.

Ammonium Stripping

The most popular technique for removing NH_3-N from landfill leachate is ammonium stripping treatment because of its high removal effectiveness. Landfill leachates often include high amounts of NH_4-N, which makes wastewater more noxious. Gases are often produced by this process from dissolved CH_4 and NH_3-N in leachate. This

Table 1: Social, economic, and environmental benefits of using nature-based remediation technology and nature-based brownfield redevelopment strategies

Remedial Technologies Based on Nature	Advantages			Disadvantages
	Social Benefits	**Economic Benefits**	**Advantages for the Environment and Ecology**	
Constructed wetland	Enhance appearance and establish a gathering/social area	Improve flood management; lessen the strain on drainage systems	Reduce the danger of flooding, establish a healthy ecosystem, and minimise the environmental impact of life cycles	Operational difficulties; a protracted rehabilitation process; and a need for space
Phytoremediation	Enhance appearance and provide health benefits	Enhance flood control	Reduce life cycle environmental impact, enhance air quality, and establish a healthy ecosystem	Removing or sequestering pollutants slowly
In-situ bioremediation	Rapid contamination elimination	Lower capital expenditure	Little environmental impact throughout life	Potential toxic byproducts
Nanoremediation via green synthesis	Minimise the usage of hazardous chemicals; remove impurities quickly	Reduced operating costs	Low influence on the environment over the course of its life cycle	Challenges in large-scale production
Biochar, compost, and green mulch stabilisation	Eliminate waste; improve aesthetics	Reduce operational costs; improve dust suppression and flood control	Reduced environmental impact over life cycle; enhanced air quality	Hazards associated with the release of pollutants as time passes and soil conditions alter
Reforestation and landscape architecture integration	Enhance aesthetics and offer chances for leisure and recreation	–	Establish a thriving ecosystem	Higher maintenance costs; longer turnaround time
Greenspace conversion on brownfield locations	Expand the area available for public recreation and enhance social interaction, aesthetics, and public health	Improve flood control and lower building envelope costs	Environmental footprint with a low life cycle; establish a robust ecosystem	Lack of income and increased maintenance costs

Source: Song et al. (2019)

approach requires high pH levels in order for the column to function well (Haslina et al. 2021). In terms of running costs, ammonium stripping was shown to be more economical than alternative treatment techniques including reverse osmosis and nanofiltration. Despite its advantages, the following are this technique's main disadvantages: (1) NH_3-N gas is released into the atmosphere; (2) the effluent's pH must be controlled prior to discharge (Alfaia et al. 2023).

Chemical Oxidation

Toxic or non-biodegradable contaminants in wastewater must undergo chemical oxidation treatment. Consequently, this method might be applied to effectively treat low BOD_5/COD (stabilised) landfill leachates (Derco et al. 2010, Mojiri et al. 2021).

Amokrane et al. (2020) reported 20–50% COD reduction from landfill leachate treated with oxidants such as calcium hypochlorite, ozone, potassium permanganate, and chlorine. Using the Fenton reagent, scientists discovered that the efficacy of reducing COD in biologically processed and mature landfill leachates was 75% and 60%, respectively.

Ion Exchange

The term "ion exchange process" refers to the reversible transfer of ions between the liquid and solid phases when there is no discernible change in the solid's structure. This treatment approach may successfully eliminate the remaining metal pollutants in order to meet the ever-stricter discharge restrictions in developed countries (Zheng et al. 2023). Ion exchange isn't often used to treat landfill leachate; however, in Germany, it was shown to be quite effective at removing chemicals that weren't biodegradable and that included humic components. Researchers have found that ion exchange resins are frequently and effectively used to recover organic compounds and ions from water and wastewater, as well as to polish landfill leachate treatment. The leachate should be biologically treated before beginning the ion exchange process (Barman et al. 2023). The ion exchange method could be used to effectively reduce or eliminate all soluble metallic elements, whether they are cationic or anionic. Ion exchange's high running costs restrict its commercial applicability. The need for an appropriate pre-treatment system, such as one that removes suspended contaminants from landfill leachate, is another restriction. However, the elimination of heavy metals from leachate is suitable for the process of ion exchange (Mojiri et al. 2021).

Electrochemical Oxidation

The electrochemical oxidation technique has garnered significant attention in wastewater treatment recently due to its straightforward and effective functioning. Refractory contaminants may be destroyed by this treatment method. Refractory pollutants might be eliminated via indirect oxidation, electrochemical oxidation, or direct anodic oxidation in this procedure. Ti, TiO_2, PbO_2/Ti, SnO_2/Ti, Fe, aluminium, iron, and graphite are among the electrode types that have been investigated for electrochemical treatment (Deng et al. 2020).

Electrochemical oxidation was used to clean up the semi-aerobic landfill leachate from Pulau Burung, Malaysia. Throughout the duration of the treatment, carbon electrodes and sodium sulphate (Na_2SO_4) were used as the electrolyte. BOD_5, COD, and colour removal efficiencies were 69.8%, 67.6%, and 83.7%, respectively, under ideal operating conditions of a 4-hour reaction time, 79.8 mA/cm^2 current density, and 1414 mg/L influent COD. Additionally, BOD, COD, and colour may be removed more effectively with the use of a graphite carbon electrode (Bashir 2007, Khoo et al. 2020, Hamid et al. 2021).

This method's drawbacks include: (1) high energy consumption, which makes it more costly than other treatment processes; (2) limited application in stabilised leachate treatment; (3) high current density, which may accelerate electrode corrosion; and (4) inefficiency in removing inorganic substances and NH_3-N.

Membrane Technology

A membrane is a substance that forms a thin barrier that may oppose some components of a fluid's movement and, therefore, influence the constituents' separation. A thin layer of material with a limited pore size domain and high surface porosity frequently affects a membrane's physical structure. The treatment of landfill leachate involves using several membrane filtering methods, including reverse osmosis, ultrafiltration, nanofiltration, and microfiltration (Dabaghian et al. 2019).

Microfiltration (MF)

Microfiltration has pore widths ranging from 0.05 to 10 microns, which can capture large colloidal particles, microscopic particles, and microbiological organisms. According to landfill leachate treatment, this strategy shouldn't be used alone. It is recommended to use it as a pre-treatment step in combination with chemical treatment methods or other membrane processes (such as reverse osmosis, nanofiltration, or ultrafiltration) to remove suspended contaminants and colloids. Piatkiewicz et al. (2001) were able to remove 25% to 35% of COD by using this technology as a pre-filtration stage (Mojiri et al. 2021).

Nanofiltration (NF)

In contrast to reverse osmosis membranes and ultrafiltration, nanofiltration has been used to remove heavy metals and refractory organic compounds from landfill leachate because of its unique properties. This technique eliminated 60% to 70% and 50% of the COD and NH_3-N, respectively, from the landfill leachate. In a different work, Trebouet et al. (2001) combined physical processes with nanofiltration to remove a fair amount of COD (70–80%) from landfill leachate (Zhou et al. 2024).

Ultrafiltration (UF)

The selective process of ultrafiltration uses pressures as high as 10 bar. This method works well for eliminating suspended materials using direct filtration or biological treatment in lieu of a sedimentation unit (Yan et al. 2022). The kind of material that makes up the membrane has a significant impact. According to Syzdek and Ahlert (1984), this procedure might work well as a pre-treatment for reverse osmosis. The leachate's larger molecular weight components, which tend to clog reverse osmosis membranes, may be eliminated using this method. A 50% elimination of COD was achieved using ultrafiltration by itself.

Reverse Osmosis (RO)

For treating mature landfill leachate, membrane techniques—in particular, reverse osmosis—seem more effective than conventional procedures. This technique uses a semi-permeable membrane to separate two liquids with different concentrations (Yan et al. 2022). To improve treatment efficiency, a landfill leachate treatment facility in Korea was upgraded using an integrated membrane technique that used reverse osmosis and membrane bioreactor technologies (Deng et al. 2020). The young landfill leachate eliminated 96% and 97% of the COD and NH_3-N, respectively. Other investigations found that the landfill leachate contained 98% less COD and NH_3-N.

Physical Treatment

Coagulation and Flocculation

One basic physical-chemical process in landfill leachate is the coagulation-flocculation approach. Although it is a successful pretreatment method for removing or reducing non-biodegradable organic matter from landfill leachate, its limited ability to remove organic pollutants makes it inappropriate for a comprehensive treatment of landfill leachate. The principal mechanism of the flocculation process' elimination, according to Duan and Gregory (2003), is the integration of colloids in an amorphous hydroxide precipitate. According to Wang et al. (2023), cationic hydrolysis products then neutralise the charge of negatively charged contaminants. However, the coagulation/flocculation process has certain disadvantages, such as the inability to remove NH_3-N, low removal efficiency for high-strength landfill leachate, the production of large amounts of sludge, and an increase in iron or aluminium concentration that is also visible in the liquid phase (Chen et al. 2023).

Combined Electrocoagulation/Fenton Process

Electrocoagulation (EC) and the Fenton-like process are commonly applied in practice for treating landfill leachate alone due to their advantages. The mineralisation rate of organic content is enhanced in the Fenton or Fenton-like process. However, the effect of COD removal with EC is effective due to charge neutralisation (of both cations and anions in the leachate) and sweep flocculation, which supports the removal of heavy metals (Testolin et al. 2020, Song et al. 2022). Thus, a combined biotreatment system with electrocoagulation and the Fenton-like process saves costs. Therefore, a comprehensive study of the combined process of biological treatment and the Fenton-like process can be performed to remove recalcitrant components adsorbed on the secondary effluent in the neutralization stage and resolve competitive cost issues for landfill leachate treatment (Galvão et al. 2020).

Nature-based Solutions

In the field of water treatment, NBS includes a wide range of purification methods based on natural and sustainable materials such as soil, clays, biochar, waste, or plants hardened in construction (dams, embankments, green roofs, etc.) (Ramírez-Agudelo et al. 2020). Phytoremediation and constructed wetlands are the most well-known

NBS techniques, which are primarily based, ideally, on the cooperation of plants and their rhizosphere microorganisms. The plants create a balanced environment where the interface between the roots, the living plant tissues—especially the roots—and the microorganisms facilitates degradation processes and the neutralisation of xenobiotics like heavy metals, synthetic organic compounds, and, to a lesser extent, radioactive substances (Greksa et al. 2024).

Constructed Wetlands

Constructed wetlands show great potential for leachate treatment, especially due to their natural and environmentally friendly approach. The different types of wetland systems are horizontal flow wetlands, vertical flow wetlands, hybrid systems, and free-water surface systems (Dell' Osbel et al. 2020). The addition of various media to intensify pollutant removal and improve performance has been a great advantage. A free-water surface system integrated with a sludge treatment system using *O. sativa* found that this integration planted zone effectively reduced TOC, SS, and Fe in filtered leachate. Moreover, a cardio plant type in vertical flow wetlands was suggested, which has a high potential removal rate of cations where fly ash was implemented as a medium in the wetland (Sharma et al. 2021). Constructed wetlands can receive dilute leachate for treatment, and the typical removal efficiencies are 57–93% for chemical oxygen demand, 53–94% for total nitrogen, 69% for total nitrogen, and 63% for ammonia, among others (Parde et al. 2021). Wetland treatment systems typically utilise surface flow (FWS, HSSF, and VSSF) and subsurface flow (HSSF and VSSF) wetland cells. Among these, HSSF is the most appropriate for treating landfill leachate as it provides better treatment efficiency and is easily developed with cost-effectiveness. They operate in shallow layers of aggregated soil and sediment and are influenced by plant roots and microbial communities. High surface flow velocity is not considered effective as it leads to hydraulic short-circuiting, reducing performance. However, moderate surface flow rates regulate the aerobic–anoxic–anaerobic zones for leachate treatment, which can enhance the transformation of pollutants (Spiniello et al. 2023).

Design and Operation

Nature-based solutions (NBS) are a widely recognized and utilized tool for combating landfilled leachate pollution through dilution, immobilization and microbial transformation of leachate pollutants. Municipal and landfill-disposed leachate is directed through a biotechnological system that processes and transforms these compounds. NBS includes specific low-impact development, green infrastructure, ecosystem-based adaptation, sustainable urban and rural drainage systems, or nature-based engineering (Pereira et al. 2021). This biological technology is widely referred to as 'experimenting with ecosystems'. Practical examples of NBS include stormwater wetlands, constructed wetlands, reed beds, vertical and horizontal subsurface flow wetlands, ponds and lakes, and floating wetlands for landfill leachate treatment. This chapter provides a focused overview and critical discussion on the ability of NBS to tackle a subset of municipal and landfill-based emerging compounds associated with the nitrogen cycle, copper, zinc, and the per- and polyfluorocarboxylic acids that leachate may contain (Ganapathi et al. 2024).

Regulatory Frameworks and Policies

A variety of governments have begun to phase in regulations to control the impacts of landfill waste, particularly in response to tightening environmental regulations and legislation for water quality. This includes substantial growth in the adoption of national policies and national plans for safely handling this waste. The introduction outlines the growing trend of new EU landfills—sanitary and inert—and a decreasing trend of controlled and unauthorised landfills. Many of these regulations have implemented a variety of requirements for health risks and environmental protection, including the role of waste. However, it must be fairly acknowledged that the operation of the landfill and the conditions of the landfills are broader concerns. Nevertheless, the Member States continue to have the unilateral authority to approve their own landfill rules. In this context, the growing resonance of landfill leachate adequately reflects the importance of continuous surveillance at existing landfills. Since it contains high concentrations of dangerous and non-biodegradable chemicals, landfill leachate generates persistent chemical contamination.

Case Studies in Landfill Leachate Management

Site necessity, site selections, legal aspects, and fundraising are the initial steps to be considered for executing a landfill project. Different research findings show that site selection is an ever-beholding land-based scientific study. Indeed, public discussion and participation are vital components in all landfill projects. It is mandatory by the laws in most of the developed and developing world to develop a viable landfill project that can be sustainable and fulfil the ever-increasing demands of society as far as the environmentally friendly management of municipal solid waste is concerned. Additionally, the design and operation of a landfill greatly depend on the site conditions and constraints. This chapter attempts to highlight different issues regarding legal aspects, site necessity, site selection, and preconditions, with reference to some cases from India, together with the funded costs and upgrading costs of the landfill (Derin et al. 2020, Balew et al. 2022, Mallick 2021).

The principle of sustainable development should be the most important consideration in waste management operations. Landfill projects are no exception. However, at present, it is difficult to identify and assess the secondary and tertiary effects involved in such actions. It is necessary to adopt the principles of sustainable development in all landfill projects. The methods of waste minimization and waste analysis, as well as incineration, recycling, and biological processes, should be integrated into the overall manufacturing plant concepts right from the beginning of all manufacturing activities. All of these actions aim to create an awareness of global protective actions and also stimulate the practice of reducing landfill usage (Fatimah et al. 2020, Chen et al. 2020).

Future Directions in Landfill Leachate Research

There are many areas within landfill leachate research that present opportunities for further investigation. Understanding the availability of priority pollutants in leachates and the nature of their interactions with the landfill matrix will help define the impact of specific pollutants once they are released from the solid state. There is also a need to communicate scientific findings to local city managers, landowners, students, and other stakeholders to facilitate their application in actual separation and sorting practices (Anand and Palani 2022, Rogers et al. 2021, Abunama et al. 2021, Rajoo et al. 2020).

Multiple disciplines related to landfills and leachate have already begun addressing these concerns in fields such as medicine (antibiotics), food security (veterinary drugs), and packaging (NOM residues). This work must be continued, and communication within the field and among stakeholder communities must be strengthened. This can be achieved through greater conference exchange between researchers, technology providers, and end users to develop sustainable solutions.

Conclusion

The use of nature-based solutions (NBS) offers a practical and sustainable way to mitigate landfill leachate and lessen the impact of new contaminants. Methods that remove toxins using biological, natural, physical, and chemical processes offer advantages over conventional techniques. In addition to improving pollution removal, NBS can provide further benefits, including habitat creation and biodiversity enhancement. Long-term environmental and public health benefits can be ensured, ecological integrity can be preserved, and leachate management techniques can be improved by further research and the widespread implementation of NBS.

References

Abdel-Shafy, H.I., Ibrahim, A.M., Al-Sulaiman, A.M. and Okasha, R.A. 2024. Landfill leachate: Sources, nature, organic composition, and treatment: An environmental overview. Ain Shams Engineering Journal, 15(1): 102293.

Abunama, T., Moodley, T., Abualqumboz, M., Kumari, S and Bux, F. 2021. Variability of leachate quality and polluting potentials in light of leachate pollution index (LPI) – A global perspective. Chemosphere, 282: 131119.

Afolabi, O.O., Wali, E., Ihunda, E.C., Orji, M.C., Emelu, V.O., Bosco-Abiahu, L.C. et al. 2022. Potential environmental pollution and human health risk assessment due to leachate contamination of groundwater from anthropogenic impacted site. Environmental Challenges, 9: 100627.

Ahmad, W., Zubair, M., Ahmed, M., Ahmad, M., Latif, S., Hameed, A. et al. 2023. Assessment of potentially toxic metal (loid) s contamination in soil near the industrial landfill and impact on human health: An evaluation of risk. Environmental Geochemistry and Health, 45(7): 4353-4369.

Alfaia, R.G.D.S.M., de Almeida, R., do Nascimento, K.S. and Campos, J.C. 2023. Landfill leachate pretreatment effects on nanofiltration and reverse osmosis membrane performance. Process Safety and Environmental Protection, 172: 273-281.

Amangabara, G.T., Amadi, C.U., Nelson-Kalu, C.T. and Ukuli, M.A. 2022. Investigation of the impact of municipal solid waste leachate on the groundwater quality of the Eliozu-Eligbolo community, Port harcourt, Nigeria. International Journal of Advanced Academic Research, 8(7): 35-53. ijaar.org

Anand, N. and Palani, S.G. 2022. A comprehensive investigation of toxicity and pollution potential of municipal solid waste landfill leachate. Science of The Total Environment, 838: 155891.

Ashiq, W., Nadeem, M., Ali, W., Zaeem, M., Wu, J., Galagedara, L. et al. 2020. Biochar amendment mitigates greenhouse gases emission and global warming potential in dairy manure based silage corn in boreal climate. Environmental Pollution, 265: 114869.

Badmus, S.O., Amusa, H.K., Oyehan, T.A. and Saleh, T.A. 2021. Environmental risks and toxicity of surfactants: overview of analysis, assessment, and remediation techniques. Environmental Science and Pollution Research, 1-20. springer.com

Bakhshoodeh, R., Alavi, N., Oldham, C., Santos, R.M., Babaei, A.A., Vymazal, J. et al. 2020. Constructed wetlands for landfill leachate treatment: A review. Ecological Engineering, 146: 105725.

Balew, A., Alemu, M., Leul, Y. and Feye, T. 2022. Suitable landfill site selection using GIS-based multi-criteria decision analysis and evaluation in Robe town, Ethiopia. Geo Journal, 87(2): 895-920.

Bandala, E.R., Liu, A., Wijesiri, B., Zeidman, A.B. and Goonetilleke, A. 2021. Emerging materials and technologies for landfill leachate treatment: A critical review. Environmental Pollution, 291: 118133. qut.edu.au

Barbosa, L.C.F.M., de Oliveira, O.J., Machado, M.C., Morais, A.C.T., Bozola, P.M. and Santos, M.G.F. 2022. Lessons learned from quality management system ISO 9001: 2015 certification: Practices and barrier identification from Brazilian industrial companies. Benchmarking: An International Journal, 29(8): 2593-2614.

Barman, M.K., Bhattarai, A. and Saha, B. 2023. Applications of ion exchange resins in environmental remediation. Vietnam Journal of Chemistry, 61(5): 533-550.

Barquilha, C.E. and Braga, M.C. 2021. Adsorption of organic and inorganic pollutants onto biochars: Challenges, operating conditions, and mechanisms. Bioresource Technology Reports, 15, 100728.

Bouaouda, S., Souabi, S., Bouyakhsass, R., Taleb, A., Madinzi, A. and Anouzla, A. 2023. Techniques for treating leachate discharges: A critical review. Euro-Mediterranean Journal for Environmental Integration, 8(3): 573-599.

Bravi, L., Santos, G., Pagano, A and Murmura, F. 2020. Environmental management system according to ISO 14001: 2015 as a driver to sustainable development. Corporate Social Responsibility and Environmental Management, 27(6): 2599-2614. academia.edu

Britz, T.J. (2020). Landfill leachate treatment. Microbiology of Landfill Sites.

Chen, H., Xu, H., Zhong, C., Liu, M., Yang, L., He, J. et al. 2023. Treatment of landfill leachate by coagulation: A review. Science of The Total Environment, 169294.

Cherni, Y., Elleuch, L., Messaoud, M., Kasmi, M., Chatti, A. and Trabelsi, T. 2021. Recent technologies for leachate treatment: A review. Euro-Mediterranean Journal for Environmental Integration, 6: 1-20. researchgate.net

Costa, A.M., Daflon, S.D.A. and Campos, J.C. 2024. Landfill leachate and ecotoxicity. *In*: A Review of Landfill Leachate: Characterization Leachate Environment Impacts and Sustainable Treatment Methods, 129-180. Cham: Springer Nature Switzerland.

Dell'Osbel, N., Colares, G.S., Oliveira, G.A., Rodrigues, L.R., da Silva, F.P., Rodriguez, A.L. et al. 2020. Hybrid constructed wetlands for the treatment of urban wastewaters:

Increased nutrient removal and landscape potential. Ecological Engineering, 158: 106072. academia.edu

Deng, Y., Zhu, X., Chen, N., Feng, C., Wang, H., Kuang, P. et al. 2020. Review on electrochemical system for landfill leachate treatment: Performance, mechanism, application, shortcoming, and improvement scheme. Science of The Total Environment, 745: 140768.

Derin, P., Yeşilnacar, M.İ. and Çullu, M.A. 2020. A study on evaluation of site selection in sanitary landfill with regard to urban growth. Environmental Research and Technology, 4(2): 117-125. dergipark.org.tr

Dixit, A., Singh, D. and Shukla, S.K. 2024. Assessment of human health risk due to leachate contaminated soil at solid waste dumpsite, Kanpur (India). International Journal of Environmental Science and Technology, 21(1): 909-924.

Dushkova, D. and Haase, D. 2020. Not simply green: Nature-based solutions as a concept and practical approach for sustainability studies and planning agendas in cities. Land, 9(1): 19.

Ergene, D., Aksoy, A. and Sanin, F.D. 2022. Comprehensive analysis and modeling of landfill leachate. Waste Management, 145: 48-59.

Essien, J.P., Ikpe, D.I., Inam, E.D., Okon, A.O., Ebong, G.A. and Benson, N.U. 2022. Occurrence and spatial distribution of heavy metals in landfill leachates and impacted freshwater ecosystem: An environmental and human health threat. Plos One, 17(2): e0263279. plos.org

Fang, X., Li, J. and Ma, Q. 2023. Integrating green infrastructure, ecosystem services and nature-based solutions for urban sustainability: A comprehensive literature review. Sustainable Cities and Society, 98: 104843.

Fatimah, Y.A., Govindan, K., Murniningsih, R. and Setiawan, A. 2020. Industry 4.0 based sustainable circular economy approach for smart waste management system to achieve sustainable development goals: A case study of Indonesia. Journal of Cleaner Production, 269: 122263.

Ganapathi, H., Awasthi, S. and Vasudevan, P. 2024. Wetlands as a nature-based solution for urban water management. *In*: Nature-based Solutions for Circular Management of Urban Water, 259-275. Cham: Springer International Publishing.

Gautam, P. and Kumar, S. 2021. Characterisation of hazardous waste landfill leachate and its reliance on landfill age and seasonal variation: A statistical approach. Journal of Environmental Chemical Engineering, 9(4): 105496.

Genethliou, C., Tatoulis, T., Charalampous, N., Dailianis, S., Tekerlekopoulou, A.G. and Vayenas, D.V. 2023. Treatment of raw sanitary landfill leachate using a hybrid pilot-scale system comprising adsorption, electrocoagulation and biological process. Journal of Environmental Management, 330: 117129.

Giri, S. 2021. Water quality prospective in Twenty First Century: Status of water quality in major river basins, contemporary strategies and impediments: A review. Environmental Pollution, 271: 116332.

Gong, H., Hu, J., Rui, X., Luo, J. and Zhu, N. 2023. Unveiling the occurrence, distribution, removal, and environmental impacts of 65 emerging contaminants in neglected fresh leachate from municipal solid waste. Journal of Hazardous Materials.

Greksa, A., Ljubojević, M. and Blagojević, B. 2024. The value of vegetation in nature-based solutions: Roles, challenges, and utilization in managing different environmental and climate-related problems. Sustainability, 16(8): 3273.

Hale, S.E., Folde, M.S., Melby, U.H., Sjødahl, E.U., Smebye, A.B. and Oen, A.M. 2022. From landfills to landscapes—Nature-based solutions for water management taking into account legacy contamination. Integrated Environmental Assessment and Management, 18(1): 99-107.

Hamid, M.A.A., Aziz, H.A. and Yusoff, M.S. 2021. Electrocoagulation process in the treatment of landfill leachate. Sustainable Solutions for Environmental Pollution: Waste Management and Value-Added Products, 1: 257-304.

Hassan, A., Periathamby, A., Ahmed, A., Innocent, O. and Hamid, F.S. 2020. Effective bioremediation of heavy metal–contaminated landfill soil through bioaugmentation using consortia of fungi. Journal of Soils and Sediments, 20: 66-80. academia.edu

Health Organization, W. 2021. Guidelines on Recreational Water Quality. Volume 1: Coastal and Fresh Waters. google.com

Hoque, M.M. and Rahman, M.T.U. 2020. Landfill area estimation based on solid waste collection prediction using ANN model and final waste disposal options. Journal of Cleaner Production, 256: 120387.

Iravanian, A. and Ravari, S.O. 2020. Types of contamination in landfills and effects on the environment: A review study. *In*: IOP Conference Series: Earth and Environmental Science, Vol. 614(1), p. 012083. IOP Publishing. iop.org

Islam, S., Bano, H., Bhat, J.I.A., Aziz, M.A., Bhat, S.U.I., Nazir, N. et al. 2023. Landfill leachate a new threat to water quality: A case study from the Temperate Himalayas. Environmental Monitoring and Assessment, 195(6): 689. researchsquare.com

Jabeen, S., Khan, T., Jaiswal, A. and Bala, S. 2024. Green nanotechnology for clean energy and environmental sustainability. *In*: Sustainable Nanomaterials: Synthesis and Environmental Applications, 1-20. Singapore: Springer Nature Singapore.

Japperi, N.S., Asri, Z.Z.M., Bakar, W.Z.W., Dollah, A., Fuad, M.F.I.A and Hussein, S.N.C.M. 2021. Review on landfill gas formation from leachate biodegradation. Malaysian Journal of Chemical Engineering and Technology (MJCET), 4(1): 39-49. mohe.gov.my

Kamal, A., Makhatova, A., Yergali, B., Baidullayeva, A., Satayeva, A., Kim, J. et al. 2022. Biological treatment, advanced oxidation and membrane separation for landfill leachate treatment: A review. Sustainability, 14(21): 14427. mdpi.com

Khurshid, H., Mustafa, M.R.U., Rashid, U., Isa, M.H., Ho, Y.C. and Shah, M.M. 2021. Adsorptive removal of COD from produced water using tea waste biochar. Environmental Technology and Innovation, 23: 101563.

Kurek, M.R., Garcia-Tigreros, F., Wickland, K.P., Frey, K.E., Dornblaser, M.M., Striegl, R.G. et al. 2023. Hydrologic and landscape controls on dissolved organic matter composition across western North American Arctic lakes. Global Biogeochemical Cycles, 37(1): e2022GB007495.

Lange, L., Connor, K.O., Arason, S., Bundgård-Jørgensen, U., Canalis, A., Carrez, D. et al. 2021. Developing a sustainable and circular bio-based economy in EU: By partnering across sectors, upscaling and using new knowledge faster, and for the benefit of climate, environment and biodiversity, and people and business. Frontiers in Bioengineering and Biotechnology, 8: 619066. frontiersin.org

Lebron, Y.A.R., Moreira, V.R., Brasil, Y.L., Silva, A.F.R., de Souza Santos, L.V., Lange, L.C. et al. 2021. A survey on experiences in leachate treatment: Common practices, differences worldwide and future perspectives. Journal of Environmental Management, 288: 112475.

Li, Y., Ren, Y., Ji, J., Li, Y.Y. and Kobayashi, T. 2023. Anaerobic membrane bioreactors for municipal wastewater treatment, sewage sludge digestion and biogas upgrading: A review. Sustainability, 15(20): 15129.

Liang, H., Jia, Y., Khanal, S. K., Huang, D., Sun, L. and Lu, H. 2024. Electrochemical-coupled sulfur-driven autotrophic denitrification for nitrogen removal from raw landfill leachate: Evaluation of performance and mechanisms. Water Research, 256: 121592.

Luo, H., Zeng, Y., Cheng, Y., He, D and Pan, X. 2020. Recent advances in municipal landfill leachate: A review focusing on its characteristics, treatment, and toxicity assessment. Science of The Total Environment, 703: 135468.

Mohammad, A., Singh, D. N., Podlasek, A., Osinski, P and Koda, E. 2022. Leachate characteristics: Potential indicators for monitoring various phases of municipal solid waste decomposition in a bioreactor landfill. Journal of Environmental Management, 309, 114683.

Mohd-Salleh, S.N.A., Shaylinda, M.Z.N., Othman, N., Azizan, M.O., Yashni, G. and Afnizan, W.M.W. 2020. Sustainability analysis on landfilling and evaluation of characteristics in landfill leachate: A case study. In IOP Conference Series: Materials Science and Engineering, Vol. 736(7), 072002. IOP Publishing.

Mojiri, A., Ziyang, L., Tajuddin, R.M., Farraji, H. and Alifar, N. 2016. Co-treatment of landfill leachate and municipal wastewater using the ZELIAC/zeolite constructed wetland system. Journal of Environmental Management, 166: 124-130.

Mojiri, A., Ziyang, L., Hui, W., Ahmad, Z., Tajuddin, R.M., Amr, S.S.A. et al. 2017. Concentrated landfill leachate treatment with a combined system including electro-ozonation and composite adsorbent augmented sequencing batch reactor process. Process Safety and Environmental Protection, 111: 253-262.

Mojiri, A., Zhou, J.L., Ratnaweera, H., Ohashi, A., Ozaki, N., Kindaichi, T. et al. 2021. Treatment of landfill leachate with different techniques: An overview. Water Reuse, 11(1): 66-96. iwaponline.com

Montvydienė, D., Šulčius, S., Jurgelėnė, Ž., Makaras, T., Kalcienė, V., Taraškevičius, R. et al. 2020. Contrasting ecotoxic effects of landfill leachate and cyanobacterial biomass on aquatic organisms. Water, Air, and Soil Pollution, 231: 1-14.

Morita, A.K., Ibelli-Bianco, C., Anache, J.A., Coutinho, J.V., Pelinson, N.S., Nobrega, J. et al. 2021. Pollution threat to water and soil quality by dumpsites and non-sanitary landfills in Brazil: A review. Waste Management, 131: 163-176.

Nath, A. and Debnath, A. 2022. A short review on landfill leachate treatment technologies. Materials Today: Proceedings, 67: 1290-1297.

Nika, M.C., Ntaiou, K., Elytis, K., Thomaidi, V.S., Gatidou, G., Kalantzi, O.I. et al. 2020. Wide-scope target analysis of emerging contaminants in landfill leachates and risk assessment using Risk Quotient methodology. Journal of Hazardous Materials, 394: 122493.

Nimonkar, Y.S., Kajale, S., Dake, M., Ranade, D.R., Yadav, K.K., Kumar, R. et al. 2022. A culture-based and culture-independent approach to the study of landfill leachate bacterial and archaeal communities. Anaerobe, 77: 102626.

Ntagia, E., Amulya, K., Le, S.L., Le, L.T., Tran, C.S., Ha, T.L. et al. (2023). Pilot and full scale applications of floating treatment wetlands for treating diffuse pollution. Science of The Total Environment, 165595.

Osazee, I.T. 2021. Landfill in a sustainable waste disposal. European Journal of Environment and Earth Sciences, 2(4): 67-74.

Parde, D., Patwa, A., Shukla, A., Vijay, R., Killedar, D.J. and Kumar, R. 2021. A review of constructed wetland on type, treatment and technology of wastewater. Environmental Technology and Innovation, 21: 101261.

Parvin, F. and Tareq, S.M. 2021. Impact of landfill leachate contamination on surface and groundwater of Bangladesh: A systematic review and possible public health risks assessment. Applied Water Science, 11(6): 100.

Pereira, P., Inácio, M., Karnauskaitė, D., Bogdzevič, K., Gomes, E., Kalinauskas, M. et al. 2021. Nature-based solutions impact on urban environment chemistry: Air, soil, and water. Nature-based Solutions for Flood Mitigation: Environmental and Socio-economic Aspects, 79-137.

Qiao, P., Wang, S., Li, J., Zhao, Q., Wei, Y., Lei, M. et al. 2023. Process, influencing factors, and simulation of the lateral transport of heavy metals in surface runoff in a mining area driven by rainfall: A review. Science of the Total Environment, 857: 159119.

Rajoo, K.S., Karam, D.S., Ismail, A. and Arifin, A. 2020. Evaluating the leachate contamination impact of landfills and open dumpsites from developing countries using the proposed Leachate Pollution Index for Developing Countries (LPIDC). Environmental Nanotechnology, Monitoring and Management, 14: 100372.

Ramírez-Agudelo, N.A., Porcar Anento, R., Villares, M. and Roca, E. 2020. Nature-based solutions for water management in peri-urban areas: Barriers and lessons learned from implementation experiences. Sustainability, 12(23): 9799.

Rogers, E.R., Zalesny Jr, R.S. and Lin, C.H. 2021. A systematic approach for prioritizing landfill pollutants based on toxicity: Applications and opportunities. Journal of Environmental Management, 284: 112031.

Saeed, T., Miah, M.J., Majed, N., Alam, M.K. and Khan, T. 2021. Effect of effluent recirculation on nutrients and organics removal performance of hybrid constructed wetlands: Landfill leachate treatment. Journal of Cleaner Production, 282: 125427.

Salam, M. and Nilza, N. 2021. Hazardous components of landfill leachates and its bioremediation. *In*: Soil Contamination – Threats and Sustainable Solutions. IntechOpen. intechopen.com

Sharma, K., Kumar, P. and Singh, J. 2024. Ecology and diversity of microbial communities involved in the removal of priority contaminants and micropollutants in waste. *In*: Development in Waste Water Treatment Research and Processes, 181-199. Elsevier.

Sharma, P., Tripathi, S., Purchase, D. and Chandra, R. 2021. Integrating phytoremediation into treatment of pulp and paper industry wastewater: Field observations of native plants for the detoxification of metals and their potential as part of a multidisciplinary strategy. Journal of Environmental Chemical Engineering, 9(4): 105547.

Shi, L., Chen, H., Meng, H., Cheng, R., Dai, J., Zheng, X. et al. 2021. How environmental policy impacts technology adoption: A case of landfill leachate. Journal of Cleaner Production, 310: 127484.

Shrestha, S. 2020. Leachate characterization and management at Sisdol landfill site in Nepal. Journal of Development Innovations, 4(1): 119-132.

Siddiqi, S.A., Al-Mamun, A., Baawain, M.S. and Sana, A. 2022. A critical review of the recently developed laboratory-scale municipal solid waste landfill leachate treatment technologies. Sustainable Energy Technologies and Assessments, 52: 102011.

Simeonov, I., Chorukova, E. and Kabaivanova, L. 2025. Two-stage anaerobic digestion for green energy production: A review. Processes, 13(2): 294.

Siyar, R., Doulati Ardejani, F., Norouzi, P., Maghsoudy, S., Yavarzadeh, M., Taherdangkoo, R. et al. 2022. Phytoremediation potential of native hyperaccumulator plants growing on heavy metal-contaminated soil of Khatunabad copper smelter and refinery, Iran. Water, 14(22): 3597.

Tałałaj, I.A., Bartkowska, I. and Biedka, P. 2021. Treatment of young and stabilized landfill leachate by integrated sequencing batch reactor (SBR) and reverse osmosis (RO) process. Environmental Nanotechnology, Monitoring and Management, 16: 100502. sciencedirect.com

Teng, C., Zhou, K., Peng, C. and Chen, W. 2021. Characterization and treatment of landfill leachate: A review. Water Research, 203: 117525.

Thakur, T.K., Barya, M.P., Dutta, J., Mukherjee, P., Thakur, A., Swamy, S.L. et al. 2023. Integrated phytobial remediation of dissolved pollutants from domestic wastewater through constructed wetlands: An interactive macrophyte-microbe-based green and low-cost decontamination technology with prospective resource recovery. Water, 15(22): 3877.

Thongyuan, S., Khantamoon, T., Aendo, P., Binot, A. and Tulayakul, P. 2021. Ecological and health risk assessment, carcinogenic and non-carcinogenic effects of heavy metals contamination in the soil from municipal solid waste landfill in Central, Thailand. Human and Ecological Risk Assessment: An International Journal, 27(4): 876-897.

Trihadiningrum, Y., Wilujeng, S.A., Tafaqury, R., Radita, D.R. and Radityaningrum, A.D. 2023. Evidence of microplastics in leachate of Randegan landfill, Mojokerto City, Indonesia, and its potential to pollute surface water. Science of The Total Environment, 874: 162207.

Vasilachi, I.C., Asiminicesei, D.M., Fertu, D.I. and Gavrilescu, M. 2021. Occurrence and fate of emerging pollutants in water environment and options for their removal. Water, 13(2): 181.

Vaverková, M.D., Elbl, J., Koda, E., Adamcová, D., Bilgin, A., Lukas, V. et al. 2020. Chemical composition and hazardous effects of leachate from the active municipal solid waste landfill surrounded by farmlands. Sustainability, 12(11): 4531.

Wang, W., Li, R., Bu, F., Gao, Y., Gao, B., Yue, Q. et al. 2023. Coagulation and membrane fouling mechanism of Al species in removing humic acid: Effect of pH and a dynamics process analysis. Separation and Purification Technology, 309: 123130.

Wijekoon, P., Koliyabandara, P.A., Cooray, A.T., Lam, S.S., Athapattu, B.C. and Vithanage, M. 2022. Progress and prospects in mitigation of landfill leachate pollution: Risk, pollution potential, treatment and challenges. Journal of Hazardous Materials, 421: 126627.

Yaashikaa, P.R., Kumar, P.S., Nhung, T.C., Hemavathy, R.V., Jawahar, M.J., Neshaanthini, J.P. et al. 2022. A review on landfill system for municipal solid wastes: Insight into leachate, gas emissions, environmental and economic analysis. Chemosphere, 309: 136627.

Yan, Z., Lu, Z., Chen, X., Jiang, Y., Huang, Z., Liu, L. et al. 2022. Membrane distillation treatment of landfill leachate: Characteristics and mechanism of membrane fouling. Separation and Purification Technology, 289: 120787.

Zhang, P., Chai, J., Cao, J., Qin, Y., Dang, M., Geng, K. et al. 2023. Landfill leachate generation mechanism study: A review. International Journal of Environmental Science and Technology, 20(8): 9271-9290.

Zhao, Z., Zhang, Y., Yu, L., Hou, D., Wu, X., Li, K. et al. 2023. Fenton pretreatment to mitigate membrane distillation fouling during treatment of landfill leachate membrane concentrate: Performance and mechanism. Water Research, 244: 120517.

Zheng, L., Lin, H., Dong, Y., Li, B. and Lu, Y. 2023. A promising approach for simultaneous removal of ammonia and multiple heavy metals from landfill leachate by carbonate precipitating bacterium. Journal of Hazardous Materials, 456: 131662.

Zhou, C., Li, N., Guo, W., Wang, R., Yan, Y. and Shao, S. 2024. Removing refractory organic substances from landfill leachate using a nanofiltration membrane bioreactor: Performance and mechanisms. Chemical Engineering Journal, 484: 149704.

Zielińska, M., Kulikowska, D. and Stańczak, M. 2020. Adsorption – Membrane process for treatment of stabilized municipal landfill leachate. Waste Management, 114: 174-182.

CHAPTER
10

Potential and Limitations of Nature-based Solutions in Wastewater Treatment

Rahul Kumar[1], Reetika Bagauli[2], Snežana Andjelković[3], Marika Pellegrini[4], Sourav Chattaraj[5], Arindam Ganguly[6], Bahman Khoshru[7], Anuprita Ray[8] and Debasis Mitra[1*]

[1] Department of Microbiology, Graphic Era Deemed to be University, 566/6, Bell Road, Clement Town, Dehradun - 248002, Uttarakhand, India
[2] Department of Chemistry, Graphic Era Deemed to be University, 566/6, Bell Road, Clement Town Dehradun - 248002, Uttarakhand, India
[3] Institute for Forage Crops, 37251, Kruševac, Serbia
[4] Department of Life, Health and Environmental Sciences, University of L'Aquila, Coppito, L'Aquila, Italy
[5] Centre for Industrial Biotechnology Research, School of Pharmaceutical Science, Siksha 'O' Anusandhan University, Kalinga Nagar, Bhubaneswar, Odisha - 751 003, India
[6] Department of Microbiology, Bankura Sammilani College, Bankura, West Bengal - 722102, India
[7] Soil and Water Research Institute (SWRI), Agricultural Research, Education and Extension Organization (AREEO), Karaj 31785-311, Iran
[8] Molecular Biology and Genetic Engineering, School of Bioengineering and Biosciences, Lovely Professional University, Phagwara, Punjab, India

Introduction

Overview of Wastewater Treatment Challenges

There are 2.1 billion people in the world with inadequate sanitation (where wastewater flows straight into surface waterways) and 2.4 billion without enhanced sanitation (i.e. sanitation facilities that hygienically isolate human excreta from human touch). Improper handling of wastewater and faeces poses a serious threat to public health and the environment, even with advancements in recent decades (United Nations 2021). According to the United Nations World Water Assessment

*Corresponding author: debasismitra.mb@geu.ac.in

DOI: 10.1201/9781003527350-10

Programme (WWAP 2009), a staggering 80% of wastewater is discharged untreated (Wan Rosely and Voulvoulis 2023). As interest in low-cost, eco-friendly treatment methods increases, wastewater utility managers often face challenges in determining the best way to integrate conventional infrastructure, such as treatment plants, with ecological alternatives, such as wetlands. The aim of this study is to investigate the implementation of nature-based solutions (NBS) for wastewater treatment and the benefits they provide for society. Cohen-Shacham et al. (2016) define NBS as "actions to protect, sustainably manage, and restore natural or modified ecosystems that address societal challenges effectively and adaptively, simultaneously providing human well-being and biodiversity benefits." This definition was provided by the International Union for the Conservation of Nature.

Access to clean water and sanitation is one of the Sustainable Development Goals established by the United Nations General Assembly in 2015. The UN has estimated that by 2030, 700 million people may be forced to relocate due to water scarcity. In 2017, it was reported that 2.2 billion people worldwide lacked access to clean drinking water. The release of large volumes of untreated wastewater into natural water bodies, exacerbated by rapid urbanisation and industrialisation, poses a serious threat to water resources (Obotey and Rathilal 2020). A wastewater treatment plant is as vital as any other facility in building a sustainable community. Several units comprise domestic wastewater treatment, which removes various pollutants from WW (Metcalf et al. 2007, Hu and Dold 2012, Hreiz et al. 2015). Toxic compounds, heavy metals, microbes, biological materials, microplastics, oil, and viruses can be found in industrial wastewater (Hakak et al. 2020). According to the United Nations Educational, Scientific, and Cultural Organization (UNESCO), over 80% of this effluent is discharged into the water without treatment (Ryder 2017).

Industries are not properly adopting wastewater treatment because of the high set-up and operating costs of wastewater treatment facilities, ample space requirements in the vicinity of the sector, lax legislative enforcement of discharge limits, and a lack of technical understanding of wastewater treatment systems. Certain nations have adopted serious strategies to reduce water pollution and conserve water. For example, China implemented the Water Pollution Control Action Plan in January 2015, emphasising the protection and conservation of water resources. To adhere to this strategy, the industry must create wastewater treatment systems that remove pollutants from water, recycle water, and enable reuse (Chen et al. 2020) (Table 1). The Water Framework Directive (2000/60/EC) governs the use of recovered wastewater in the EU and is a part of the circular economy plan (WFD). Although 88% of EU wastewater is now secondarily treated owing to WFD, reuse rates still need to be higher (Mesa-Perez and Berbel 2020). The World Bank raised awareness among governing parties in Latin America and the Caribbean through its "Wastewater: From Waste to Resource" campaign on the possibility of turning wastewater into valuable resources, such as energy, reusable water, fertilizers, and biosolids (Rodriguez et al. 2020) (Figure 1).

One of the main components of the effluent released by industry is wastewater. The food and beverage industries produce a large amount of sewage, but most biocides mixed into water come from petrochemical and metal processing industries. These petroleum fractions include diesel oil, petrol, diesel oil-derived fuels,

Table 1: Water pollution by chemical

Pollutant Name	Pollutant Sources	Pollutant Types	Source Type	Major Challenges	Water Quality Problems
DDT, PCDFs, PCDDs, PCBs, PAHs, PBDEs	Numerous (combustion, agricultural, waste, sites, spills, and others)	Persistent organic pollutants (POPs)	Globally dispersed point and diffuse	Eliminate current POPs, contain current sources, and forbid using newly developed POPs.	Food chain biomagnification and a range of health implications
DDT, triazines, chloraceanilides, lindane	Agriculture	Pesticides	Diffuse	Agricultural land control of pesticide discharge, pesticide abuse	Biologically active chemical contamination of surface and groundwater; unintentional poisoning (especially in underdeveloped nations)
U, F, As, Se, geosmin, microcystins	Natural pollutants Geogenic pollutants Biogenic pollutants	Taste and odour compounds, cyanotoxins, and inorganic pollutants	Diffuse	Creation of efficient methods for treating households, regulation, eutrophication, and consumer acceptance	Fluorosis, cancer, human health, and aesthetics (taste and odour)
Cyanide, mercury, sulfuric acid and copper	Mining	Heavy metals, leaching, and acid	Mostly point	Metal removal, acid neutralisation, and the addition of coefficient nontoxic reagents	Acute poisoning, persistent neurotoxicity, and metal remobilization
Nitroaromatic explosives, chlorinated solvents, chromium, U, technetium	Hazardous source	Diverse	Point	Pollutant containment and oversight of mitigation measures, such as natural attenuation	Long-term pollution of sources of drinking water

Dengue, cholera, hepatitis A and B, typhoid fever, schistosomiasis, diarrhoea	Urban wastewater in nations that are developing or emerging	Viruses and microorganisms	Point	Enhancing cleanliness and hygiene, providing safe drinking water, and using affordable, effective drinking water disinfection methods	Malnutrition, child mortality, and human health
17 α-ethinylestradiol, diclophenac	Urban wastewater in develop nations	Hormones, pharmaceuticals	Point	Reduction of wastewater's micropollutant contents with polishing treatment	Effects of ecotoxicology in rivers on fish feminization

Note: U – Uranium; PCDFs – Polychlorinated dibenzofurans; PAHs – Polycyclic aromatic hydrocarbons; As – Arsenic; PCDDs – Polychlorinated dibenzo-p-dioxins; PBDEs – Polybrominated diphenyl ethers; PCBs – Polychlorinated biphenyls; DDT – Dichlorodiphenyltrichloroethane

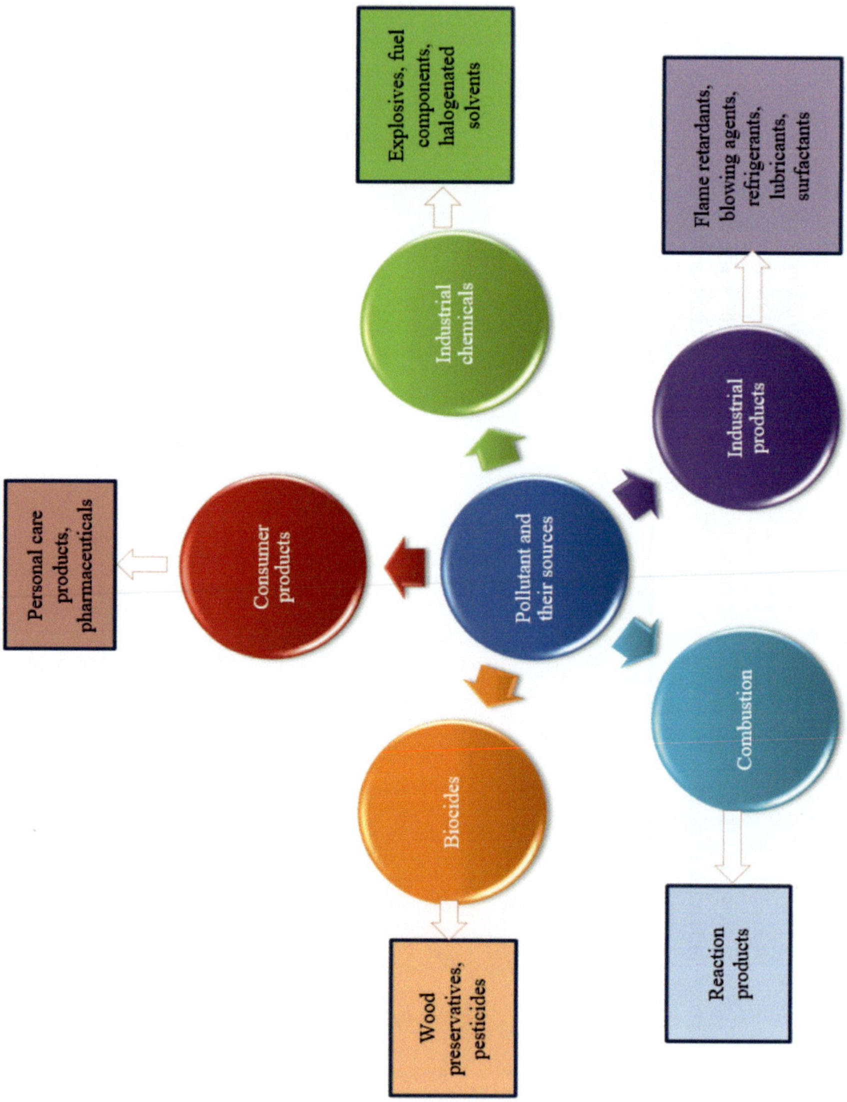

Figure 1: Pollutants and their sources.

antibiotics, biocides, heavy metals, and manure. These ingredients can be found in wastewater (Cai et al. 2020). It consists of physical, chemical, and biological treatment procedures that can minimize the amount of sludge and sewage water, create recycled water, and shield the receiving water bodies from pollution (Wakelin et al. 2008). Benefits to the environment from WW include healthier, more stable, and safer aquatic ecosystems, and a decrease in the quantity of untreated wastewater discharged into the atmosphere (Hasan and Muhammad 2020). Every wastewater treatment method offers an opportunity to find a long-term solution to some of the issues related to water scarcity (Antalová et al. 2018). An ecosystem that is both sustainable and healthy can be developed through effective wastewater management (Hasan and Muhammad 2020) (Table 2).

Growing Global Water Pollution Concerns

The 21st century has seen the emergence of numerous significant challenges for humanity, many of which are connected to issues surrounding the availability and quality of water (UN-Water 2009). Climate change is expected to exacerbate these problems by intensifying the water cycle, melting glaciers, and increasing water temperatures, which may result in more frequent droughts and floods (Huntington 2006, Oki and Kanae 2006). The lack of proper sanitation has the most immediate and severe consequences for human health, particularly in relation to the problem of unsafe drinking water, which currently affects over a third of the global population. Other potential hazards include exposure to pathogens or toxic chemicals while engaging in leisure activities e.g. swimming in polluted surface water or through the food chain e.g. consuming fish and seafood that have accumulated toxic chemicals because of irrigating plants with contaminated water. A substantial portion of the Earth's readily accessible freshwater, approximately 36% of the total, is used extensively for domestic, industrial, and agricultural purposes (Schwarzenbach et al. 2006). Consequently, water pollution with synthetic and geogenic natural chemicals has become a significant public concern in virtually every country around the globe, as most of these activities lead to the contamination of water resources (Saad 2009).

The Need for Sustainable and Effective Wastewater Treatment Solutions

Chemical water pollutants can be separated into two groups: the relatively limited number of micropollutants, including nutrients such as nitrogen, usually at a milligram per litre level (Gruber and Galloway 2008) as well as phosphorus species (Filippelli 2008), in addition to organic natural components (Jorgenson 2009). Although the causes and effects of these classical pollutants are well recognized, there is still a scientific challenge in developing sustainable treatment systems for them (Larsen et al. 2007). For instance, elevated nutrient loading may result in harmful algal blooms, oxygen depletion, and increased primary biomass production (Lohse et al. 2009, Heisler et al. 2008). Another long-term issue is the rising salt load that surface water receives from road salt and intensive irrigation. Elevated salinity levels hinder crop development in agriculture and prohibit their direct use as drinking water (Kaushal et al. 2005). Owing to the overuse of aquifers and rising sea levels, marine

Table 2: Nature-based solutions (NBS) – benefits and processes for environmental monitoring

Benefits	Definition	Source
Sequestration of carbon	The process of using physical or biological processes, like photosynthesis, to take carbon out of the atmosphere and place it in a reservoir or carbon sinks (like oceans, forests, or soils).	(Hickmann et al. 2021)
Mitigation of flooding	The control of water flows due to ecosystems' chemical and physical features, which helps people manage and use hydrological systems and reduces or eliminates the possibility of harm to human use, health, or safety (e.g. damage mitigation because of fewer and smaller flood/storm events).	(Haines and Potschin 2012)
Minimization of storm peaks	Rainfall volume during storms may occasionally surpass the capacity of the drainage systems, resulting in sudden overflowing; however, infiltration, retention, and detention features of the majority of NBS will stop this from happening. For instance, vegetation increases friction along the rain flow path to prolong the runoff process and lessen the peak flow, while the permeability and porosity of the ground where NBS are installed enable infiltration during the peak event.	(Brears 2020, Huang et al. 2020)
Biosolids	Wastewater treatment facilities create biosolids, which are organic materials rich in nutrients that are obtained from treated wastewater sludge. These leftovers can be recycled and used as fertiliser to maintain and enhance productive soils and promote plant development after they have been cleaned up and processed.	(US EPA 2021s)
Fauna (Biodiversity)	Diversity within and between species as well as within ecosystems is a measure of the variability found in all living things, including terrestrial, marine, and other aquatic ecosystems, as well as the ecological complexes of which they are a part. Every mammal (kingdom Animalia), every type of fungus, and every group of bacteria.	(Cragg et al. 2012)
Production of biomass	The gathering of plant material above ground by means of routine removal and harvesting. In certain circumstances, harvesting biomass may result in a greater removal of phosphate and nitrogen. The biomass material that has been harvested can then be used for other economically beneficial uses.	(Kim and Geary 2001)
Leisure	Individuals frequently make their decisions about where to spend their free time partly on the features of the local natural or artificial landscapes. People may use the environment for sport and recreation in the context of NBS being used for wastewater, depending on the degree of treatment as well as the technology and design employed at a site.	(Haines and Potschin 2012, Carpenter et al. 2009)

Beauty value	Many individuals find aesthetic worth or beauty in many components of ecosystems; this is shown in their support for parks, "scenic drives," and the places where they choose to live. The biophysical traits or attributes of species or ecosystems (settings, landscapes, cultural spaces) that people value for reasons other than utilitarianism could apply to NBS employed for wastewater treatment.	(Haines and Potschin 2012, Carpenter et al. 2009)
Flora biodiversity	Diversity within and between species as well as within ecosystems is a measure of the variability found in all living things, including terrestrial, marine, and other aquatic ecosystems, as well as the ecological complexes of which they are a part. Everything that belongs to the kingdom Plantae.	(Cragg et al. 2012)
Reusing water	The utilisation of treated wastewater, such as that produced by NBS, in a variety of applications, including industrial processes, environmental restoration, potable water supply, groundwater replenishment, and agricultural irrigation, is referred to as water reuse. This practice has the potential to enhance resilience, sustainability, and water security by providing alternatives to existing water sources.	(ISO 2018, Meeks 2022)
Food supply	Food derived from undomesticated flora and fauna. This includes the by-products of wild animal species that can be used as raw materials for food production, as well as the parts of cultivated plant species that can be harvested and used for food production.	(Brears 2020, Haines and Potschin 2012)
Controlling the temperature	The control of localised temperatures and humidity during hot weather, particularly through transpiration and ventilation.	(Brears 2020, Haines and Potschin 2012)

salt intrusion into groundwater is becoming worse in many coastal areas, including China and India (Post 2005).

Nature-based Solutions

The world's population is constantly growing, which has increased the amount of water required for industrial, agricultural, and residential use. As a result, the significant amount of wastewater generated has an impact on the quality of water. However, various uses can be derived from this wastewater (Onu et al. 2023, Zvimba et al. 2021, Sawunyama et al. 2024a). Sustainable growth requires effective wastewater management. Eleven of the 17 Sustainable Development Goals were effectively achieved with the aid of wastewater treatment. Wastewater contains a variety of contaminants such as microorganisms, insecticides, dyes, heavy metals, medications, and surfactants (Saxena et al. 2020, Kamboj et al. 2024). Even at low concentrations, these contaminants have a substantial negative impact on aquatic ecosystems, and human and animal health. In recent years, significant efforts have been made to address the problems associated with wastewater treatment, including the development of treatment techniques, policies, and implementation (Sangamnere et al. 2023, Geng et al. 2009, Samantaray et al. 2024). The following text may be classified as either advanced treatment techniques or conventional methods. Advanced treatment methods include constructed wetlands, bioelectrical systems, membrane filtration technologies, adsorption, advanced oxidation processes such as Fenton oxidation processes, photocatalysis, electrochemical oxidation, and enzymatic treatment (Ahmed et al. 2021a, Shindhal et al. 2021, Priyadarshini et al. 2023).

Ozonation and coagulation-flocculation are examples of the conventional techniques. However, only a small portion of the various treatment approaches listed for wastewater treatment are used for home and commercial reasons owing to budgetary and technological limitations (Crini and Lichtfouse 2019). To address the unresolved issues of global water shortages and pollution, water treatment technologies that are multifunctional, scalable, robust, chemical-free, low-impact on the environment, and energy-efficient are required (Sun et al. 2021). A new class of pollutants, known as emerging contaminants, has recently been found in the environment because of inadequate wastewater treatment techniques; these pollutants have not yet been completely removed from the ecosystem (El Moukhtari et al. 2023).

Further studies are therefore required to improve current wastewater treatment systems to make them fully compliant with established criteria for the removal of contaminants. Membrane technology has gained international attention owing to its easy, ecologically friendly, energy-efficient, and effective removal of impurities with minimal chemical use (Singh and Hankins 2016, Kotobuki et al. 2021, Sawunyama et al. 2024b). Membranes function as selective barriers that allow certain substances to pass while preventing others from doing so (Singh and Hankins 2016, Ahmed et al. 2022a). The basic idea behind using membranes in wastewater treatment is to generate a driving force across the membrane to capture impurities and allow water molecules to pass selectively. The two main types of membranes are inorganic

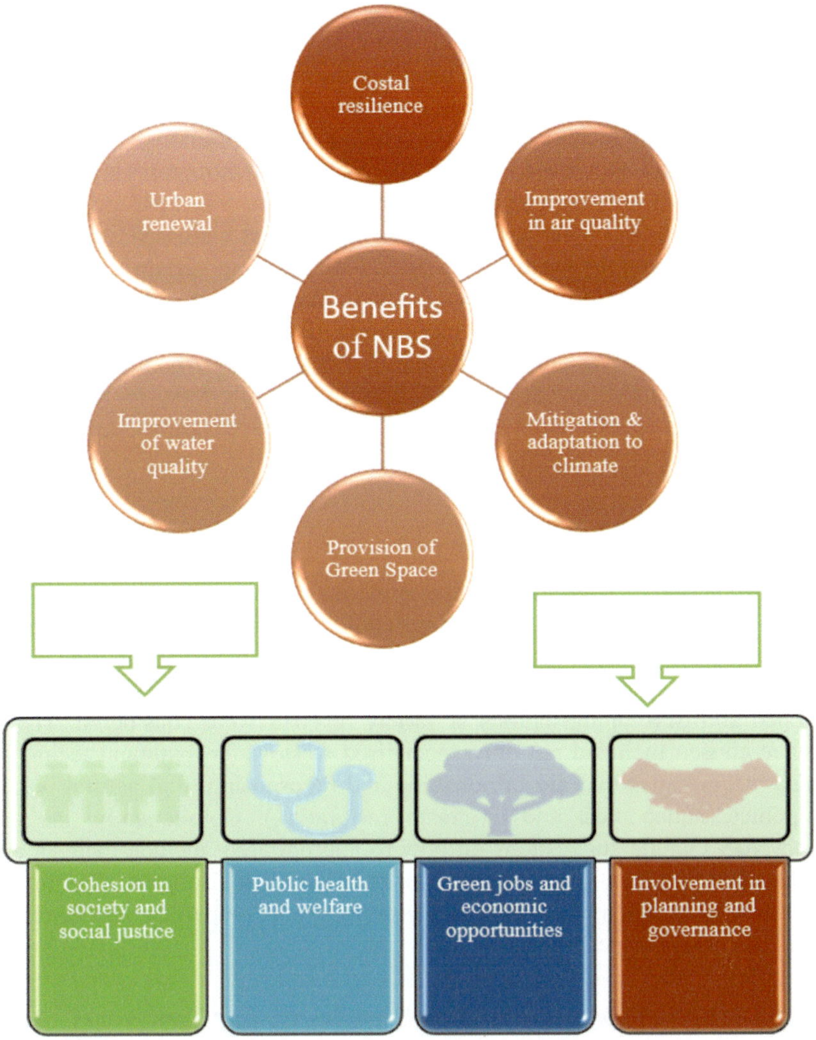

Figure 2: Co-benefits of natured-based solutions.

(ceramic) and organic (polymeric) (Kotobuki et al. 2021). Nonporous polymeric materials like cellulose acetate, polyethersulphone, polyimide, polysulphone, polymethylpentene, polycarbonate, polydimethylsiloxane, and polyphenylene oxide are the building blocks of organic membranes (Binazadeh et al. 2023, Ntone et al. 2023).

Intended Audience

The book's main target audience consists of managers and operators of wastewater utilities, municipal governments, authorities, and local governments. These populations can use this publication to learn about the NBS choices that they can include in their treatment plans and to obtain an idea of any possible co-benefits. These readers may be further empowered by the information supplied to conduct a preliminary cost-benefit analysis that considers the functionality and operation of various NBS in their locality. Aside from financial organisations, other significant audiences are those having a say in how urban infrastructure is planned and developed, such as land developers and urban planners. The NBS outlined in this book can support other objectives in urban design, such as increased liveability using green areas. Likewise, information is provided regarding certain co-benefits that NBS can assist with and that correspond with the broader socioeconomic objectives of financing organisations, such as international development banks. Furthermore, by using the information in this book, environmental groups and allied associations can better understand how various NBS apply to specific local project situations. Essential details and references for various NBS have been consolidated to benefit students and scholars.

Techniques

The NatureSan working group used the guidelines provided in the scoping section to select NBS for the treatment of household wastewater. Through a series of workshops, the types of NBS were decided, and the material required for each factsheet and case study was outlined. To consider those that can be used in both developed and developing nations, a variety of NBS were considered. The NBS that were chosen included not only specialised wastewater treatment technologies but also technologies that aid in polishing and repurposing (e.g. natural wetlands, hydroponics, aquaponics, and in-stream restoration). Consequently, the details of each NBS include suggestions on how to appropriately include them in an all-encompassing treatment plan. Members of the IWA Task Group on NBS for Water and Sanitation, as well as the NatureSan working group, prepared and peer-reviewed NBS factsheets and case studies.

Nature-based Solution Applications for Treating Wastewater

Wastewater treatment systems can be utilised to treat various types of wastewater, including stormwater, leachates, and industrial, agricultural, and municipal wastewater. The objective of employing NBS in wastewater treatment is to develop engineered systems that mimic and benefit from healthy ecosystems with minimal reliance on mechanical components. NBS eliminates contaminants from wastewater, such as suspended particles, organics, nitrogen, phosphorus, and pathogens, using natural elements and processes, including plants, soil, porous media, microbes, and other materials (Kadlec and Wallace 2009). Additionally, NBS can eliminate new contaminants, such as biocides and steroid hormones (Chen et al. 2019), pesticides (Vymazal and Březinová 2015), and personal care items (Ilyas et al. 2020). It is possible to combine various types of NBS to obtain the required level of therapeutic effectiveness. Enhancing the natural environment and adjacent habitats, as well as increasing water quality, are some of the ways in which using NBS for wastewater treatment can lead to healthier settings. Natural places and NBS can support physical and mental health, pure air, and clean water, and help boost human health. Moreover, NBS has therapeutic qualities and an attractive aesthetic, bringing people together and fostering a sense of community. Less expensive water treatment, reduced flood damage, healthier fisheries, better leisure options, increased tourism, and greater economic growth are among the financial gains. A comprehensive cost-benefit analysis is required when evaluating NBS solutions to take these benefits into account (El Zein et al. 2016, WWAP 2017). According to the European Investment Bank (2020), wastewater treatment operators can enhance customer engagement, reduce operating costs, generate new revenue streams, and offer public environmental goods and services by investing in NBS. Depending on the cost of land, the technologies employed, the availability of resources, and operation and maintenance expenses, initial expenditures are frequently less than those of conventional activated sludge (CAS) systems (El Zein et al. 2016, Vymazal 2010).

United Nations Sustainable Development Goals

Because they offer many advantages, such as food and energy security, human health and livelihoods, sustainable economic growth, and ecosystem restoration (Kumar et al. 2024), NBS is becoming increasingly recognized as creative approaches to managing water-related risks and supporting the 2030 Agenda for Sustainable Development (Martín et al. 2020). Several NBS services can help meet several Sustainable Development Goal (SDG) targets, including lowering greenhouse gas emissions and environmental pollutants, preserving a steady groundwater level, and even slowing global warming (Seifollahi et al. 2019). NBS for wastewater treatment is closely related to SDG 6: Sanitation and Clean Water. Notwithstanding, the advantages that NBS provide can differ among social groups and across different temporal and spatial scales; therefore, their contribution to different SDGs depends on the circumstances (Martín et al. 2020). Wetlands have the capacity to impact various ecosystem processes connected to multiple Sustainable Development Goals (SDGs), including those related to No Poverty (SDG 1), Zero Hunger (SDG 2), Clean Water and Sanitation (SDG 6), Responsible Consumption and Production (SDG 12), and Climate Action (SDG 13) (Seifollahi et al. 2019). Additionally, the location and application of Nature-Based Solutions (NBS) may contribute to achieving SDGs 3 (Good Health and Well-Being), 7 (Affordable and Clean Energy), 11 (Sustainable Cities and Communities), 14 (Life Below Water), and 15 (Life on Land) (Seifollahi et al. 2019).

Co-benefits of Treating Wastewater with NBS

Solutions based on nature (NBS) for wastewater treatment offer a range of benefits that extend beyond the primary objective of enhancing water quality. These benefits span the environmental, social, and economic dimensions. Below is a comprehensive breakdown of the co-benefits associated with utilising NBS for wastewater treatment. NBS, including constructed wetlands and riparian buffers, create habitats for a diverse range of species, support wildlife, and enhance ecological resilience. Vegetated systems, such as green roofs and wetlands, sequester carbon dioxide and moderate urban temperatures, mitigate the impacts of climate change, and reduce the urban heat island effect. NBS effectively remove nutrients, contaminants, and pathogens from wastewater, improve water quality, and safeguard aquatic ecosystems (Figure 2). Techniques such as riparian buffers stabilise soil, reduce erosion, protect water bodies from sedimentation, and maintain water quality. Green infrastructure, such as bioswales, manages stormwater runoff, reduces flood risk, and promotes groundwater recharge.

Water Treatment and Techniques and their Used Cases

Scientific Wastewater Treatment

Reusing wastewater for land application and irrigation significantly reduces possible pollutants that can contaminate the incoming water bodies, resulting from sewage being dumped on land instead of being released into the receiving water bodies. The water used for irrigation can be successfully replaced by wastewater applied to land (Fedler 1999, Fedler et al. 2008). Therefore, the burden of agricultural irrigation on natural water supplies can be reduced by applying wastewater to the land (Fedler et al. 2008). Furthermore, when wastewater is recycled as agricultural irrigation water, it can provide the soil with organic and inorganic elements, such as nitrogen and phosphate, which are employed as sources of fertilisers (Fedler et al. 2008). According to a contract signed in the 1930s, the City of Lubbock was required to pump all sewage effluents into Grey Farm (USEPA 1986). This involved applying 200 acres (or 80 hectares) of land to an average daily flow of 1 million gallons (MGD) of secondary processed effluent (Fedler 1999). This contract was set up because there is not enough rainfall in this area to sustain crops and groundwater is not always readily available. Furthermore, treating and disposing of the city's wastewater more cheaply is another benefit of this choice.

Wetlands Under Vertical-flow Treatment

With a population of 428,000, the Pingshan District is situated northeast of Shenzhen City in Guangzhou Province. The Pingshan River Basin comprises 77% of the district's total land area (129.4 km^2). Low water levels and sediment deposition in rivers characterise the subtropical oceanic climate of Shenzhen. The Pingshan River Basin is highly polluted because of the industrial and domestic effluent dumped into it by the surrounding industries. Beginning in 2011, industries began leaving the Pingshan River Basin. Between 2014 and 2018, eight vertical-flow treatment wetlands (VFTWs) were constructed to revive and restore the biological functions of Pingshan. Watershed Basin VFTWs, which have a 50-hectare total capacity, were built and used to treat wastewater from the Shangyang Wastewater Treatment Plant (WWTP), which has a daily treatment capacity of 1,365,000 m^3. The Pingshan District and other regions, such as Longgang District, are included in the service area of the Shangyang WWTP. Approximately 340,000 people are believed to live in the Shangyang WWTP catchment area. To satisfy the Grade VI requirement of the National "Environmental Quality Standards for Surface Water (GB3838-2002)," VFTWs were created as a polishing step. The Grade VI standard has the following limits: 1.5 mg/L for NH_4-N, 0.3 mg/L for total phosphorus (TP), 5 mg/L for dissolved oxygen, 1.5 mg/L for chemical oxygen demand (COD), and 6 mg/L for biochemical oxygen demand (BOD5).

Wetland with French Vertical-flow Treatment in Orhei Municipality, Mazovia

Because of its location on a hill, where wastewater from the city had to be piped, the original wastewater treatment plant (WWTP) in Orhei was fitted with a high-rate percolating filter, which proved exceedingly expensive. It was no longer sufficiently effective to treat the entire city. As a result, the Moldovan government chose to replace it with a French vertical-flow treatment wetland (French VFTW) through a feasibility study and World Bank finance project. After comparing treatment wetlands (TWs) with other technologies (activated sludge, sequencing batch reactors, and percolating filters), World Bank consultants selected a French VFTW to minimise operating costs using the most affordable water tariff available in the local economic context. The World Bank supported and provided funding for the design of the Orhei French VFTW, overseen during construction by a multinational joint venture made up of Posch and Partners (Austria), SWS Consulting, Iridra, and Hydea (Italy). The World Bank, Moldovan Environmental Ministry, and European Union contributed to the realisation of the plant. The Project Implementation Unit awarded the German joint venture Heilit-BioPlanta the construction contract for the system.

Combined Sewer Overflow Treatment Wetland in Kenten, Germany

The capacities of combined sewage systems and linked wastewater treatment plants (WWTPs) are constrained by specific design parameters, such as double the flow that occurs on a typical day without rain, known as "dry weather flow." If this capacity is surpassed, specific locations in the sewer network must release untreated mixed sewage and stormwater into the surface water bodies. Storage basins that catch the sewer leak and divert it to the WWTP following a rainy event are conventional solutions. If their volume is surpassed, pre-settled diluted wastewater is released into the surface water. Treatment wetlands for combined sewer overflows (CSO-TWs) offer additional storage capacity and quick sewer spill treatment, which helps lessen this issue. The CSO-TW described in this study is situated across the Bergheim-Kenten WWTP in a peri-urban area outside Bergheim. Before the implementation of the CSO-TW, surplus water from the sewage network was collected in two stormwater basins at the Bergheim-Kenten WWTP site and directed for treatment following a rainstorm event. When precipitation events persisted, the overflow of the storage basins was released into the River Erft. The biological state of rivers is greatly affected by pollution from CSO discharge, which also interferes with the objectives of the European Water Framework Directive. For these reasons, in

2003, the "Erftverband" implemented more than 30 CSO-TWs, including the one in Kenten. To enhance the quality of the river water, the public water association oversees the 1,900 km^2 catchment area that runs along 106.6 km of the Erft River. In the German state of North-Rhine Westphalia, where this case study is situated, the Ministry of Environment has provided financial support for installing CSO-TWs for more than ten years.

Tidal-flow Treatment Reciprocating Wetland Demonstration, Hawaii, USA

Through the development of diverse microbial biofilms and a broad continuum of biologically mediated treatment environments, reciprocating wetlands, a specific subset and precursor of tidal flow and fill-and-drain wetlands, have been shown to accelerate and enhance wastewater treatment processes (Austin and Nivala 2009, Nivala et al. 2019, Behrends and Choperena 2012). The benefits of this technology include decentralisation, low cost, optimised nitrogen removal, and reuse of treated wastewater. Among the reuse alternatives are the aquaculture of baitfish, irrigation of fodder crops, subsurface irrigation of landscape plants, and toilet flushing. Researchers at the Tennessee Valley Authority have created modular, scalable, and energy-efficient reciprocating subsurface-flow treatment wetland (TW) systems that improve both aerobic (with oxygen) and anoxic (without oxygen) environments. Based on this plan, the U.S. Department of Defense funded a commercial-scale reciprocating wetland demonstration that was run and observed for two years to assess the suitability of the reciprocating technology for the decentralised treatment of municipal sanitary wastewater. A system design was created based on a wastewater loading rate of 227 m^3 per day (60,000 gallons per day) or a 3-day hydraulic retention time. On the Hawaiian island of Oahu, the treatment facility is situated to the north of the city of Wahiawa. Wastewater treatment operations started in December 2000, and the effectiveness of the treatment was checked weekly for 114 weeks.

Forced Aeration in Intensified Treatment Wetlands in Tarcenay, France

Domestic wastewater is effectively treated by treatment wetlands (TWs). As there is typically no problem with space, they have taken over as the primary technology used in rural areas of France. However, the issue of space constraints for the construction of a new treatment plant emerges when considering larger treatment capacities or in the case of facility retrofitting. Strict outflow regulations make this worse, necessitating several types and phases of treatment wetlands. In this situation, enhanced TWs appear to be a good substitute for saving building expenses because they require less material. Upscaling and retrofitting the Tarcenay (old pond) wastewater treatment plant (WWTP) is necessary to comply with the increased outflow requirements. In this case, a phosphorus removal filter utilising apatite was used after implementation of a one-stage TW system with forced aeration (Rhizosph'air). Two parts make up the Rhizosph'air method (patented by Syntea, Naturally Wallace, and Rietland): a horizontally saturated filter with forced aeration after a vertically unsaturated filter that receives raw wastewater. With a nominal daily flow of 293 m^3, it is a single-stage

TW that accepts raw wastewater and is built for 1,400 population equivalents (p.e.). 15 mg of biological oxygen demand (BOD5)/L, 90 mg of chemical oxygen demand (COD)/L, 20 mg of total suspended solids (TSS)/L, 15 mg of Kjeldahl nitrogen/L, and 1.5 mg of phosphorus/L are the requirements for the outlet. Total nitrogen (TN) is not necessary, although INRAE (previously Irstea) conducted monitoring in 2018 and 2019, with the goal of optimising aeration cycles for better TN performance.

Conclusion

Nature-based solutions (NBS) in wastewater treatment offer a promising alternative to conventional methods, leveraging natural processes and ecosystems to purify water. These solutions, such as constructed wetlands, green roofs, and bioreactors, have demonstrated considerable potential in providing cost-effective, sustainable, and environmentally friendly wastewater treatment options. They can enhance biodiversity, improve water quality, and mitigate climate change impacts by reducing the greenhouse gas emissions associated with traditional wastewater treatment plants. However, the limitations of NBS should also be considered. Various factors, including local climatic conditions, the scale of the treatment system, and the characteristics of the wastewater, can influence their effectiveness. Additionally, implementing an NBS requires significant land area and ongoing maintenance to ensure optimal performance. Integrating NBS into urban settings can also present space constraints and challenges for public acceptance. In conclusion, although nature-based solutions offer a viable and ecologically sustainable approach to wastewater treatment, they are not a one-size-fits-all solution. Successful implementation of NBS requires careful planning, site-specific considerations, and often a hybrid approach that combines natural and conventional treatment methods. By addressing these limitations and optimising their design and management, NBS can be crucial for advancing sustainable water management practices.

References

Ahmed, S.F., Mehejabin, F., Momtahin, A., Tasannum, N., Faria, N.T., Mofijur, M. et al 2022a. Strategies to improve membrane performance in wastewater treatment. Chemosphere, 306: 135527.

Ahmed, S.F., Mofijur, M., Nuzhat, S., Chowdhury, A.T., Rafa, N., Uddin, M.A. et al. 2021b. Recent developments in physical, biological, chemical, and hybrid treatment techniques for removing emerging contaminants from wastewater. Journal of Hazardous Materials, 416: 125912.

Antalová, V., Slučiaková, S. and Haluš, M. 2018. Estimating environmental benefits of wastewater treatment in Slovakia. Economic Analysis, 3.

Austin, D. and Nivala, J. 2009. Energy requirements for nitrification and biological nitrogen removal in engineered wetlands. Ecological Engineering, 35(2): 184-192.

Baker, F., Smith, G.R., Marsden, S.J. and Cavan, G. 2021. Mapping regulating ecosystem service deprivation in urban areas: A transferable high-spatial resolution uncertainty aware approach. Ecological Indicators, 121: 107058.

Behrends, L.L. and Choperena, J. 2012. Tidal-flow constructed wetlands (TFW), for treating and reusing high strength animal production wastewater. *In*: 2012 Dallas, Texas, July 29-August 1, 2012 (p. 1). American Society of Agricultural and Biological Engineers.

Binazadeh, M., Rasouli, J., Sabbaghi, S., Mousavi, S.M., Hashemi, S.A. and Lai, C.W. 2023. An overview of photocatalytic membrane degradation development. Materials, 16(9): 3526.

Brears, R.C. 2020. Nature-based Solutions to 21st Century Challenges. Routledge.

Cai, Y., Chen, D., Li, N., Xu, Q., Li, H., He, J. et al. 2020. A self-cleaning heterostructured membrane for efficient oil-in-water emulsion separation with stable flux. Advanced Materials, 32(25): 2001265.

Carpenter, S.R., Mooney, H.A., Agard, J., Capistrano, D., DeFries, R.S., Díaz, S. et al. 2009. Science for managing ecosystem services: Beyond the millennium ecosystem assessment. Proceedings of the National Academy of Sciences, 106(5): 1305-1312.

Chen, B., Yang, S., Cao, Q. and Qian, Y. 2020. Life cycle economic assessment of coal chemical wastewater treatment facing the 'Zero liquid discharge' industrial water policies in China: Discharge or reuse? Energy Policy, 137: 111107.

Chen, J., Liu, Y.S., Deng, W.J. and Ying, G.G. 2019. Removal of steroid hormones and biocides from rural wastewater by an integrated constructed wetland. Science of the Total Environment, 660: 358-365.

Cohen-Shacham, E., Walters, G., Janzen, C. and Maginnis, S. 2016. Nature-based solutions to address global societal challenges. IUCN: Gland, Switzerland, 97: 2016-2036.

Cossio, C., McConville, J., Rauch, S., Wilén, B.M., Dalahmeh, S. and Mercado, A. 2018. Wastewater management in small towns – Understanding the failure of small treatment plants in Bolivia. Environmental Technology, 39(11): 1393-1403.

Cragg, G.M., Katz, F., Newman, D.J. and Rosenthal, J. 2012. The impact of the United Nations Convention on biological diversity on natural products research. Natural Product Reports, 29(12): 1407-1423.

Crini, G. and Lichtfouse, E. 2019. Advantages and disadvantages of techniques used for wastewater treatment. Environmental Chemistry Letters, 17: 145-155.

El Moukhtari, F., Martín-Pozo, L. and Zafra-Gómez, A. 2023. Strategies based on the use of microorganisms for the elimination of pollutants with endocrine-disrupting activity in the environment. Journal of Environmental Chemical Engineering, 11(1): 109268.

El Zein, Z.A.K.A.A., Abdou, A. and Abd ElGawad, I. 2016. Constructed wetlands as a sustainable wastewater treatment method in communities. Procedia Environmental Sciences, 34: 605-617.

Fedler, C.B. (1999, August). Long-term land application of municipal wastewater—A case study. *In*: Proceedings of the ASCE International Water Resources Engineering Conference, Seattle, WA, USA (pp. 8-11).

Fedler, C.B., Duan, R., Borrelli, J. and Green, C. 2008. Design & Operation of Land Application Systems from a Water, Nitrogen & Salt Balance Approach. https://shorturl.at/jXnnj

Filippelli, G.M. 2008. The global phosphorus cycle: Past, present, and future. Elements, 4(2): 89-95.

Geng, Y., Zhu, Q., Doberstein, B. and Fujita, T. 2009. Implementing China's circular economy concept at the regional level: A review of progress in Dalian, China. Waste Management, 29(2): 996-1002.

Gruber, N. and Galloway, J.N. 2008. An Earth-system perspective of the global nitrogen cycle. Nature, 451(7176): 293-296.

Haines-Young, R. and Potschin, M. 2012. Common international classification of ecosystem services. Nottingham: Centre for Environmental Management, University of Nottingham.

Hakak, S., Khan, W.Z., Gilkar, G.A., Haider, N., Imran, M. and Alkatheiri, M.S. 2020. Industrial wastewater management using blockchain technology: Architecture, requirements, and future directions. IEEE Internet of Things Magazine, 3(2): 38-43.

Hasan, H.A. and Muhammad, M.H. 2020. A review of biological drinking water treatment technologies for contaminants removal from polluted water resources. Journal of Water Process Engineering, 33: 101035.

Heisler, J., Glibert, P.M., Burkholder, J.M., Anderson, D.M., Cochlan, W., Dennison, et al. 2008. Eutrophication and harmful algal blooms: A scientific consensus. Harmful Algae, 8(1): 3-13.

Hickmann, T., Widerberg, O., Lederer, M. and Pattberg, P. 2021. The United Nations Framework Convention on Climate Change Secretariat as an orchestrator in global climate policymaking. International Review of Administrative Sciences, 87(1): 21-38.

Hreiz, R., Latifi, M.A. and Roche, N. 2015. Optimal design and operation of activated sludge processes: State-of-the-art. Chemical Engineering Journal, 281: 900-920.

Hu, Z., Houweling, D. and Dold, P. 2012. Biological nutrient removal in municipal wastewater treatment: New directions in sustainability. Journal of Environmental Engineering, 138(3): 307-317.

Huang, Y., Tian, Z., Ke, Q., Liu, J., Irannezhad, M. and Fan, D. 2020. Nature-based solutions for urban pluvial flood risk management. Wiley Interdisciplinary Reviews: Water, 7(3): e1421.

Huntington, T.G. 2006. Evidence for intensification of the global water cycle: Review and synthesis. Journal of Hydrology, 319(1-4): 83-95.

Ilyas, H., Masih, I. and van Hullebusch, E.D. 2020. Pharmaceuticals' removal by constructed wetlands: A critical evaluation and meta-analysis on performance, risk reduction, and role of physicochemical properties on removal mechanisms. Journal of Water and Health, 18(3): 253-291.

International Organization for Standardization 2018. ISO 2670 Water reuse—Vocabulary. https://www.iso.org/ obp/ui/#iso:std:iso:20670:ed-1:v1:en (accessed March 15th, 2021).

Jorgenson, A.K. 2009. Political-economic integration, industrial pollution and human health: A panel study of less-developed countries, 1980-2000. International Sociology, 24(1): 115-143.

Kadlec, R.H. and Wallace, S.D. 2009. Treatment Wetlands. 2nd Edition. CRC Press.

Kamboj, N., Kumar, R., Kumar, N. and Gautam, P. 2024. Potential and limitation of peptides from native plants of Uttarakhand. International Journal of Peptide Research and Therapeutics, 30(5): 55.

Kaushal, S.S., Groffman, P.M., Likens, G.E., Belt, K.T., Stack, W.P., Kelly, V.R. et al. 2005. Increased salinization of fresh water in the northeastern United States. Proceedings of the National Academy of Sciences, 102(38): 13517-13520.

Kim, S.Y. and Geary, P.M. 2001. The impact of biomass harvesting on phosphorus uptake by wetland plants. Water Science and Technology, 44(11-12): 61-67.

Kotobuki, M., Gu, Q., Zhang, L. and Wang, J. 2021. Ceramic-polymer composite membranes for water and wastewater treatment: Bridging the big gap between ceramics and polymers. Molecules, 26(11): 3331.

Kumar, R., Kamboj, N. and Mitra, D. 2024. Microbial innovations in agriculture: Interdisciplinary approaches to leveraging microbes for food sustainability and security. Indian J Microbiol Res, 11(3): 129-139.

Larsen, T.A., Maurer, M., Udert, K.M. and Lienert, J. 2007. Nutrient cycles and resource management: Implications for the choice of wastewater treatment technology. Water Science and Technology, 56(5): 229-237.

Lohse, K.A., Brooks, P.D., McIntosh, J.C., Meixner, T. and Huxman, T.E. 2009. Interactions between biogeochemistry and hydrologic systems. Annual Review of Environment and Resources, 34: 65-96.

Martín, E.G., Giordano, R., Pagano, A., Van Der Keur, P. and Costa, M.M. 2020. Using a system thinking approach to assess the contribution of nature based solutions to sustainable development goals. Science of the Total Environment, 738: 139693.

Meeks, D. 2022. A Survey of Regulatory, Technical, and Public Outreach Challenges and Opportunities for Direct Potable Reuse with an Emphasis on Tucson and Pima County, Arizona.

Mesa-Pérez, E. and Berbel, J. 2020. Analysis of barriers and opportunities for reclaimed wastewater use for agriculture in Europe. Water, 12(8): 2308.

Metcalf and Eddy, Inc, Asano, T., Burton, F.L., Leverenz, H., Tsuchihashi, R. and Tchobanoglous, G. 2007. Water Reuse (pp. 6-15). United States of America: McGraw-Hill Professional Publishing.

Nivala, J., Kahl, S., Boog, J., van Afferden, M., Reemtsma, T. and Müller, R.A. 2019. Dynamics of emerging organic contaminant removal in conventional and intensified subsurface flow treatment wetlands. Science of the Total Environment, 649: 1144-1156.

Ntone, E.P.N., Rahman, S.A., Abdul Wahab, M.S., Samah, R.A. and Ahmad, A.L. 2023. A mini review on membrane potential for pharmaceutical and personal care product (PPCP) removal from water. Water, Air, & Soil Pollution, 234(7): 412.

Obotey Ezugbe, E. and Rathilal, S. 2020. Membrane technologies in wastewater treatment: A review. Membranes, 10(5): 89.

Oki, T. and Kanae, S. 2006. Global hydrological cycles and world water resources. Science, 313(5790): 1068-1072.

Onu, M.A., Ayeleru, O.O., Oboirien, B. and Olubambi, P.A. 2023. Challenges of wastewater generation and management in sub-Saharan Africa: A review. Environmental Challenges, 11: 100686.

Post, V.E.A. 2005. Fresh and saline groundwater interaction in coastal aquifers: Is our technology ready for the problems ahead? Hydrogeology Journal, 13, 120-123.

Priyadarshini, M., Sathe, S.M. and Ghangrekar, M.M. 2023. Hybrid treatment solutions for removal of micropollutant from wastewaters. Microconstituents in the Environment: Occurrence, Fate, Removal and Management, 491-512.

Rodriguez, D.J., Serrano, H.A., Delgado, A., Nolasco, D. and Saltiel, G. 2020. From waste to resource. Water Papers. World Bank https://doi. org/10.1596/33436.

Ryder, G. 2017. The United Nations World Water Development Report, 2017. Wastewater: The Untapped Resource.

Saad, L. 2009. Water pollution Americans' top green concern. Retrieved September, 23, 2011.

Samantaray, A., Chattaraj, S., Mitra, D., Ganguly, A., Kumar, R., Gaur, A. et al. 2024. Advances in microbial based bio-inoculum for amelioration of soil health and sustainable crop production. Current Research in Microbial Sciences, 100251.

Sangamnere, R., Misra, T., Bherwani, H., Kapley, A. and Kumar, R. 2023. A critical review of conventional and emerging wastewater treatment technologies. Sustainable Water Resources Management, 9(2): 58.

Sawunyama, L., Ajiboye, T.O., Oyewo, O. and Onwudiwe, D.C. 2024a. Ceramic-polymer composite membranes: Synthesis methods and environmental applications. Ceramics International, 50(3): 5067-5079.

Sawunyama, L., Oyewo, O., Onwudiwe, D.C. and Makgato, S.S. 2024b. Photocatalytic degradation of tetracycline using surface defective black TiO_2–ZnO heterojunction photocatalyst under visible light. Heliyon, 9(11).

Saxena, R., Saxena, M. and Lochab, A. 2020. Recent progress in nanomaterials for adsorptive removal of organic contaminants from wastewater. ChemistrySelect, 5(1): 335-353.

Schwarzenbach, R.P., Escher, B.I., Fenner, K., Hofstetter, T.B., Johnson, C.A., Von Gunten, U. et al. 2006. The challenge of micropollutants in aquatic systems. Science, 313(5790): 1072-1077.

Seifollahi-Aghmiuni, S., Nockrach, M. and Kalantari, Z. 2019. The potential of wetlands in achieving the sustainable development goals of the 2030 Agenda. Water, 11(3): 609.

Shindhal, T., Rakholiya, P., Varjani, S., Pandey, A., Ngo, H.H., Guo, W. et al. 2021. A critical review on advances in the practices and perspectives for the treatment of dye industry wastewater. Bioengineered, 12(1): 70-87.

Singh, R. and Hankins, N. (Eds.). 2016. Emerging Membrane Technology for Sustainable Water Treatment. Elsevier.

Sun, M., Wang, X., Winter, L.R., Zhao, Y., Ma, W., Hedtke, T. et al. 2021. Electrified membranes for water treatment applications. ACS ES&T Engineering, 1(4): 725-752.

US Environmental Protection Agency. 2021. Basic Information about Biosolids. https://www.epa.gov/ biosolids/basic-information-about-biosolids (accessed March 2nd, 2021).

Vymazal, J. 2010. Constructed wetlands for wastewater treatment. Water, 2(3): 530-549.

Vymazal, J. and Březinová, T. 2015. The use of constructed wetlands for removal of pesticides from agricultural runoff and drainage: A review. Environment International, 75: 11-20.

Wakelin, S.A., Colloff, M.J. and Kookana, R.S. 2008. Effect of wastewater treatment plant effluent on microbial function and community structure in the sediment of a freshwater stream with variable seasonal flow. Applied and Environmental Microbiology, 74(9): 2659-2668.

Wan Rosely, W.I.H. and Voulvoulis, N. 2023. Systems thinking for the sustainability transformation of urban water systems. Critical Reviews in Environmental Science and Technology, 53(11): 1127-1147.

World Water Assessment Programme (United Nations), UN-Water. 2009. Water in a Changing World. https://shorturl.at/W8eXy

WWAP, U. 2017. WWAP (United Nations World Water Assessment Programme).

Zvimba, J.N., Musvoto, E.V., Nhamo, L., Mabhaudhi, T., Nyambiya, I., Chapungu, L. et al. 2021. Energy pathway for transitioning to a circular economy within wastewater services. Case Studies in Chemical and Environmental Engineering, 4: 100144.

CHAPTER

11

Omitting Antibiotic Pollution via Novel Ecofriendly Approaches: A Green Waste Water Management

Himani Deepak* **and Verinder Virk**

Department of Microbiology, Kanya Gurukula Campus, Gurukula Kangri (Deemed to be University) Haridwar, Uttarakhand, India

Introduction

Antibiotics are crucial in medicine, agriculture, aquaculture, and animal husbandry. However, both prescribed and non-prescribed, global antibiotic consumption has been steadily increasing. As a result, large quantities of antibiotics and their byproducts end up in wastewater, posing risks to human and aquatic life. Conventional water treatment plants may not completely remove these pollutants due to their unique properties. Various techniques like physicochemical processes, advanced oxidation processes, and bioremediation involving fungi, algae, and bacteria are being explored to address this issue. Recent advancements in omics, nanotechnology, electrokinetics, and electrochemical processes are also being considered for antibiotic degradation in water sources (Gul et al. 2022). Approximately, 700,000 annual deaths occur due to antibiotic-resistant infections worldwide as a result of antibiotic-resistant bacteria (Duarte et al. 2022). In parallel, antibiotic resistance genes (ARGs) and antibiotic-resistant bacteria (ARBs) can have environmental dispersion, which is why they have been recognized as emerging environmental contaminants (Zhang et al. 2021). Antibiotics are commonly found in surface water in East Asian countries such as China and South Korea, with concentrations ranging from <1 ng/L to hundreds of μg/L. Wastewater treatment plants, aquaculture, livestock production, and urban sewage are the main sources of antibiotic pollution (Anh et al. 2021). Antibiotics are now recognized as significant pollutants found in high concentrations in various environments worldwide, including water sources and soil. This presence can lead to ecological risks for aquatic life and contribute to the development of antibiotic

*Corresponding author: hdeep04@gmail.com

DOI: 10.1201/9781003527350-11

resistance in bacteria. Developing countries face particular challenges in managing this issue effectively, raising concerns about the impact on aquatic ecosystems (Lyu et al. 2020, Kovalakova et al. 2020).

The presence of antibiotic residues in surface waters can impact microbial communities by increasing mutation rates, promoting horizontal gene transfer, and selecting antibiotic-resistant bacteria. Various sectors like human medicine, agriculture, aquaculture, and veterinary medicine contribute to this selective pressure, developing drug resistance in bacteria (Atterby et al. 2021). Antimicrobial resistance determinants, often found in wastewater, pose a significant environmental threat by potentially spreading drug resistance. The overuse and misuse of antibiotics contribute to the development of antibiotic-resistant bacteria and genes (Vaughn et al. 2022). Rivers polluted with these bacteria and genes increase the risk of antibiotic resistance transmission to humans and disrupt the ecological balance of aquatic environments (Mounzer et al. 2024). Antibiotic resistance genes, both extracellular and intracellular, play a key role in spreading resistance. Horizontal gene transfer enables the movement of these genes between different bacterial species, with water acting as a major pathway for dissemination, potentially impacting human health (Koniuszewska et al. 2020).

Horizontal transfer of antibiotic resistance genes (ARGs) is a key mechanism for spreading antibiotic resistance. Pathogenic cells can acquire resistance from benign cells via conjugation, transduction, and natural transformation (Miller et al. 2016). Natural transformation plays a significant role in the dissemination of antibiotic resistance genes. Monitoring the frequency of horizontal gene transfer (HGT) mechanisms is crucial to assess the risk posed by intra- and environment-derived ARGs in receiving environments (Zhang et al. 2018). Antibiotics infiltrate groundwater through leaching, surface water-groundwater interaction, and pipe network leakage. In urban areas, population density, road complexity, and aging infrastructure worsen pollution, making it more complex and difficult to manage (Liu et al. 2024).

The overuse of antibiotics has resulted in the rise of antibiotic-resistant bacteria (ARB) and genes (ARGs) in the environment, posing a threat to ecosystems and human health. Even minimal exposure to antibiotics can lead to resistance in pathogenic bacteria, causing harm to disease prevention efforts (Li et al. 2023a). In addition to posing a threat to humankind directly, antibiotics taken above the recommended dosage can result in allergies, poisoning, and even death. On the other hand, excessive intake of antibiotics imbalances the gut microbes, which can lead to intestinal diseases (Owusu-Doubreh et al. 2023). Antibiotic concentrations found in China's coastal region range from 33.14 to 417.78 ng L^{-1} in rivers and 0.90 to 86.33 ng L^{-1} at particular locations. The major antibiotics found in the coastal regions include roxithromycin, lincomycin, azithromycin, and sulfamethoxazole. The increased antibiotic levels directly correlate to chemical oxygen demand, total phosphorous and nitrogen concentrations (Xu et al. 2023).

Conventional methods for removing antibiotics from the environment include physical (filtration and sedimentation), chemical (chlorination and ozonation), and biological techniques (activated sludge and constructed wetlands). However, conventional methods impose some disadvantages, including high cost, ineffectiveness, incomplete antibiotic degradation and generation of antibiotic resistance bacteria (Michael et al. 2013, Ahmed et al. 2017, Luo et al. 2024). Innovative methods for removing antibiotic pollution offer promising alternatives to traditional techniques. Utilizing advanced technologies and natural processes, techniques like advanced oxidation processes (AOPs) and bioremediation are achieving high degradation rates of antibiotics (Fabragat and Pagan 2024). Adsorption techniques with special adsorbents, membrane technologies, and electrochemical methods also play key roles in effectively capturing and removing antibiotic contaminants from water, ultimately reducing environmental impact (Najafi 2024).

Conventional Methods for Antibiotic Removal

Physical Methods

The most common physical treatment methods for antibiotic removal include adsorption, ionic resin, and membrane filtration. Adsorption involves physical and chemical interactions like van der Waals forces, π-π interactions, hydrophobic interactions, and hydrogen bonds. Chemical adsorption occurs when electrons are transferred, exchanged, or shared between adsorbate molecules and adsorbent surface atoms (or molecules), resulting in adsorption chemical bonds (Huang et al. 2022). Antibiotic resistance genes (ARGs) in swine wastewater are a growing environmental concern. Integrated membrane filtration has been used to treat swine wastewater, reducing conventional pollutants and 99.79% of ARGs. Quantitative real-time PCR was used to track ARG levels during filtration, showing a significant reduction (Liang et al. 2021). Conventional wastewater treatment technologies like activated sludge and chlorination have low removal efficiency for β-lactams and macrolides (less than 50%). In comparison, tetracyclines can reach 80%, while sulfonamides are only about 20% (Sousa et al. 2018). Tetracycline (TC) is a common antibiotic used to treat bacterial infections, but its residues are difficult to degrade in wastewater. Peroxymonosulfate (PMS) activation-based advanced oxidation processes (AOPs) have shown promise in removing TC due to their cost-effectiveness and high oxidation efficiency (Chen et al. 2024).

Factors such as solution pH, initial antibiotic concentration, and adsorbent dose have a major impact on antibiotic removal because they affect the charges of the adsorbent and adsorbate. Higher starting antibiotic concentrations reduced removal efficiency, but more adsorbent boosted it (Sharma et al. 2017). Adsorption is a useful method to remove antibiotics from water using solid materials that attract and bind pollutants through physical or weak chemical interactions. The selection of the right adsorbent depends on water conditions, types of antibiotics, and desired water quality. More research is needed to develop new adsorbents for the effective removal of pharmaceutical products under different conditions (Pourhakkak et al. 2021).

Chemical Method

Chemical treatments for antibiotic removal involve various oxidative and reductive processes designed to degrade and eliminate antibiotic compounds from water (Fig. 1). Chlorination is one of the most widely used methods, where chlorine is

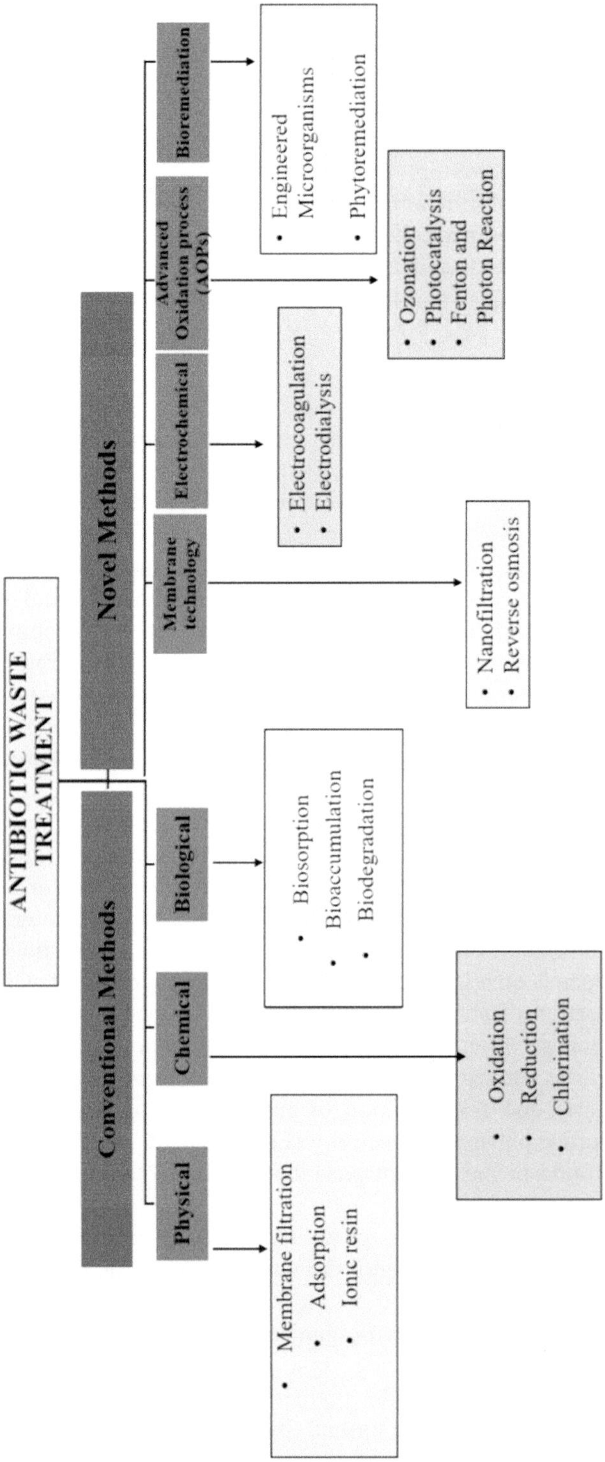

Figure 1: Methods for removing antibiotic pollution.

used to oxidize and break down antibiotics (Wiesch et al. 2017). However, this process can lead to the formation of harmful disinfection byproducts and may not be effective against all types of antibiotics. Ozone (O_3) is an effective method to remove fishery antibiotics, including florfenicol, oxytetracycline, sulfadimethoxine, sulfamethoxazole, and trimethoprim from aquaculture effluents to prevent their harmful effects on receiving waters (Gorito et al. 2022). Although chemical methods, including ozonation and chlorination, have promising results in eliminating degraded antibiotics from the environment, they possess several disadvantages, including the generation of toxic byproducts and disruption of the local ecosystem. Also, chemical methods were found to reduce soil fertility and soil structure and disrupt beneficial soil microbiota.

Biological Methods

Biological treatment is a technology that uses microorganisms to break down organic pollutants in water into harmless substances. It relies on biodegradation and biosorption mechanisms to remove antibiotics from water. This process mimics environmental self-purification, enhancing water quality (Khalil and Liu 2021). The main mechanism for antibiotic removal includes biodegradation and bioadsorption. The main genera found in degrading sulfonamides and tetracyclines are Paenarthrobacter and Hydrogenophaga, respectively, efficiently removing sulfonamides and tetracyclines at 16 mg/L (Li et al. 2024b). Antibiotic pollution in surface waters is common and can bioaccumulate in phytoplankton, inhibiting algal growth (Wang et al. 2017). High doses of antibiotics can impact metabolism, inhibiting photosynthesis through various mechanisms. Erythromycin and chlortetracycline have been shown to inhibit photosynthetic activity in specific microalgae species. Antibiotics can also induce oxidative stress in microalgae and inhibit protein synthesis in chloroplasts. Some antibiotics, at lower concentrations, have been shown to promote algae growth through cytoprotection and antioxidant enzyme enhancement (Aderemi et al. 2018). Antibiotic removal through adsorption relies on hydrophobic and electrostatic interactions, while antibiotic removal via biodegradation depends on releasing extracellular enzymes. Algae play a crucial role in the biomagnification of antibiotics and other organic pollutants in aquatic food chains through bioadsorption, bioaccumulation, and biodegradation processes. A study investigated the behavior of norfloxacin, sulfamethazine, and roxithromycin in Chlorella vulgaris; the concentration of antibiotics in the extracellular matter increased with alkaline phosphatase activity (Long et al. 2024). The main antibiotic resistance genes found in the effluent are *intI1*, *sul1*, *sul2*, *tetA*, *tetW* and *tetX* (Li et al. 2024a).

Although conventional methods for antibiotic pollution offer eco-friendly and cost-effective solutions, they have major drawbacks. One of the major challenges is the incomplete degradation of certain antibiotics resistant to microbial breakdown. Antibiotics can hinder beneficial microorganisms' growth and treatment efficiency. Treatment methods like activated sludge systems and wetlands require large areas and time. Environmental factors like temperature and pH can affect treatment effectiveness. Phytoremediation is limited by plant types and their ability to accumulate antibiotics. Careful monitoring and integration with other technologies are necessary for reliable antibiotic removal.

Novel Approaches to Antibiotic Removal

Recently, various innovative technologies have been developed to address antibiotic pollution. Advanced oxidation processes like photocatalysis and Fenton reactions are effective in degrading antibiotics rapidly. Bioremediation with engineered microorganisms and phytoremediation with genetically modified plants offer targeted solutions. Meanwhile, adsorption techniques using nanomaterials ensure efficient capture of antibiotics. Membrane technologies like nanofiltration and reverse osmosis enable precise separation, while electrochemical methods like electrocoagulation and electrodialysis offer energy-efficient solutions. These advancements improve efficiency and sustainability in combating antibiotic pollution.

Antibiotics are a major concern in aquatic environments due to antibiotic resistance and bioaccumulation toxicity. Magnetic nanocomposites (BC-$NiFe_2O_4$) using nickel ferrite nanoparticles on activated biochar derived from banana peel effectively degrade ciprofloxacin antibiotic pollution from the pharmaceutical effluent. The adsorption capacity of the BC-$NiFe_2O_4$ heterostructure indicates that magnetically loading $NiFe_2O_4$ nanoparticles onto the surface of porous biochar improved its stability and adsorption performance for CIP removal, with the metal-antibiotic complex having a significant effect on CIP removal (Azam et al. 2023). The major interactions involved in bioadsorption include π-π stacking and H-bonding interactions.

Membrane Technologies for Treatment of Antibiotic Pollution

Antibiotics play a crucial role in improving public health, but their presence in wastewater from pharmaceutical industries is a major environmental concern. Conventional wastewater treatment methods are not effective in completely removing antibiotics. Unmetabolized antibiotics in wastewater pose a threat to the environment. A recent study explores various techniques, including membrane technology, to effectively remove antibiotics from wastewater, highlighting the importance of hybrid systems for optimal removal. Various treatment techniques have been tested for removing antibiotics, including biological aerated filters, anaerobic digestion, biologically active carbon filters, etc. (Nasrollahi et al. 2022). Recently, membrane technologies, e.g., nanofiltration and reverse osmosis, are gaining significant attention in eliminating 80–100% of antibiotic waste pollution from the natural environment (de Illurdoz et al. 2022).

Nanofiltration

One of the most important membrane technologies is nanofiltration and reverse osmosis (NF/RO), which are used for wastewater treatment due to their high efficiency in removing trace contaminants. The membranes show promise in controlling antibiotics and estrogens. While some studies suggest rejection is based on hydrophobic adsorption followed by size exclusion, recent findings show that rejection mechanisms are complex and may vary depending on factors such as molecular size and pH levels. NF/RO membranes play a crucial role in effectively treating water by removing contaminants such as antibiotics and estrogens (Yang et al. 2024). Nanostructured materials possess unique properties like increased adsorption capacity, chemical stability, reactivity, mobility, and antimicrobial activity and are versatile for contaminant removal (Moreira et al. 2022). A pilot-scale double membrane system consisting of a membrane bioreactor (MBR) and nanofiltration (NF) membrane effectively treated antibiotic wastewater at a pharmaceutical company in Wuxi, China. Antibiotic removal rates for spiramycin (SPM) and new spiramycin exceeded 95%. At the same time, organics analysis identified proteins, soluble microbial by-products, fulvic acid-like, and humic substances as the main components in the biological effluent, which were effectively rejected by the NF membrane (Wang et al. 2014). Nanofiltration (NF) is an efficient membrane process known for its molecular separation abilities and low energy consumption (Cao et al. 2020, Wang et al. 2023). NF membranes can retain antibiotics while allowing ions and water to pass through, reducing energy costs in treating complex wastewater (Bai et al. 2023).

Reverse Osmosis

Reverse osmosis (RO) is one of the most effective membrane technologies for removing antibiotics from water because it can filter out very small molecules and ions. RO membranes are commonly used for water reclamation systems, particularly for industrial consumers needing high-quality reclaimed wastewater like boiler feedwater and semiconductor process water (Nayak et al. 2022). The membranes used in reverse osmosis can reject organic molecules and ions, making them suitable for removing contaminants in wastewater. A recent study found that certain antibiotics, such as caffeine, omeprazole, and sulfamethoxazole, can be efficiently treated using nanofiltration (NF) and reverse osmosis (RO) membranes. The RO membranes effectively removed the compounds with rejection rates exceeding 99%. The NF membranes showed varying retention capabilities, influenced by compound and membrane characteristics (Matin et al. 2023).

A study evaluated the effectiveness of seawater reverse osmosis membrane in removing antibiotics from seawater using high ionic strength water. The RO membranes efficiently removed a significant amount of antibiotic ciprofloxacin (Alonso et al. 2018).

Fluoroquinolones, like ciprofloxacin (CIP), are commonly prescribed antibiotics, with CIP concentrations rising in water sources, which can harm aquatic life due to their low biodegradability and toxicity. The excessive use of antibiotics, especially CIP, poses a challenge for managing these residues in water reuse for environmental safety. Reverse osmosis plays a critical role in water reuse, but antibiotics in the concentrate raise public health concerns (Igwegbe et al. 2021). Granular activated carbon (GAC) has shown promising results due to its high adsorption capacity. Metal impregnations can enhance GAC's adsorption capabilities, making it more effective in capturing specific contaminants like antibiotics. Research has demonstrated that iron-doped activated carbon improves CIP removal efficiency. However, the disposal of exhausted carbon remains a limitation, and *in-situ* regeneration techniques are still lacking (Bednarek et al. 2024).

Electrochemical Methods

Various water treatment methods, such as Fenton's process, electrochemical degradation, photodegradation, and advanced oxidation processes, are available (Akhil et al. 2021). While traditional methods have limitations like lower efficacy and high energy needs, electrochemical approaches stand out for their success in removing organic contaminants, including disposed antibiotics. Electrooxidation, in particular, is a cost-effective, efficient, and environmentally friendly technique for water treatment, with the added benefit of selectively controlling electrode reactions to minimize secondary pollutants (Nagban et al. 2022). The two major electrochemical methods for the treatment of antibiotic pollution include electrocoagulation and electrodialysis.

Electrocoagulation

Electrocoagulation (EC) is an electrochemical process that removes pollutants, pesticides, metals, and other impurities from water and eliminates harmful microorganisms. By applying direct current, electrode plates dissolve into the solution, increasing metal concentration, which precipitates as oxides and hydroxides. EC is cost-effective and widely used due to its design and materials (Segura et al. 2017). Veterinary antibiotics from wastewater can be efficiently removed using the electrocoagulation method. The major antibiotics that can be removed via electrocoagulation include ampicillin, doxycycline, sulfathiazole, and tylosin, with doxycycline showing the highest removal rate at ~100% (Baran et al. 2018). Despite the effectiveness of electrocoagulation (EC), limitations such as passivation of electrodes and low yield of strong oxidizing substances like •OH are evident. For most antibiotics, EC only separates them from water without degradation. Moreover, EC may not effectively remove individual refractory antibiotics, as shown by EC(Al) with only 4.8% removal of sulfamethoxazole (Leng et al. 2022). Electrocoagulation with an iron electrode removed ciprofloxacin from hospital effluent via chemisorption mechanism. Electrode consumption was 0.0625 g per run, and electrical energy usage was 0.522 $kWh \cdot m^{-3}$ under ideal circumstances, which showed that electrocoagulation can be a useful method for effectively removing ciprofloxacin from wastewater (Yoosefian et al. 2017).

Electrodialysis

Another promising, innovative electrochemical membrane approach for recovering and concentrating nutrients from wastewater is electrodialysis (Arola et al. 2019). Electrodialysis with ultrafiltration membrane (EDUF) technology combines electrodialysis (ED) and UF processes, utilizing UF membranes to separate compounds with molecular weights above the UFM molecular weight cut-off. EDUF eliminates transmembrane pressure, preventing UFM fouling. This hybrid technology offers benefits for both ED and UF without their drawbacks. Applications include isolating polyphenols from tobacco extract, fractionating bioactive peptides, and separating anthocyanins from cranberry juice (Lu et al. 2016). Electrochemical advanced oxidation processes have emerged as a promising solution, producing reactive radicals that efficiently degrade antibiotics. Three major EAOPs—electro-oxidation, photoelectro-catalytic oxidation, and electro-Fenton—are analyzed based on their performance and different influencing factors. While EAOPs offer eco-friendly treatment options, some drawbacks include toxic intermediates and high costs. Combining EAOPs with membrane separation and biological treatments can enhance the efficiency of waste removal (Mahmud et al. 2024). Additionally, EAOPs have a few technological benefits, such as reduced chemical reagent use, automation potential, and simplicity of use (Xiong et al. 2019). Electrodialysis technology has the potential for nutrient recovery from anaerobic digestate. A recent study focused on removing 17 β-estradiol (E2), a natural estrogen found in manure digestate. Membrane adsorption and anodic oxidation were found to be effective for the removal of E2. It was found that the anion exchange membrane had a higher E2 adsorption capacity than the cation exchange. Anode-ED is a promising solution for E2 removal during nutrient recovery from digestate (Cui et al. 2024).

Advanced Oxidation Processes (AOP)

Advanced oxidation processes (AOPs) are novel treatment methods to remove contaminants, including antibiotics, from water by generating highly reactive species such as hydroxyl radicals. These radicals can effectively break down complex organic molecules into simpler or less harmful substances. Common AOPs include ozonation, photocatalysis, Fenton reactions, and ultraviolet (UV) radiation combined with hydrogen peroxide.

Ozonation

Ozonation, a well-known technology, is widely utilized for disinfection and antibiotic pollution degradation in water treatment. Studies confirm its cost-effective ability to eliminate pathogens and degrade various micropollutants in water treatment. One of the types of ozonation methods involves the electro-peroxone (EP) process, which includes generating hydrogen peroxide (H_2O_2) from cathodic oxygen (O_2) reduction during ozonation to oxidize contaminants (Srinivasan 2024). A study comparing the removal of antibiotic-resistant bacteria (ARB) and antibiotic-resistance genes (ARGs) in groundwater found that the EP process resulted in higher exposure to hydroxyl radicals (•OH) but lower exposure to ozone (O_3) compared to ozonation alone. While this decrease in O_3 exposure moderately affected ARB inactivation and ARG degradation, the EP process effectively cleaved plasmid DNA into shorter fragments, reducing the risk of ARG transformation (Zheng et al. 2023). Ozonation utilizes ozone gas to oxidize and degrade antibiotic molecules. While effective, it is energy-intensive and can also produce toxic byproducts. Another chemical treatment for antibiotic removal includes advanced oxidation processes (AOPs) utilizing H_2O_2, which have become a popular method for removing antibiotics from wastewater due to their efficiency (Wang et al. 2023). Hydrogen peroxide-based AOPs are popular for their strong oxidation properties, eco-friendliness, and high degradation efficiency, gaining widespread acceptance involving various treatments like chemical, photochemical, electrochemical, and photoelectrochemical methods. They produce reactive oxygen species (ROS) such as the powerful oxidant hydroxyl radical (•OH) with a high standard reduction potential of 2.72 V/SCE (Brillas 2022).

Photocatalysis

Photocatalysis is a novel advanced oxidation process that involves reactions in the presence of light and a semiconductor photocatalyst, which boosts reaction rates. Photocatalysis offers various applications including antibacterial, deodorizing, air purifying, and water purification, addressing the rising demand for clean water due to population growth (Ameta et al. 2018). Photocatalysis is a promising method for removing antibiotics from water because it is eco-friendly and efficient. However, traditional photocatalysts have limitations like UV light absorption and quick charge recombination. Recent research focuses on developing visible light-driven photocatalysts to overcome these challenges. Strategies include doping metals and non-metals, creating new semiconductor photocatalysts, forming heterojunctions, and enhancing systems with surface plasmon resonance for better degradation of antibiotic wastewater (Li and Shi 2016). $ZnIn_2S_4$ is a promising photocatalyst for antibiotic degradation due to its visible light absorption, band gap, and affordability (Zhang et al. 2024). Spherical metal oxides (SMOs) are effective photocatalysts for degrading antibiotics due to their unique properties like strong light absorption, high separation efficiency of electron-hole pairs, and good catalytic activity (Zhu et al. 2024). A recent advancement involves the development of Co-doped CuWO4/BN quantum dots to improve peroxymonosulfate activation for pollutant removal. CW/4Co/2BNQDs degraded 94.8% of tetracycline within 30 min, with higher rate constants than other catalysts (Chen et al. 2024). A TiO_2 photocatalyst co-precipitated with terbium and manganese using the liquid phase plasma (LPP) method effectively decomposed tetracycline into oxides such as Tb_2O_3, TbO_2, Mn_2O_3, and MnO_2 on the TiO_2 surface (You and Jung 2024).

Fenton and Photon Reactions

The Photo-Fenton reaction is an enhancement of the advanced oxidation process which involves the addition of UV or visible light to improve the generation of hydroxyl radicals further. The UV light helps regenerate Fe^{2+} from Fe^{3+}, sustaining the reaction and increasing the overall efficiency of antibiotic removal. Recently, a new eco-friendly bimetallic catalyst, AgO/FeO/OSBC, synthesized from Artocarpus heterophyllus Lam leaves, efficiently degrades sulfamethoxazole and amoxicillin in water under various conditions and exhibits high removal rates. The catalyst demonstrates stability and effectiveness in repeated cycles, offering a sustainable solution for antibiotic removal (Lalhriatpuia et al. 2024). Although the Fenton reaction shows possibilities for antibiotic elimination, it has limitations due to slower Fe(III) reduction and partial degradation (Liu et al. 2023). However, the use of a rotating cylinder electrode (RCE) reactor is effective for efficient degradation of the antibiotic ampicillin (AMP) through the photoelectro-Fenton (PEF) process. O_2 generated from water oxidation at anodic plates is transported to the RCE, where H_2O_2 is continuously produced and can operate under sunlight (Cornejo et al. 2023).

Traditional Fenton reactions have various limitations, including the formation of toxic byproducts, hydrogen peroxide consumption, high energy requirements, and limited pH environments. These limitations can be overcome by applying modern Fenton reactions, including metal-organic frameworks. Metal-organic frameworks (MOFs) are ideal templates for constructing bimetallic oxides, overcoming the shortcomings of low exposure of active sites in traditional photon catalysts. One-dimensional hollow structures, like hollow tubular structures, show promise in enhancing catalytic performance by enriching pollutants and oxidants, increasing specific surface area, and aiding electron transfer. Researchers have developed MOF-derived Fe-doped indium oxide hollow nanotubes to improve the efficient removal of organic pollutants and antibiotic-resistant bacteria from wastewater using continuous flow reactors for photo-Fenton sterilization and disinfection (Li et al. 2023b).

Bioremediation

Bioremediation utilizes microbial metabolism with optimal environmental conditions and nutrients to break down complex contaminants like petroleum hydrocarbons into simpler ones. Biotechnological approaches for remediation have gained attention recently, with studies on biostimulation and bioaugmentation being prominent in the literature (Adams et al. 2015). Microorganisms used for bioremediation are called bioremediators and can be used *in situ* or *ex situ*. Examples of bioremediation technologies include phytoremediation, bioventing, and composting. This natural biological process offers an effective way to eliminate or neutralize contaminants (Sharma 2012). Bioremediation is favored over physical and chemical removal methods due to the generation of toxic byproducts. However, antibiotic residues in wastewater remain a global threat. Advancements in molecular and synthetic biology may lead to the engineering of microbe-metabolite biodevices and the development of new strains with desired properties for sustainable antibiotic elimination from the environment (Apreja et al. 2022). It is well documented that bioaugmentation with the bacterium *Burkholderia cepacia* immobilized on sugarcane bagasse (SCB) showed promise in soil degrading tetracycline antibiotics (TCAs). However, certain factors, including a 15% inoculum dose, temperature of 28–43 °C, and acidic pH enhance the bacterium's capacity to degrade tetracycline from the soil (Hong et al. 2020). Recently, researchers explored response surface methodology to investigate the removal of sulfamethoxazole and ofloxacin using the marine diatom *Chaetoceros muelleri* (Mojiri et al. 2021). A recent study focused on antibiotic degradation using immobilized HRP on functionalized cellulose support. Immobilized enzymes showed higher Km and Vmax values than free HRP, leading to enhanced degradation rates. LC-MS analysis revealed the degradation mechanism of antibiotics involving hydroxylation, oxidation, dehydration, and decarbonylation reactions (Galodia and Chakma 2024).

Engineered Microorganisms

Engineered microorganisms play an important role in bioremediation, where microorganisms like bacteria and fungi neutralize the toxins from the environment (Adetunji et al. 2024). Genetically modified organisms (GMOs) with enhanced metabolic pathways can help accelerate bioremediation by overproducing beneficial proteins and enzymes (Table 1). Factors such as soil contamination levels, site conditions, and adoption of remediation methods are carefully examined (Patel et al. 2024). The bioremediation process can convert toxic substances to non-toxic ones by a variety of engineered bacteria species such as *Achromobacter, Pseudomonas, Dehalococcoides, Rhodococcus, Comamonas, Burkholderia, Alcaligenes, Bacillus subtillus, Aspergilus niger, Deinococcus radioduran, Acidithiobacillus ferrooxidans, Mesorhizobium huakuii, Pseudomonas K-62, Ralstonia,* and *Rhodopseudomonas palustris* (Zhang et al. 2020, Rafeeq et al. 2023).

Table 1: List of microorganisms reported for removal of antibiotic pollution

Microorganisms	Treated Pollutant	Mode of Action	Reference
Phanerochaete chrysosporium	Tetracycline	Manganese peroxidase; Demethylation, oxidation of dimethyl amino, decarboxylation, hydroxylation, oxidative degeneration	Sun et al. 2021
Bradyrhizobium, Gemmatimenas, Burkholderia	Sulfonamides	Hydroxylation; Sad A gene, NADH Dependent monooxygenase	Chen et al. 2023
Geobacter, Xanthobacter, Bradyrhizobium, Achromobacter	Sulfadiazine	Antibiotic resistance gene reduction	Yang et al. 2019
Saccharaomyces cerevisae	Cephalexin	*S. cerevisae* immobilized in calcium alginate as biosorbent	Rusu et al. 2021
Trametes versicolor (White Rot Fungi)	Azure B, Sulfacetamide	Ligninolytic enzymes; laccase	Zhang et al. 2023
Chlorella vulgaris, Clanostachys rosea, Bacillus licheniformis	Tetracycline, Oxytetracycline Chlortetracycline	–	Huang et al. 2024
Pseudomonas stutzeri DLY-21	Sulfonamides	Degradation of sulfonamide antibiotics to sulfoxide (SDM) ≈ sulfachloropyridazine (SCP) > sulfa quinoxaline (SQ) > sulfadiazine (SQ)	Li et al. 2024b

Phytoremediation

Phytoremediation is an effective and cost-efficient method to remove antibiotics from water, generating fewer harmful by-products. Aquatic plants aid in phytoremediation by accumulating, transforming, and degrading contaminants (Yan et al. 2021). Phytoremediation in constructed wetlands shows promise in removing antibiotics through plant uptake, microbial degradation, and sorption. Recent studies have found that sulfonamide, macrolide, tetracycline, and fluoroquinolone antibiotics can be effectively removed through the phytoremediation method (McCorquodale-Bauer et al. 2023). Research on phytoremediation for antibiotic removal in wastewater treatment has revealed valuable insights, including plant uptake, wetland removal, rhizodegradation, and plant effects on antibiotic resistance genes. A recent study investigated the phytotoxicity of penicillin-3, a common antibiotic in agriculture, that can be eradicated from antibiotic-contaminated soils with the use of watercress, industrial hemp, rice, and *Elodea canadensis* (Timofeeva and Tyukalova 2023). The coontail *Ceratophyllum demersum* effectively removed tetracycline from antibiotic-contaminated wastewater treatment (Ali et al. 2023). On the other hand, vetiver grass (*Chrysopogon zizanioides*) from wetlands was found effective in eliminating tetracycline, ciprofloxacin, and nutrients (nitrogen and phosphorus) from wastewater, demonstrating its potential as an environmentally beneficial wastewater treatment solution (Panja et al. 2021).

Challenges and Future Studies

Antibiotics are widely used and have a tenacious nature, although eliminating them from natural environmental sources is still considered a challenging task. Because of their stability and effectiveness, antibiotics may withstand natural degradation processes and are therefore found in the environment for extended periods. Due to their long persistence in nature and the different chemical structures of various antibiotics, it is challenging to design universal removal techniques, which makes the treatment procedures expensive and time-consuming. Furthermore, there is a serious risk to public health since antibiotic residues in the environment encourage the growth of bacteria that are resistant to antibiotics. Antibiotics are not efficiently removed by conventional wastewater treatment plants, which results in their release into untreated natural water bodies even after treatment. To address these challenges, several future directions have been explored. Advanced oxidation processes, including ozonation and photocatalysis, are being optimized for large-scale applications to convert antibiotics into less harmful substances. Biological processes, including bioremediation using microorganisms capable of degrading antibiotics and genetic engineering to enhance their effectiveness, offer promising solutions. Adsorption techniques using materials like activated carbon and biochar and advanced membrane filtration technologies such as nanofiltration and reverse osmosis are being developed to improve antibiotic removal efficiency. The field of green chemistry focuses on creating environmentally friendly antibiotics that degrade more easily after use. However, policy measures, stricter regulations on antibiotic discharge, public awareness, and education campaigns are essential to promote responsible antibiotic use and proper disposal. Continued research and innovation are necessary to develop sustainable strategies for mitigating the environmental impact of antibiotics. Future directions in antibiotic removal must emphasize advanced and sustainable methods to combat environmental contamination. The major key areas for antibiotic removal include advanced oxidation processes such as photocatalysis, nanotechnology with carbon nanotubes, engineered microorganisms, etc. Apart from this, more focus should be placed on the development of antibiotic-degrading plants through genetic modification and advanced membrane filtration hybrid systems.

References

Abel zur Wiesch, P., Clarelli, F. and Cohen, T. 2017. Using chemical reaction kinetics to predict optimal antibiotic treatment strategies. PLoS Comput. Biol., 13(1): e1005321.

Adams, G.O., Fufeyin, P.T., Okoro, S.E. and Ehinomen, I. 2015. Bioremediation, biostimulation and bioaugmention: A review. Int. J. Environ. Bioremediat. Biodegrad, 3(1): 28-39.

Aderemi, A.O., Novais, S.C., Lemos, M.F., Alves, L.M., Hunter, C. and Pahl, O. 2018. Oxidative stress responses and cellular energy allocation changes in microalgae following exposure to widely used human antibiotics. Aquat. Toxicol., 203: 130-139.

Adetunji, C.O., Ahamed, M.I. and Altalhi, T. (eds.). 2024. Genetically Engineered Organisms in Bioremediation. CRC Press.

Ahmed, M.B., Zhou, J.L., Ngo, H.H., Guo, W., Thomaidis, N.S. and Xu, J. 2017. Progress in the biological and chemical treatment technologies for emerging contaminant removal from wastewater: A critical review. J. Hazard. Mater., 323: 274-298.

Akhil, D., Lakshmi, D., Senthil Kumar, P., Vo, D.V. and Kartik, A. 2021. Occurrence and removal of antibiotics from industrial wastewater. Environ Chem Lett., 19(2): 1477-1507.

Akhil, P., Kharwar, Y.P., Rath, S., Krishnan, R., Roy, S.C., Shanmugam, S. et al. 2024. $BaTaO_2N$ and quantum dots-based CuO nanocomposites for HER by solar electrochemical water splitting. Inorg. Chem. Commun., 168: 112828.

Ali, M.F., Ali, Q.A., M-Ridha, M.J., Mohammed, S.J. and Bohan, H.R. 2023. Phytoremediation of tetracycline via the coontail Ceratophyllum demersum in antibiotics-contaminated water. Biocatal. Agric. Biotechnol., 53: 102887.

Alonso, J.J.S., El Kori, N., Melián-Martel, N. and Del Río-Gamero, B. 2018. Removal of ciprofloxacin from seawater by reverse osmosis. J. Environ. Manage., 217: 337-345.

Ameta, R., Solanki, M.S., Benjamin, S. and Ameta, S.C. 2018. Photocatalysis. *In*: Advanced Oxidation Processes for Waste Water Treatment, 135-175. Academic Press.

Anh, H.Q., Le, T.P.Q., Da Le, N., Lu, X.X., Duong, T.T., Garnier, J. et al. 2021. Antibiotics in surface water of East and Southeast Asian countries: A focused review on contamination status, pollution sources, potential risks, and future perspectives. Sci. Total Environ., 764: 142865.

Apreja, M., Sharma, A., Balda, S., Kataria, K., Capalash, N. and Sharma, P. 2022. Antibiotic residues in environment: Antimicrobial resistance development, ecological risks, and bioremediation. Environ. Sci. Pollution. Res., 1-17.

Arola, K., Ward, A., Mänttäri, M., Kallioinen, M. and Batstone, D. 2019. Transport of pharmaceuticals during electrodialysis treatment of wastewater. Water. Res., 161: 496-504.

Atterby, C., Nykvist, M., Lustig, U., Andersson, D.I., Järhult, J.D. and Sandegren, L. 2021. Selection of resistant bacteria in mallards exposed to subinhibitory concentrations of ciprofloxacin in their water environment. Antimicrob. Agents. Chemother., 65(3): 10-1128.

Bai, Y., Liu, B., Li, J., Li, M., Yao, Z., Dong, L. et al. 2023. Microstructure optimization of bioderived polyester nanofilms for antibiotic desalination via nanofiltration. Sci. Adv., 9(18): 6134.

Baran, W., Adamek, E., Jajko, M. and Sobczak, A. 2018. Removal of veterinary antibiotics from wastewater by electrocoagulation. Chemosphere, 194: 381-389.

Bednárek, J., Matějová, L., Koutník, I., Vráblová, M., Cruz, G.J.F., Strašák, T. et al. 2024. Revelation of high-adsorption-performance activated carbon for removal of fluoroquinolone antibiotics from water. Biomass. Convers. Bioref., 14(2): 2585-2599.

Brillas, E. 2022. Progress of homogeneous and heterogeneous electro-Fenton treatments of antibiotics in synthetic and real wastewaters. A critical review on the period 2017–2021. Sci. Toral. Environ., 819: 153102.

Cao, Y., Chen, X., Feng, S., Wan, Y. and Luo, J. 2020. Nanofiltration for decolorization: Membrane fabrication, applications and challenges. Ind. Eng. Chem. Res., 59(45): 19858-19875.

Chen, J., Ke, Y., Zhu, Y., Chen, X. and Xie, S. 2023. Deciphering of sulfonamide biodegradation mechanism in wetland sediments: From microbial community and individual populations to pathway and functional genes. Water. Res., 240: 120132.

Chen, R., Zhang, H., Dong, Y. and Shi, H. 2024. Dual metal ions/BNQDs boost PMS activation over copper tungstate photocatalyst for antibiotic removal: Intermediate, toxicity assessment and mechanism. J. Mater. Sci. Trechnol., 170: 11-24.

Cornejo, O.M., Sirés, I. and Nava, J.L. 2023. Continuous H_2O_2 production sustained by anodic O_2 for the destruction of the antibiotic ampicillin by photoelectro-Fenton process in a rotating cylinder electrode reactor. J. Environ. Chem. Eng., 11(2): 109326.

Cui, W., Meng, J., Wang, S., Hu, Z., Liu, G. and Zhan, X. 2024. 17β-estradiol (E2) removal in anode-electrodialysis (anode-ED) during nutrient recovery from pig manure digestate J. Hazard. Mater., 462: 132754.

de Ilurdoz, M.S., Sadhwani, J.J. and Reboso, J.V. 2022. Antibiotic removal processes from water & wastewater for the protection of the aquatic environment – A review. J. Water. Process. Eng., 45: 102474.

Duarte, A.C., Rodrigues, S., Afonso, A., Nogueira, A. and Coutinho, P. 2022. Antibiotic resistance in the drinking water: Old and new strategies to remove antibiotics, resistant bacteria, and resistance genes. Pharmaceuticals, 15(4): 393.

Fabregat, V. and Pagán, J.M. 2024. Technical–economic feasibility of a new method of adsorbent materials and advanced oxidation techniques to remove emerging pollutants in treated wastewater. Water, 16(6): 814.

Galodiya, M.N. and Chakma, S. 2024. Immobilization of enzymes on functionalized cellulose nanofibrils for bioremediation of antibiotics: Degradation mechanism, kinetics, and thermodynamic study. Chemosphere, 349: 140803.

Garcia-Segura, S., Eiband, M.M.S., de Melo, J.V. and Martínez-Huitle, C.A. 2017. Electrocoagulation and advanced electrocoagulation processes: A general review about the fundamentals, emerging applications and its association with other technologies. J. Electroanal. Chem., 801: 267-299.

Gul, S., Hussain, S., Khan, H., Khan, K.I., Khan, S., Ullah, S. et al. 2022. Advances in bioremediation of antibiotic pollution in the environment. Biol. Approaches. Control. Pollut., 49-78. Woodhead Publishing.

Hong, X., Zhao, Y., Zhuang, R., Liu, J., Guo, G., Chen, J. et al. 2020. Bioremediation of tetracycline antibiotics-contaminated soil by bioaugmentation. RSC Adv., 10(55): 33086-33102.

Huang, J., Wang, Z., Zhao, C., Yang, H. and Niu, L. 2024. Performance of four different microalgae-based technologies in antibiotics removal under multiple concentrations of antibiotics and strigolactone analogue GR24 administration. Sci. Rep., 14(1):16004.

Huang, S., Yu, J., Li, C., Zhu, Q., Zhang, Y., Lichtfouse, E. et al. 2022. The effect review of various biological, physical and chemical methods on the removal of antibiotics. Water, 14(19): 3138.

Igwegbe, C.A., Oba, S.N., Aniagor, C.O., Adeniyi, A.G. and Ighalo, J.O. 2021. Adsorption of ciprofloxacin from water: A comprehensive review. J. Ind. Eng. Chem., 93: 57-77.

Khalil, M. and Liu, Y. 2021. Greywater biodegradability and biological treatment technologies: A critical review. Int. Biodeterior. Biodegrad., 161: 105211.

Koniuszewska, I., Korzeniewska, E., Harnisz, M., Kiedrzyńska, E., Kiedrzyński, M., Czatzkowska, M. et al. 2020. The occurrence of antibiotic-resistance genes in the Pilica River, Poland. Ecohydrol. Hydrobiol., 20(1): 1-11.

Kovalakova, P., Cizmas, L., McDonald, T.J., Marsalek, B., Feng, M. and Sharma, V.K. 2020. Occurrence and toxicity of antibiotics in the aquatic environment: A review. Chemosphere, 251: 126351.

Lalhriatpuia, C., Jeon, C. and Tiwari, D. 2024. Highly efficient novel heterojunction catalyst in the elimination of antibiotics under the photo-Fenton-like process. Colloids and Surfaces A: Physicochemical and Engineering Aspects, 699: 134651.

Leng, X., Tegladza, I.D., Kadier, A., Dai, H. and Lu, J. 2022. In-situ generation of both hydroxyl radical and adsorptive flocs in electro-coagulation process with air breathing cathode. Process. Saf. Environ. Prot., 164: 345-353.

Li, D. and Shi, W. 2016. Recent developments in visible-light photocatalytic degradation of antibiotics. Chin. J. Catal., 37(6): 792-799.

Li, F., Wang, P., Li, M., Zhang, T., Li, Y. and Zhan, S. 2023a. Efficient photo-Fenton reaction for tetracycline and antibiotic resistant bacteria removal using hollow Fe-doped In_2O_3 nanotubes: From theoretical research to practical application. Water. Res., 240: 120088.

Li, S., Hofstra, N., van de Schans, M.G., Yang, J., Li, Y., Zhang, Q. et al. 2023b. Riverine antibiotics from animal production and wastewater. Environ. Sci. Technol. Lett., 10(11): 1059-1067.

Li, X., Lu, Z., Wu, B., Xie, H. and Liu, G. 2024a. Antibiotics and antibiotic resistance genes removal in biological aerated filter. Bioresour. Technol., 395: 130392.

Li, Y., Yang, T., Lin, X., Huang, J., Zeng, J., Cai, Q. et al. 2024b. Isolation, identification, and optimization of conditions for the degradation of four sulfonamide antibiotics and their metabolic pathways in *Pseudomonas stutzeri* strain DLY-21. Heliyon, 10(7).

Liang, C., Wei, D., Zhang, S., Ren, Q., Shi, J. and Liu, L. 2021. Removal of antibiotic resistance genes from swine wastewater by membrane filtration treatment. Ecotoxicol. Environ. Saf., 210, 111885.

Liu, C., Wang, P., Huang, P., Yang, Z. and Zhou, G. 2023. Photo-induced heterogeneous regeneration of Fe (II) in Fenton reaction for efficient polycyclic antibiotics removal and in-depth charge transfer mechanism. J. Colloid. Interface. Sci., 638: 768-777.

Liu, K., Gan, C., Gan, Y., He, J., Du, Y., Tong, L. et al. 2024. Occurrence and source identification of antibiotics and antibiotic resistance genes in groundwater surrounding urban hospitals. J. Hazard. Mater., 465: 133368.

Lu, H., Zou, W., Chai, P., Wang, J. and Bazinet, L. 2016. Feasibility of antibiotic and sulfate ions separation from wastewater using electrodialysis with ultrafiltration membrane. J. Clean. Prod., 112: 3097-3105.

Luo, Y., Sun, Y., Wei, X., He, Y., Wang, H., Cui, Z. et al. 2024. Detection methods for antibiotics in wastewater: A review. Bioprocess. Biosyst. Eng., 47(9): 1433-1451.

Lyu, J., Yang, L., Zhang, L., Ye, B. and Wang, L. 2020. Antibiotics in soil and water in China – A systematic review and source analysis. Environ. Pollut., 266: 115147.

Mahmud, F., Banhi, T.S., Roy, H., Dihan, M.R., Islam, S., Cai, Y. et al. 2024. Antibiotic-contaminated wastewater treatment and remediation by electro-advanced oxidation processes (EAOPs). Groundw. Sustain. Dev., 101181.

Matin, A., Jillani, S.M.S., Baig, U., Ihsanullah, I. and Alhooshani, K. 2023. Removal of pharmaceutically active compounds from water sources using nanofiltration and reverse osmosis membranes: Comparison of removal efficiencies and in-depth analysis of rejection mechanisms. J. Environ. Manage., 338: 117682.

McCorquodale-Bauer, K., Grosshans, R., Zvomuya, F. and Cicek, N. 2023. Critical review of phytoremediation for the removal of antibiotics and antibiotic resistance genes in wastewater. Sci. Total. Environ., 870: 161876.

Michael, I., Rizzo, L., McArdell, C.S., Manaia, C.M., Merlin, C., Schwartz, T. et al. 2013. Urban wastewater treatment plants as hotspots for the release of antibiotics in the environment: A review. Water. Res., 47(3): 957-995.

Mojiri, A., Baharlooeian, M. and Zahed, M.A. 2021. The potential of Chaetoceros muelleri in bioremediation of antibiotics: Performance and optimization. Int. J. Environ. Res. Public Health, 18(3): 977.

Moreira, J.B., Santos, T.D., Zaparoli, M., de Almeida, A.C.A., Costa, J.A.V. and de Morais, M.G. 2022. An overview of nanofiltration and nanoadsorption technologies to emerging pollutants treatment. Appl. Sci., 12(16): 8352.

Mounzer, C.K., Baydoun, S.A., Amer, R.A. and Borjac, J.M. 2024. Occurrence of antibiotics and antibiotic-resistant bacteria in the Lebanese polluted Litani River. Environ. Monit. Assess., 196(1): 90.

Nabgan, W., Saeed, M., Jalil, A.A., Nabgan, B., Gambo, Y., Ali, M.W. et al. 2022. A state of the art review on electrochemical technique for the remediation of pharmaceuticals containing wastewater. Environ. Res., 210: 112975.

Najafi, M. 2024. Adsorption process of antibiotics by novel adsorbents. *In*: Traditional and Novel Adsorbents for Antibiotics Removal from Wastewater, 301-367. Elsevier.

Nasrollahi, N., Vatanpour, V. and Khataee, A. (2022). Removal of antibiotics from wastewaters by membrane technology: Limitations, successes, and future improvements. Sci. Total. Environ., 838: 156010.

Owusu-Doubreh, B., Appaw, W.O. and Abe-Inge, V. 2023. Antibiotic residues in poultry eggs and its implications on public health: A review. Sci. Afr., 19: 01456.

Panja, S., Sarkar, D., Zhang, Z. and Datta, R. 2021. Removal of antibiotics and nutrients by vetiver grass (Chrysopogon zizanioides) from a plug flow reactor based constructed wetland model. Toxics, 9(4): 84.

Patel, D., Solanki, J., Azagury, A. and Subramanian, R.B. 2024. Molecular cloning for the production of genetically modified organisms for bioremediation. Genet. Eng. Org. Bioremed.,, 119-135. CRC Press.

Pourhakkak, P., Taghizadeh, A., Taghizadeh, M., Ghaedi, M. and Haghdoust, S. 2021. Fundamentals of adsorption technology. Interface. Sci. Technol., 33: 1-70. Elsevier.

Rafeeq, H., Afsheen, N., Rafique, S., Arshad, A., Intisar, M., Hussain, A. et al. 2023. Genetically engineered microorganisms for environmental remediation. Chemosphere, 310: 136751.

Rusu, L., Grigoraș, C.G., Simion, A.I., Suceveanu, E.M., Șuteu, D. and Harja, M. 2021. Application of Saccharomyces cerevisiae/calcium alginate composite beads for cephalexin antibiotic biosorption from aqueous solutions. Materials, 14(16): 4728.

Sharma, S. 2012. Bioremediation: Features, strategies and applications. Asian. J. Pharm. Life. Sci., 2231: 4423.

Sharma, V., Kumar, R.V., Pakshirajan, K. and Pugazhenthi, G. 2017. Integrated adsorption-membrane filtration process for antibiotic removal from aqueous solution. Powder. Technol., 321: 259-269.

Sousa, M.R.S., Lora-Garcia, J. and López-Pérez, M.F. 2018. Modelling approach to an ultrafiltration process for the removal of dissolved and colloidal substances from treated wastewater for reuse in recycled paper manufacturing. J. Water. Process. Eng., 21: 96-106.

Srinivasan, R. 2024. Sustainable degradation of ciprofloxacin in water by the electro-peroxone process via a graphite felt electrode system. Discov. Water., 4(1): 4.

Sun, X., Leng, Y., Wan, D., Chang, F., Huang, Y., Li, Z. et al. 2021. Transformation of tetracycline by manganese peroxidase from Phanerochaete chrysosporium. Molecules, 26(22): 6803.

Timofeeva, S.S. and Tyukalova, O.V. 2023, September. Penicillin Antibiotics and Their Phytotoxicity. *In*: International Conference on Construction, Architecture and Technosphere Safety, 569-577. Cham: Springer Nature Switzerland.

Vaughn, V.M., Hersh, A.L. and Spivak, E.S. 2022. Antibiotic overuse and stewardship at hospital discharge: The reducing overuse of antibiotics at discharge home framework. Clin. Infect. Dis., 74(9): 1696-1702.

Wang, J., Wei, Y., Li, K., Cheng, Y., Li, M. and Xu, J. 2014. Fate of organic pollutants in a pilot-scale membrane bioreactor-nanofiltration membrane system at high water yield in antibiotic wastewater treatment. Water. Sci. Technol., 69(4): 876-881.

Wang, X.L., Wang, Q., Xue, Y.X., Zhang, B., Han, S.R., Zhang, H. et al. 2023. Preparation of composite nanofiltration membrane with interlayer for pharmaceutical rejection. Sep. Purif. Technol., 312: 123411.

Wang, Z., Du, Y., Yang, C., Liu, X., Zhang, J., Li, E. et al. 2017. Occurrence and ecological hazard assessment of selected antibiotics in the surface waters in and around Lake Honghu, China. Sci. Total. Environ., 609: 1423-1432.

Xu, N., Shen, Y., Jiang, L., Jiang, B., Li, Y., Yuan, Q. et al. 2023. Occurrence and risk levels of antibiotic pollution in the coastal waters of eastern China. Environ. Sci. Pollut. Res., 30(27): 71371-71381.

Yan, Y., Deng, Y., Li, W., Du, W., Gu, Y., Li, J. et al. 2021. Phytoremediation of antibiotic-contaminated wastewater: Insight into the comparison of ciprofloxacin absorption, migration, and transformation process at different growth stages of E. crassipes. Chemosphere, 283: 131192.

Yang, L., Xia, C., Jiang, J., Chen, X., Zhou, Y., Yuan, C. et al. 2024. Removal of antibiotics and estrogens by nanofiltration and reverse osmosis membranes. J. Hazard. Mater., 461: 132628.

Yang, Y., Liu, G., Song, W., Ye, C., Lin, H., Li, Z. and Liu, W. 2019. Plastics in the marine environment are reservoirs for antibiotic and metal resistance genes. Environ Int., 123: 79-86.

Yang, Z., Li, H., Li, N., Sardar, M.F., Song, T., Zhu, H. et al. 2022. Dynamics of a bacterial community in the anode and cathode of microbial fuel cells under sulfadiazine pressure. International Journal of Environ. Res. Public Health, 19(10): 6253.

Yoosefian, M., Ahmadzadeh, S., Aghasi, M. and Dolatabadi, M. 2017. Optimization of electrocoagulation process for efficient removal of ciprofloxacin antibiotic using iron electrode; kinetic and isotherm studies of adsorption. J. Mol. Liquid., 225: 544-553.

You, C.S. and Jung, S.C. 2024. Photo-catalytic destruction of tetracycline antibiotics using terbium and manganese co-precipitated TiO_2 photocatalyst. J. Environ. Chem. Eng., 12(1): 111666.

Zhang, B., Xu, R., Feng, Y. and Wang, J. 2024. Photocatalytic degradation of antibiotics in municipal wastewater over $ZnIn_2S_4$. Ionics, 30(3): 1291-1306.

Zhang, H., Liu, X., Liu, B., Sun, F., Jing, L., Shao, L. et al. 2023. Synergistic degradation of Azure B and sulfanilamide antibiotics by the white-rot fungus Trametes versicolor with an activated ligninolytic enzyme system. J. Hazard Mater., 458: 131939.

Zhang, J., Sui, Q., Tong, J., Zhong, H., Wang, Y., Chen, M. and Wei, Y. 2018. Soil types influence the fate of antibiotic-resistant bacteria and antibiotic resistance genes following the land application of sludge composts. Environ Int., 118: 34-43.

Zhang, T., Lv, K., Lu, Q., Wang, L. and Liu, X. 2021. Removal of antibiotic-resistant genes during drinking water treatment: A review. J. Environ. Sci., 104: 415-429.

Zhang, W., Lin, Z., Pang, S., Bhatt, P. and Chen, S. 2020. Insights into the biodegradation of lindane (γ-hexachlorocyclohexane) using a microbial system. Front. Microbiol., 11: 522.

Zheng, Q., Zhang, Y., Wang, Y. and Yu, G. 2023. Removal of antibiotic resistant bacteria and plasmid-encoded antibiotic resistance genes in water by ozonation and electro-peroxone process. Chemosphere, 319: 138039.

Zhu, K., Li, X., Chen, Y., Huang, Y., Yang, Z., Guan, G. et al. 2024. Recent advances on the spherical metal oxides for sustainable degradation of antibiotics. Coord. Chem. Rev., 510: 215813.

CHAPTER

12

Microbial Enzymes Mediated Bioremediation of Wastewater: An Overview

Aiman Tanveer[1], Supriya Gupta[1], Shruti Dwivedi[1], Kanchan Yadav[1], Varsha Rani[1,2] and Dinesh Yadav[1*]

[1] Gene Cloning and Expression Lab, Department of Biotechnology, Deen Dayal Upadhyaya Gorakhpur University, Gorakhpur - 273009, Uttar Pradesh, India

[2] Department of Biotechnology, School of Engineering and Technology, Sandip University, Nashik - 422213, Maharashtra, India

Introduction

Water serves as one of the integral environmental resources. Despite this, the water supply faces contamination after being used by various industries, institutes, residential areas, commercial establishments, etc. This leads to water pollution as the untreated wastewater from multiple sources contains decomposed organic waste that can produce gases, resulting in an awful odour. Also, the wastewater comprises a vast variety of toxic substances. Thus, wastewater must be properly treated as the toxic waste present in it might be dangerous to the environment. Different approaches have been applied for the remediation of wastewater. Conventionally, physical and chemical treatment has been used but biological treatment has come up as a sustainable alternative for waste water remediation. It has proved to be a promising technology owing to reduced handling cost and eco-friendly processing. The various bioremediation techniques include phycoremediation, mycoremediation, nanoremediation, phytoremediation, rhizoremediation and enzymatic remediation. Among them, the enzymatic approach is one of the most extensively used approach. As the enzymes are natural biocatalysts, they can efficiently act on the pollutants with high specificity and efficacy. They can specifically target contaminants and may have no effect on the various components of the effluent. They may also be utilized to act as biocatalysts by operating at a certain temperature and pH, even under mild reaction circumstances. Microbial enzymes are preferred over the enzymes from

*Corresponding author: dinesh.biotech@ddugu.ac.in

DOI: 10.1201/9781003527350-12

plant and animal sources as they can be produced with ease in a short time under cost effective conditions. Different molecular biology techniques such as mutagenesis, recombinant DNA techniques, and directed evolution can be used to modify the enzymes for enhanced catalytic activity. This makes enzymes highly acceptable as green technology for wastewater remediation. Microbes are rich sources of several enzymes, cellulases, ligninase, lipases, peroxidase, laccases, proteases etc., which play a potential role in the bioremediation process. Several commercial enzymes are also available that can be used in the bioremediation of wastewater. In the present chapter, we have discussed the different types of remediation techniques for wastewater management. We have also discussed in detail about the different types of pollutants contaminating water, recent reports on the usage of different enzymes in wastewater bioremediation, the molecular tools applied for the modification of enzymes used in wastewater treatment, and some recent examples of the same. We have also thoroughly discussed the intervention of *in silico* tools for wastewater remediation and the bioinformatics approach for the identification of enzymes used for remediation of the wastewater. Figure 1 illustrates a schematic representation of the entire process.

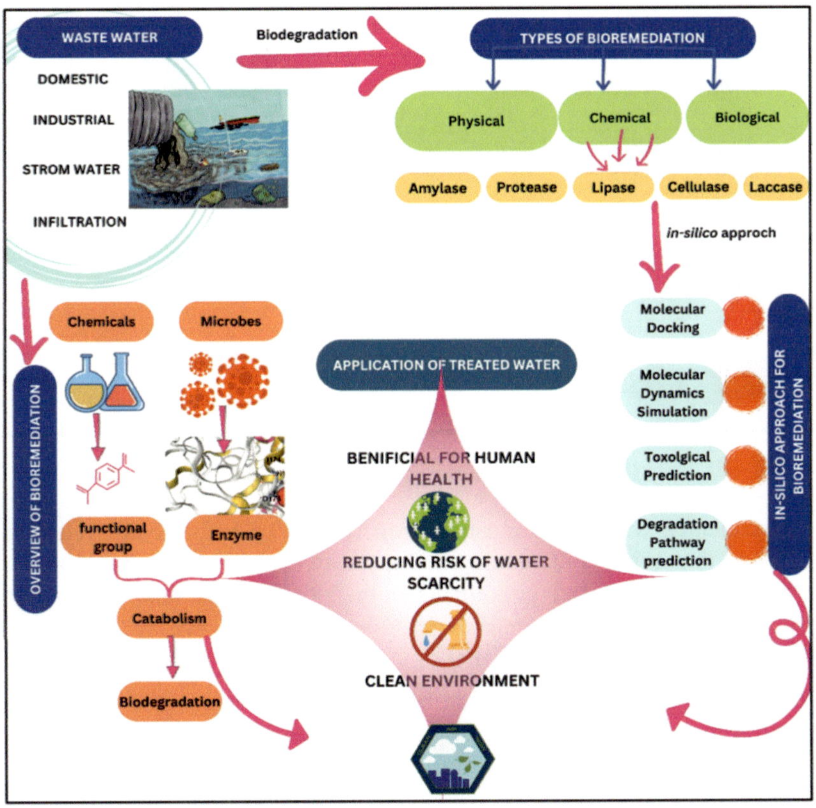

Figure 1: Overview of waste water, types of bioremediations, different omics intervention for treatment of waste water and their application.

Pollutants in Wastewater

Water is considered polluted when it contains too many dangerous contaminants, making it unfit for bathing, cooking, or drinking. Human activity is usually the cause of water pollution. The main sources of pollutants released into the environment are industrial waste sites, oil spills, heavy metals, animal and chemical wastes, deforestation, eroded sediments, trash, fertilisers, herbicides, pesticides, etc. Approximately one-third of the available renewable freshwater is used by these industries, and the pollutants they produce include a variety of man-made and natural chemical contaminants. A major factor in the reduction of both the amount and quality of drinkable water is the growing water pollution caused by waste discharged from various sources, including households, industry, agriculture, and hospitals. Figure 2 shows sources of pollutants in wastewater.

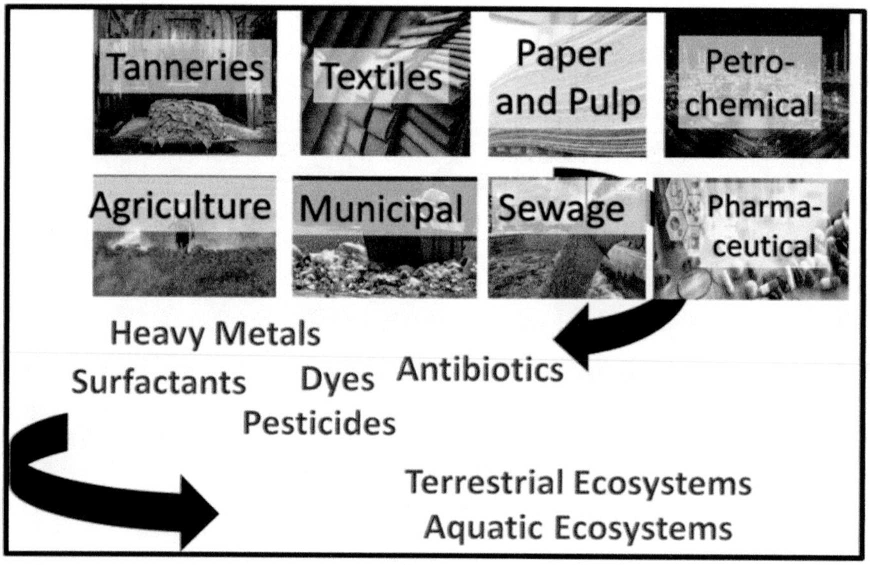

Figure 2: The source of waste water pollutants.

Consequently, wastewater must be properly treated before being released into water bodies to maximise the quality and quantity of drinkable water. Some of the major pollutants and their adverse effects on the biosphere are summarised below.

Pollutants of Domestic Origin

Many different types of pollutants can be found in domestic wastewater, including salts, pathogens, soluble and particulate organic matter, oils and greases, biowastes, home chemicals, heavy metals, nutrients, and bath and kitchen waste. The contaminants are responsible for changing the water's pH, hardness, inorganic components, pathogen load, and biological and chemical oxygen demand (BOD and COD) levels. In addition to threatening aquatic and marine life, untreated home waste water lowers crop yields and is the source of numerous waterborne and water-related illnesses. Because less than 5% of wastewater from homes and urban settings is pre-treated before being released into the environment, these water-related illnesses are remain common throughout the nation (Koul et al. 2021, Rout et al. 2021).

Pollutants of Industrial Origin

Industrial wastewater is the waste product of several production processes, such as petrochemical, textile, electroplating, pharmaceutical, and food industries. Thus, this waste includes a range of dangerous and poisonous materials, dyes, metal ions, and organic and inorganic substances (Bakr et al. 2021). The effluent generated contains high concentrations of oil, volatile chemicals, lipids, and cleaning agents that are harmful to the environment and human health. The petrochemical, iron, and steel sectors produce a sizable amount of wastewater that contains heavy metals (such as nickel, copper, zinc, chromium, cadmium, and lead) (Fawzy et al. 2019). These pollutants lead to disorders of the nervous system, kidney and liver damage, skin symptoms, dermatitis, rheumatoid arthritis, chronic asthma, and visceral malignancies. The composition and quantity of industrial discharges vary greatly, therefore, in order to produce a final effluent that meets both national and international standards, modern treatment methods are fundamentally needed. Strict guidelines have also been proposed for the handling, treatment, and repurposing of this highly contaminated wastewater (Nasr 2021).

Pollutants such as suspended particles, infectious bacteria, synthetic colours and chemicals, high organic matter content, heavy metals (Cd, Ni, Pb, Hg, As, Cu, Cr), and suspended particles are characteristic of industrial wastewaters. Nearly 80% of wastewater is processed and discharged into the environment untreated (Amin et al. 2022). Phosphorus (P) and nitrogen (N) are the two nutrients most prevalent in wastewater. The N/P ratio was comparatively greater in wastewaters from livestock, agricultural, and urban sources (Diaz et al. 2022). High BOD, COD, total organic carbon, and variable pH are typical characteristics of industrial effluents. Agricultural and food processing wastewaters have significant levels of BOD, suspended particles, and ammonia nitrogen, and their chemical composition varies depending on the industry. The meat processing industry's wastewater contains body fluids, hormones, antibiotics, and other organic pollutants, wastewater from the confectionery industry contains simple sugars. Effluents from the pulp and paper industries are tainted with high levels of alkali, inorganic acid salts, and lignin. Gasification byproducts from the steel industry, such as benzene, phenols, anthracene, cyanide, naphthalene, ammonia, and cresols, and complex organic compounds such as polycyclic aromatic hydrocarbons (PAHs), pollute the waterways. Besides PAHs, xenobiotic components from industrial effluents in the pharmaceutical and agrochemical sectors—including pesticides, dye effluents, organochlorides, plasticisers (phthalates, bisphenol), and dioxins—are also found in large quantities. These xenobiotics are referred to as persistent organic pollutants (POPs), as they are resistant to standard treatment procedures (Ahmed et al. 2022, Kumar et al. 2023). Figure 3 shows the major pollutants and their major effects on the biosphere. When these industrial and nutrient-rich effluents are released into waterbodies untreated, they lead to eutrophication, which is harmful to aquatic ecosystems and human health by causing organic compounds and heavy metals to accumulate (Nagarajan et al. 2022, Kumar et al. 2023). The most prevalent and persistent industrial pollutants are:

Heavy Metals

Although heavy metals are present in the earth's crust, the majority of environmental contamination originates from human activities such as industrial production, smelting, mining, and the use of metals. Metal refineries, coal-burning power plants, petroleum combustion, nuclear power plants, textiles, and plastics are among the industrial sources. Several studies have demonstrated that heavy metals affect cellular organelles and its components. Interactions between metal ions and nuclear proteins lead to conformational changes and DNA damage. This may cause cell death, cancer, or alteration in cell cycle. A major contributor to the carcinogenicity and toxicity of metals such as mercury (Hg), lead (Pb), arsenic (As), chromium (Cr) and cadmium (Cd) is oxidative stress and the generation of reactive oxygen species (ROS). The high level of toxicity of these five elements has a significant impact on public health, making them priority metals. Despite their low exposure levels, all of these are known to be systemic toxicants that damage a variety of organs (Tchounwou et al. 2012, Mitra et al. 2022, Briffa et al. 2020).

Synthetic Dyes

In both conventional treatment plants and natural ecosystems, dyes are stable and exhibit a high degree of diversity, with around 100,000 different varieties. They are also considered xenobiotics. These contaminants impose varying levels of toxicity and mutagenicity and cause long-term environmental problems due to their strong structural integrity and reactive nature (Yaseen et al. 2019, Tkaczyk et al. 2020). The physical appearance of dye-induced colour in water is unappealing. Red and purple in particular, being artificial, raise more concern than the more naturally occurring colours of blue, green, and brown. Colour also blocks sunlight from penetrating water, reducing the aquatic environment's gross biological productivity (photosynthesis). These substances chemically affect water quality due to the rising concentration of contaminants such as TDS, COD, and BOD. Oxygenic stress, caused by low oxygen levels, restricts the respiratory physiology of living things. On the other hand, excessive TDS increases electric conductivity, which can increase toxicity. Aquatic plants and animals eventually die and deteriorate as a result of these conditions (Al-Tohamy et al. 2022, Rafaqat et al. 2022).

According to their chemical structures, dyes fall into eight major types. Application-based dyes rely on chromophore groupings that facilitate dying according to dye solubility in dye bath, affinity for various fibres, and fixation type. Anionic, nonionic, and cationic are the general terms used to describe the solubility of dyes and pigments. Because of their vivid colours and water solubility in traditional wastewater treatments, reactive and acidic anionic dyes—among direct, acidic, and reactive—pose the greatest challenges. Anthraquinone, azoic, reactive, and disperse dyes are cationic dyes, nonionic dyes, like disperse, do not ionise in an aqueous solution. Reductive cleavage of the azo and nitro groups in dispersed dyes results in hazardous aromatic compounds in soil sediments, and few dispersion dyes efficiently bioaccumulate (Srivastava et al. 2021, Slama et al. 2020).

All 'colouring substances' combined, synthetic organic dyes are the largest class and are extensively used in the paper, textile, tanning, and printing industries. A small number of synthetic organic dyes have also found usage in the food processing, pharmaceutical, and cosmetics sectors. Toxic substances in wastewater are transported over great distances. The prolonged presence of these substance in the water and soil poses serious health concerns for aquatic life, and depletes soil fertility and the ability of aquatic plants to photosynthesize, which consequently results in anoxic conditions for aquatic plants and animals (Srivastava et al. 2021, Nair et al.2023).

Textile dyes reduce photosynthesis, prevent plant growth and development, penetrate the food chain, and cause bioaccumulation. This potentially deteriorates the aesthetic value of water bodies and raises biochemical and chemical oxygen demand. Consumption or inhalation of textile dyes can result in acute toxicity, including bladder cancer, dermatitis, neurological disorders, reduced enzymatic activity, and skin and eye irritation (Slama et al. 2020, Tomar et al. 2023, Nair et al. 2023).

Pesticides from Agricultural Origin

In compliance with the Stockholm Convention, organochlorine pesticides, widely employed in agriculture, are currently prohibited in the majority of countries and classified as persistent organic pollutants (POPs). However, even with their extended half-lives and high persistence, these pesticides and their degradation products (TPs) and transformation products (TPs) are still found in agricultural soils at high concentrations, including dichlorodiphenyltrichloroethane (DDT). Organophosphorus insecticides and neonicotinoids, which have quicker rates of biodegradation than outmoded organochlorine pesticides, have gradually replaced them in recent years (Kalyabina et al. 2020, Syafrudin et al. 2020).

Current-use pesticides (CUPs) are formulations designed to prevent PBT (persistent, bioaccumulative, and toxic) characteristics. CUPs have been found in Arctic air and surface water samples. Emerging pollutants (EPs) are compounds not yet subject to regulations but have the potential to impact the environment. Toxicological evaluations of these chemicals are needed to determine their characteristics, metabolism, and ecological fate. CUPs are frequently identified as EPs. An additional issue is the equating of the harmful effects of commercial pesticides with the effects of their active components, potentially leading to an inaccurate evaluation of their safety. The identification of hidden risks and the separation of effects have received a great deal of attention lately (Campos-Mañas et al. 2019).

Non-ionic surfactants, as well as co-solvents like N-methyl-2-pyrrolidone (NMP) and organosilicone polyethoxylates, are the most often utilised auxiliary components in pesticides. They guard against quick deterioration and increase the active ingredient's solubility. Research has demonstrated that the use of surfactants and seed oil additives prolongs the half-life of metazachlor and slows down its degradation in soil. Additionally, formulants change the physicochemical characteristics of the active components in pesticide formulations by acting as drift retardants, spreaders, dyes, and (anti)foaming agents (Alhalili et al. 2022).

The primary adverse impacts of pesticides are their toxicity to non-targeted species, their long-term effects, and their persistence in the ecosystem. But another problem is that there is no systematic, science-based method for creating adjuvant formulations, many of which have not had their toxicity thoroughly studied. Formulants, also known as co-formulants, are generic terms for inert substances typically added to pesticide formulations to improve their stability and effectiveness. For example, adding co-solvents, penetrant boosters, and inert surfactants increases the herbicide's mobility and uptake by weed leaves. Inert agrochemicals that protect crops from herbicide toxicity by improving their biochemical response are yet to be thoroughly explored (Metcalfe et al. 2016, Syafrudin et al. 2020).

Farmers and other sensitive populations, such as agricultural workers, are more likely to have the active ingredient penetrate both the target organism and their own skin when using auxiliary compounds. The composition of ingredients in commercial formulations can alter the rate of percutaneous penetration of pesticides. For example, methiocarb and pirimicarb have been found to have significantly different penetration rates, lag times, and amounts of penetrating substance. For the hazardous effects of pesticides on farmers and workers to be accurately assessed, this is a crucial component (Kalyabina et al. 2020, Dehghani et al. 2024).

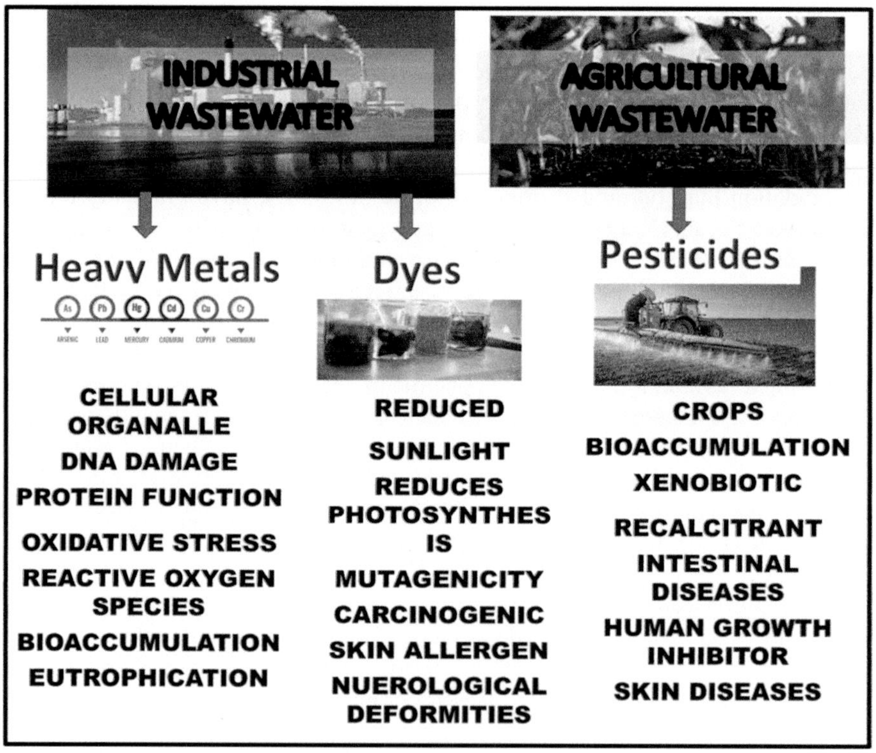

Figure 3: Major wastewater pollutants and their effects on biosphere.

Pollutants of Pharmaceutical Origin

The primary sources of pharmaceutical pollutants in the environment are hospitals, animal faeces, pharmaceutical ventures, research projects using medicinal substances, and the release of outdated medications into the environment. According to several reports, hospitals are the main contributors to pharmaceutical releases into the environment. Hospital wastewater includes radioactive elements, drug conjugates, pathogens, pharmaceutical residues and their metabolites, and other substances. Hospital effluent discharge into municipal treatment slows down the organic contaminant's biodegradation process, even at low medication concentrations (Gadipelly et al. 2014, Taharani et al. 2016). Gas production is reduced and the microbial community's capacity to denitrify effluents is diminished when diclofenac is continuously added to the anoxic sludge-treatment process. Concerns were raised about the potential impact on aquatic ecology when untreated wastewater from a wastewater treatment plant was directly discharged into natural water bodies (Massima Mouele et al. 2021). The aquatic flora and fauna are disturbed, and human health is at risk, when these pharmaceutical pollutants are present in the receiving environment. Due to their low concentrations, drug molecules have been shown in numerous short-term toxicity tests not to have an acute toxic effect on aquatic organisms, nevertheless, long-term (chronic) consequences result from their continual release and exposure to aquatic biota. Aquatic organisms' traits and behaviours alter as a result of prolonged exposure to low concentrations of drugs (Massima Mouele et al. 2021, Samal et al. 2022). The most popular and durable medicinal items include:

Antibiotics

The constant introduction and presence of antibiotics in the environment makes them a pseudo-persistent chemical. Because antibiotics are meant to kill and impede the growth of microorganisms, their presence and release are likely to be especially concerning. This is because they prevent helpful microbes from acting in treatment plants and play a role in their removal. A further consequence of continuous exposure to antibiotics is the rapid development of resistance mechanisms by the microbiological population living in wastewater. Many antibiotic compounds are found in the aqueous and solid phases of untreated wastewater. Both surface water and wastewater frequently contain sulfonamides, macrolides, and fluoroquinolone antibiotics. Treated and untreated wastewater contain tetracyclines, which are commonly used broad-spectrum antibiotics (4-epitertracyline). Antibiotics belonging to the fluoroquinolone and tetracycline families interact with metal cations, present in wastewater, resulting in the formation of complex compounds that accumulate in sewage sludge. Since 90% of all antibiotics used by humans are eliminated through urine and faeces, the presence and persistence of antibiotics in water bodies pose significant concern (Liu et al. 2023, Massima Mouele et al. 2021).

Therapeutic Harmones

Therapeutic hormones are synthetic versions of natural hormones found in plants and animals. They have a positive impact on both human and animal health. The most commonly detected hormone in the environment is oestrogen—a synthetic oestrogenic steroid used in hormone replacement therapy and birth control. As a result, oestrogen and its metabolites are a common class of newly discovered pharmaceutical pollutants. Their presence in river ecosystems has a negative impact on the development and reproduction of non-target organisms. Numerous investigations have confirmed the presence of oestrogen, at concerning concentrations, in the influents and effluents of wastewater treatment plants (Taharani et al. 2016, Samal et al. 2022).

Analgesics

Analgesics are the most common medications for treating inflammation and relieving pain. Meprobamate, ibuprofen, diclofenac, naproxen, acetaminophen, and other drugs in this class are considered significant environmental contaminants due to their persistence in aquatic environments. There is a significant health risk associated with these xenobiotic substances in natural water bodies due to the lack of knowledge regarding the long-term effects of their ingestion through drinking water. Therefore, it is advised that medications be completely and effectively removed before final effluent is released into water bodies (Xu et al. 2022).

By-products and Metabolites

During the metabolism of pharmaceutical chemicals, hydrophilic, physiologically active, and polar metabolites are excreted through urine and faeces by humans and animals. These active metabolites accumulate in the tissues of aquatic creatures and include N4-acetylsulfamethoxazole, 10,11-dihydro-10,11-epoxy carbamazepine, and 4-hydroxydiclofenac. In addition to having the ability to bind covalently to cellular proteins, they may also cause immune responses or harm to the body. It is stated that, compared to their parent chemicals, these metabolites are 50% more hazardous. Because of their greater concentration and toxicity, studies and analyses of pharmaceutical metabolites are particularly important for determining the fate of the parent compounds. The parental medicinal chemicals that are not well metabolised change and affect the activity of the microbial community. Because of their greater mobility and lower sorption potential, these metabolites are persistent and can be found in environmental samples (Frascaroli et al. 2020, Xu et al. 2021).

Types of Remediation

The wastewater remediation can be carried out by physical, chemical, or biological methods.

Physical Water Treatment

Physical wastewater treatment involves screening, sedimentation, and skimming to remove solid waste without chemicals. Sedimentation suspends insoluble particles, separating pure water. Aeration circulates air to provide oxygen to water, while filtration removes contaminants using special filters. These methods help separate contaminants and insoluble particles from wastewater, ensuring a clean and healthy environment. During the course of a cleanup operation, booms are used to collect oil pollutants and prevent them from spreading across the surface (Vocciante et al. 2019). After booms, contaminants are further absorbed and adsorbed using skimmers and sorbents (Kumari et al. 2019). The filter most frequently used is the sand filter. Combinations of engineering and physical treatment processes are also used for bioremediation. These include mechanical aeration and algae removal, dilution of water, and the building of hydraulic structures, etc. A number of engineering processes are capable of effectively restoring the quality of river water but may lead to the destruction of natural ecosystems along with high investment and maintenance costs (Peilin et al. 2019).

Chemical Remediation

Chemical remediation is carried out by the addition of chemicals to remove pollutants from wastewater. Adsorption, reduction, oxidation, complexation, flocculation, precipitation, and oxidation are the common mechanisms for chemical-mediated treatment. Algaecides are used to remove suspended solids (SS) and algae. The chemical remediation method is simple and fast, however, the chemicals themselves can pollute the environment (Xu et al. 2022). Therefore, chemicals used for flocculation or precipitation should be environmentally safe.

Biological Remediation

Bioremediation is the most reliable, safe, affordable and sustainable method of pollution treatment that involves the use of biological agents (Patel et al. 2020). To determine the viability of this method, it is important to take into account the location, nature, and level of pollution. Microbes are highly effective in the remediation of environmental pollutants. Due to their ease of growth and manipulation, they are preferred over plants for bioremediation. For a sustainable environment, it is therefore necessary to improve the use of microorganisms as agents of bioremediation.

Microbial Enzymes as Bioremediation Agents

It has been discovered that a wide variety of microbial enzymes are helpful in environmental restoration. Oxidation, ring-opening, reduction, etc., are among the enzyme-mediated mechanisms used in the bioremediation processes (Bhandari et al. 2021). Many parameters, including pH, temperature, contact duration, and concentration, control the activity of enzymes (Bhandari et al. 2021). Even though enzymatic bioremediation is a sustainable method, it is an expensive and time-consuming process. Therefore, it is not practical in settings where bioremediation needs to happen immediately (Narayanan et al. 2023). The type of enzyme and persistence of the pollutant determine the enzyme's effectiveness in the remediation process.

Amylase

Amylases are classified into (a) α-amylases (E.C. 3.2.1.2), (b) β-amylases (E.C. 3.2.1.3), and glucoamylases (E.C. 3.2.1.1). Also, they can be categorised as exo-

and endo-amylases (Fleuri et al. 2015). Microorganisms are extracellular producers of amylase. Microorganisms such as *Aspergillus niger*, *Aspergillus oryzae*, *Bacillus amyloliquefaciens*, *Bacillus licheniformis* and *Bacillus circulans* are used for α-amylase production.

Five blue-green algae species isolated from the Iraqi aquatic environment have been reported for the reduction of hydrocarbon pollution (Al-Hussieny et al. 2020). Three algal enzymes—namely amylase, protease, and beta-lactamase—were detected from these isolates and were effective in reducing hydrocarbon concentrations. The study explores the use of microorganisms in wastewater bioremediation, specifically starch bioremediation with *Aspergillus oryzae* (Costa et al. 2021). The amylase-producing fungus was tested using submerged fermentation technology and solid-state fermentation on rice hulls, demonstrating its ability to grow and express amylase activity. α-Amylase is a potential contender to digest low-density polyethylene, a hazardous contaminant generated from petroleum, in addition to n-alkanes, a subclass of alkanes thought to be the most prevalent petroleum-derived molecules (Pinto et al. 2020). Based on growth on wastewater agar medium, Sonune and Garode isolated five bacterial isolates from 34 municipal wastewater and sludge samples (Sonune and Garode 2018). All isolates produced lipase and protease, with one producing amylase. The study found that isolate NW6 was the most efficient for bioremediation, reducing BOD, COD, NO_3^-, and PO_4^{3-} by approximately 50%, 20%, 56%, and 31%, respectively, after 72 hours. The isolate was optimised for inoculum size, retention time, and agitation speed. Using a 10% inoculum size, NW6 reduced BOD, COD, nitrate, and phosphate by approximately 54%, 36%, 24%, and 26%. NW6 could be useful as an enzyme producer and in municipal wastewater treatment.

Lipase

Lipase enzymes are effective in breaking down fats and oils, particularly useful in wastewater from restaurants and food industries.

The wide range of catalytic processes that microbial lipases may catalyze in both aqueous and non-aqueous conditions makes them an indispensable biocatalyst. Because of their excellent stability and specificity, lipases are favoured for catalysis over chemical mediated process. The production of meat, dairy, fats and oils, surfactants, tanneries, cosmetics, and medicines, among other activities, is the primary driver of the worldwide need for microbial lipases, particularly bacterial and fungal lipases. Potential uses for lipases include the treatment of wastewater that is high in lipids. Overly fattening wastewater results in major environmental issues, such as clogged sewage lines and the formation of oily layers on the water's surface that obstruct the entry of sunlight and oxygen, harming aquatic life. The physical-chemical treatment techniques now in use for wastewater with high lipid content are expensive, unfriendly, and create secondary pollution issues whereas the enzyme lipase mediated process provides a sustainable approach towards waste water remediation (Nimkande and Bafana 2022).

Ortellado et al. have assessed the mycoremediation of wastewater which is rich in fats and oils using lipase enzymes obtained from the supernatants of fungal c ulture (Ortellado et al. 2022). An enzyme mixture containing enzymes, protease, amylase and lipase obtained by fermentation of jaggery, fruit wastes and water mixture in the ratio of 1:3:10 was reported to be sustainable solution for domestic wastewater treatment (Joseph et al. 2021). Crude Enzymatic Extract (CEE) of *Yarrowia lipolytica* has been reported to degrade the fat-containing effluent (Posso Mendoza et al. 2020). This was evident by the decrease in the fat content as well as reduction of BOD and COD of the effluent. Lipase from *Aspergillus niger* was also found to be effective for the treatment of vegetable oil containing effluents (Vincent et al. 2022).

Cellulase

Cellulases decomposes cellulose from plant materials, helping to reduce solid waste in wastewater. Cellulose is present in considerable amount in the black liquor released from the paper and pulp based sector. According to An et al. (2021b), the pulp and paper sector is ranked 41st, and the wastewater it produces makes up 13.0% of the total volume of wastewater released by China's diverse industries (An et al. 2021). Pulp made from plant materials requires a variety of chemicals, a lot of water, and generates a lot of effluent that is poisonous and intensely coloured (Haq et al. 2016). Chemical/biochemical oxygen demand (COD/BOD) and intense colour are typical characteristics of generated wastewater. Black liquor has significantly contaminated terrestrial and aquatic environments due to its massive pollutant load. Furthermore, a great deal of research indicates that black liquor poses a substantial environmental risk (An et al. 2022).

Thus, it is imperative to treat black liquor before discarding it. 33% of the total organic pollutants in black liquor are made up of cellulose, which is a significant impermeable pollutant. It is one of the primary sources of BOD and COD in black liquor (Azadi Aghdam et al. 2016). Therefore, it is crucial to treat cellulose effectively in order to lower the high pollutant load of black liquor.

The synergistic interaction of *Serrati*a sp. AXJ-M and *Arthrobacter* sp. AXJ-M1 may be potentially useful for bioremediation of black liquor generated from the papermaking industries (An et al. 2022). Cellulases can also be indirectly used in the bioremediation process. Maize stover can be effectively modified by cellulases to remove sulfamethazine from aqueous solutions (Zhang et al. 2020).

Laccase

Laccases degrades phenolic compounds and other complex organic pollutants, often found in industrial wastewater.

Laccases are polyphenol oxidases that contain copper and catalyse the reduction of two pairs of reducing substrate molecules by single electron transfer. This process is linked to the reduction of molecular oxygen to water by four electrons. Furthermore, laccases differ from lignin peroxidase and manganese peroxidase in that they do not require an external supply of H_2O_2 or Mn^{2+}, respectively, or other cofactors for the oxidative cleavage of a wide variety of substrates, particularly phenolic compounds. However, they do have some affinity for metal ions and certain inorganic and organometallic complexes (Unuofin et al. 2019). The review by Unifin et al. has extensively described the role of laccases and other oxidative enzymes in the bioremediation process (Unuofin et al. 2019).

Pleurotus ostreatus has been used for laccase production using cassava waste as substrate (Kumar et al. 2022). This laccase enzyme has been used in the removal of phenolic contaminants and polycyclic aromatic hydrocarbons (PAHs) from lignocellulosic biorefinery wastewater. This suggests that laccases can bioremediate waste generated from different industries. Fluoroquinolone antibiotic levofloxacin in water and pharmaceutical wastewater was reported to be effectively removed by laccases isolated from *P. eryngii*, *P. florida*, and *P. sajor caju* (Mathur et al. 2021). Their results were comparable to commercial laccase, hence can considerably reduce bioremediation costs.

Protease

Proteases are catalysts that break down peptide bonds of proteins. They assist in the degradation of contaminating proteins like keratin, casein, etc., present in wastewater. Proteases can be effectively used for the treatment of hazardous protein-rich waste in water bodies. Alkaline protease pre-treatment facilitates the disintegration of waste-activated sludge flocs, which releases biodegradable organic matter. Activated sludge can be disposed of in this manner while also recovering carbon. Short-chain fatty acids may close 60% of the carbon gap in wastewater, making it an effective and carbon-beneficial solution (Pang et al. 2020, 2022).

Sludge dewaterability is needed for the treatment and disposal of waste-activated sludge in wastewater treatment facilities. The conditioning of sludge with enzymes enhances its dewaterability (Kang et al. 2023). Several enzymes were screened for this application, but among them, α-amylase and neutral protease showed good performance in conditional optimization experiments suggesting eco-friendly conditioning of the activated sludge. In another study, protease and lipase enzymes were used for the hydrolysis and degradation of primary sludge obtained from three wastewater treatment plants in Korea. A total of 33.3% of the volatile suspended solids in the sludge were reduced by enzymatic hydrolysis when these enzymes were produced from microorganisms taken from eight different digester sites in Korea (Tongco et al. 2020). Proteases are used for eco-friendly cleaning of the equipment. As they are biodegradable, they do not cause environmental damage after usage. However, the enzyme-mediated processes are costly, adding to its demerit. Microbiological proteolytic proteases are widely studied enzymes that play crucial roles in metabolic processes and are also extensively utilized by industry (Younes and Rinaudo 2015). Proteases produced by microbial sources are classified based on their acidic or basic properties. Another parameter for their classification is on the basis of presence of functional groups and the position of peptide bond (Panda et al. 2013).

In addition to poultry waste, seawater samples contain alkaline protease-producing microorganisms. By analysis of 16S rRNA partial gene sequences, these microorganisms have been identified as *Bacillus aryabhattai P1, Bacillus sp. P4, Bacillus firmus P2, Bacillus clausii P6, , Bacillus cereus P5, Halobacillus alkaliphilus MGR92, Alkalibacterium kapii MGR70, Bacillus halodurans* MGR125 and *Alteromonas halophila MGR98* (Rathod and Pathak 2018). All these strains exhibited the property to decolourise the textile dye. These bacteria were used in two consortia to decolourise textile dyes to the maximum extent possible. Among

Table 1: Recent reports for the use of microbial enzymes in bioremediation

Enzyme	Source Organism	Application	Reference
Amylase, protease, and beta-lactamase	*Westiellopsis prolific, Anabaena variabilia, Oscillatoria pranceps, Phormidium mucicola,* and *Lyngbya digueti*	Hydrocarbon concentrations reduction	(Al-Hussieny et al. 2020)
Amylase	*Aspergillus oryzae*	Starch bioremediation of paper mill wastewater	(Costa et al. 2021)
α-amylase and neutral protease	-	Conditioning of waste activated sludge for enhanced dewatering	(Kang et al. 2023)
Lipase and protease	*Bacillus licheniformis*	Biodegradation of municipal wastewater in reasonable residence time of 96 h and 200 rpm in terms of reduction in BOD, COD, nitrate and phosphate	(Sonune and Garode 2018)
Lipase and protease	-	Enzymatic hydrolysis and degradation of primary sludge obtained from three wastewater treatment plants	(Tongco et al. 2020)
Protease, amylase and lipase	Garbage enzyme solution of jaggery, fruit wastes and water mixture in the ratio of 1:3:10	Low-cost alternative for domestic wastewater treatment	(Joseph et al. 2021)
Crude Enzymatic Extract (CEE)	*Yarrowia lipolytica*	Industrial fat effluent	(Posso Mendoza et al. 2020)
Lipase	*Aspergillus niger*	Treatment of vegetable oil effluent	(Vincent et al. 2022)
Cellulases	*Serratia sp.* and *Arthrobacter sp. AXJ-M1*	Co-culture had detoxifying effect on black liquor	(An et al. 2022)
Laccase	*Pleurotus ostreatus*	Removing phenolic contaminants and polycyclic aromatic hydrocarbons (PAHs) from lignocellulosic biorefinery wastewater	(Kumar et al. 2022)
Laccase	*P. eryngii, P. florida* and *P. sajor caju*	Fluoroquinolone antibiotic, levofloxacin, in water and pharmaceutical wastewater	(Mathur et al. 2021)
Alkaline protease	Bacterial consortia	Decolourisation of Direct Red-81 (68.88%) and Direct Orange-34 (70.78%)	(Rathod and Pathak 2018)

them, one consortia, named as SW2 has shown highest decolorisation of Direct Red-81 (68.88%) and Direct Orange-34 (70.78%) after 144 h. Recent reports of the enzymatic intervention in wastewater processing are mentioned in Table 1.

Techniques Used in Enhancing Catalytic Activity of Enzymes

The chemical industry is undergoing a transformation that presents a new challenge in terms of expanding the use of microbes and enzymes in chemical synthesis (Hazeena et al. 2019). Enzyme activity takes place under optimal conditions of pressure, temperature, and pH, both in aqueous solutions and in organic solvents. Enzymes directly isolated from microbes may not have the attributes that are needed, prompting researchers to seek alternative enzymes. With advances in biotechnology techniques, researchers may manufacture and develop new enzymes with better and novel uses. The advent of recent techniques has allowed for the improvement of the catalytic properties of existing enzymes and the development of new enzymes with enhanced properties. Some recent technologies that are being used for enzyme improvement are fusion proteins, site-directed mutagenesis, metagenomics, and bioinformatics-based tools.

Metagenomics for Novel Enzyme

Instead of using traditional microbe culture procedures, this approach uses direct sequencing of the DNA isolated from environmental samples to identify the target gene. The extracted metagenomic DNA is used to generate metagenomic libraries in appropriate vectors. Several vector alternatives have been considered, depending on the size of the insert, such as λ phage, plasmids, fosmids, cosmids, and bacterial artificial chromosomes (BAC). Following vector translation, the expression systems are adapted (Lorenz and Eck 2005).

A metagenomic study can be conducted through sequence-based screening as well as function-based screening. Sequence-based metagenomics involves sequencing the metagenomic DNA, assembling the genome, identifying the gene, searching the entire metabolic pathway, and comparing the results to other communities.

Next generation sequencing (NGS) enables the collection of massive amounts of data at a fraction of the expense of previous sequencing methods (e.g. Sanger) (Koutsandreas et al. 2019). To optimise enzyme screening and selection, metagenomics should be combined with other omics approaches such as metatranscriptomics and metaproteomics. Metatranscriptomics may be used to screen for actively expressed proteins quickly (Shakya et al. 2019). A bioinformatics approach is also employed for analysis of protein sequence, structure , Pfam analysis, and understanding phylogeny of the protein (Prayogo et al. 2020).

Engineering for Enhanced Enzyme Production

Genetic material comprises the entire information that determines its structure, function, secretion, and control. Thus, approaches including genetic structural alterations have proven to be successful in increasing production.

Optimizing promoters: The initial step in designing for increased enzyme synthesis might be at the promoter level. The promoter region is responsible for binding to RNA polymerase, which starts transcription. The use of a strong, constitutively expressing promoter with high affinity should be ideal for the construction of large-scale biotechnological products (Qin et al. 2011).

Increased expression by transcription regulation: The key molecules that control metabolic pathways are activators, inducers, and repressors, whereas transcriptional factors are required to regulate enzyme expression and production.

Increased expression by translation regulation: Translation processes encompasses the synthesis of protein from mRNA, their folding into a 3-dimensional structure, and secretion to the appropriate location. The translation process can be regulated for enhanced expression.

Increased expression by codon optimization: Codon optimisation assists in gene alteration in such a way as to enhance the level of protein expression without altering the amino acid content of the protein (Şen et al. 2020).

Directed Evolution

Directed evolution simulates natural evolution by using iterative rounds of protein diversification and selection, and screening of the library to discover proteins with optimal functionality.

In this technology, a mutant library is prepared and screened for the appearance of the desirable trait. This process is repeated until the desired trait is obtained. Higher mutants in the library can be detected if there is a likelihood of a larger number of mutations. Error-prone PCR, site saturation, chemical mutagenesis, and DNA shuffling are some of the common methods to induce genetic mutation to obtain altered peptides with desired functions (Bhatia et al. 2021).

Site Directed Mutagenesis

Using this method, an amino acid can be altered at a specific location within the protein that will significantly influence the protein's activity, allowing for rational or semi-rational mutagenesis. By inserting or deleting a specified spot within enzymes with known sequences, this can be accomplished, resulting in recombinant proteins with enhanced specificity, stability, activity, and solubility. By using site-directed mutagenesis to alter a mature protein, it is possible to produce more attractive enzymes with improved functionality (Bhatia et al. 2021).

Enzyme Immobilization

The binding of an enzyme to an inert and insoluble support material that reduces or eliminates the enzyme's mobility is known as enzyme immobilisation. Its various benefits include higher pH and temperature resistance, stability, reusability, and the capacity to recover enzymes (Homaei et al. 2013). However, because of steric hindrance, mass transfer limits, and conformational changes in the enzyme, enzyme immobilization may lead to reduction in enzymatic activity (Ding et al. 2015). To address these issues, new techniques and materials that deal with the shortcomings of existing immobilization techniques need to be developed.

To immobilize enzymes, four methods exist, namely entrapment, adsorption, covalent attachment, and crosslinking. Physical immobilization methods such as adsorption and entrapment rely on weak interactions between the enzyme and the support, such as van der Waals forces and ionic bonds. This method preserves the enzyme activity as the native enzyme conformation remains unchanged (Hernandez and Fernandez-Lafuente 2011). In chemical immobilization, covalent bonds are used to connect enzymes to support materials by using agents such as glutaraldehyde in order to form covalent bonds. By altering the enzyme structure, this may impair the enzyme's activity by blocking the active site (Al-Maqdi et al. 2021).

Enhanced Bioremediation by Catalytically Improved Protein Through Molecular Approaches

Although enzymes have been a promising tool in bioremediation, there is a constant surge for new and enhanced approaches for obtaining better results. The same applies to the bioremediation techniques whereby several modifications and alterations in the existing enzymes have been performed to remove various contaminants present in the wastewater. Here we have discussed recent reports on enzyme alteration for their intervention in bioremediation. The recent reports on the diverse enzymatic modifications and their respective impact on the pollutant are tabulated in Table 2.

As mentioned in the previous section, metagenomics serves as an important technology for fishing out significant genes directly from environmental samples. A thermo-halo-alkali-stable protease termed PersiProtease1 has been obtained from the tannery wastewater metagenome. Owing to its thermostability and resistance against salinity and several inhibitors, it can serve as a potential candidate for the biodegradation of protein-rich industrial effluents (Ariaeenejad et al. 2022).

A halotolerant protease-assisted nanoporous carbon catalyst containing halotolerant protease from the halophile *Lysinibacillus macrolides* has been used to remove proteins from effluents generated by tanneries. This wastewater contained high total dissolved solids (Pounsamy et al. 2019). Ethylenediamine (EDA)/glutaraldehyde-functionalised nanoporous activated carbon was used for the immobilisation of lipase and confirmed by techniques such as SEM, FTIR, TGA-DSC, and XRD. This protease was found to be effective in degrading the proteins in tannery wastewater with high TDS in 90 min at pH 6 and 30 °C, and further confirmed by circular dichroic studies, UV-visible and UV-fluorescence spectroscopy, and ESI-mass spectrometric analysis. It was also effective in reducing the BOD and COD values in wastewater.

Lipase enzyme isolated from Porcine Pancreas and immobilized on graphene oxide was found to be effective in removal of AZO dyes from industrial wastewater effluent (Yao et al. 2022). The characterisation results confirmed the immobilisation of the lipase enzyme. The dye removal was investigated by adsorption isotherm and kinetic studies, and it was found that the azo dyes were removed by 89.47% after 240 minutes of treatment. Also, the immobilised lipase enzyme could be reused up to four cycles. Modified lipase from *Rhizopus oligosporus* (Strain-3) is useful in olive mill wastewater treatment. This is highly useful as the olive mill effluent

Table 2: Recent reports on the use of modified and enzymatically enhanced proteins used in bioremediation of wastewater

Enzyme	Modification	Substrate Pollutant	Treatment Conditions	Impact	Reference
Laccase from *Trametes versicolor*	Insolubilized laccase as cross linked enzyme aggregates (CLEAs)	Treatment of trace organic contaminants (TrOCs)	pH 7.0 and 20 °C, 1000 U/L of CLEAs	Laccase removed 11 TrOCs in the range of about 20–99%	(George et al. 2022)
Laccase from *Trametes versicolor*	Immobilized over a magnetic amino-functionalized metal–organic framework Fe_3O_4-NH_2@MIL-101(Cr)	Biodegradation of Reactive Black 5 (RB) and Alizarin Red S (AR) dyes in water	5 mg/L dye concentration, temperature of 25 °C, and a pH of 4	Laccase system caused the total removal of AR and 81% of the RB	(Amari et al. 2021)
Laccase from *Trametes versicolor*	Laccase immobilized activated carbon derived from pomegranate peels	A range of emerging pollutants, namely diclofenac, amoxicillin, carbamazepine, and ciprofloxacin from water and wastewater	pH 4, 35 °C, and a laccase concentration of 2.5 mg/mL	Target pollutants (50 mg/L) were eliminated in 2 h	(Al-sareji et al. 2023)
Laccase from *Trametes versicolo*r	Immobilization on Cu/ Zn ZIF	Dye wastewater	–	Exhibited decolourisation efficiency for dye wastewater	(Yang et al. 2023)
Laccase from *Weissella viridescens* LB37	Immobilized to the prepared magnetic chitosan NPs (MAG-CTS NPs)	Azo dyes from wastewater	pH – 6.0, temperature – 30 °C, concentration of glutaraldehyde – 1% and immobilization time – 1 h	High removal capacity were determined for Direct blue 15, Evans blue, Reactive black 5 and Acid red 37 azo dyes using immobilized laccase	(Nadaroglu et al. 2019)
Laccase	Dopamine (DA) functionalized sodium alginate (SA)/polyethylene glycol (PEG) composite hydrogels	Decolourisation of Reactive Blue 19	-	High reutilisation rate, and immense potential in waste water treatment	(Su et al. 2024)

(*Contd.*)

(*Contd.*)

Enzyme	Modification	Substrate Pollutant	Treatment Conditions	Impact	Reference
Laccase	α-cyclodextrin-laccase nanocomposite	Degradation of diclofenac	Room temperature and at pH 5	Micropollutant removal of pharmaceutical origins from municipal wastewater	(Meiczinger et al. 2022)
Protease	Protease-immobilized nanoporous carbon catalyst	Tannery saline wastewater	2 h, pH 6–8, and 30–40 °C.	Generation and extraction of protein hydrolysates from tannery wastewater were both performed effectively	(Pounsamy et al. 2024)

is rich in lipids and oils, which may pose consequences on the environment if not managed by proper treatment. Lipase immobilisation using chitosan beads is a cost-effective and environmentally friendly solution for treating wastewater with high lipid content (Waseem et al. 2024). Cell-bound lipases (CBLs) produced by the co-culture of Magnusiomyces spicifer AW2 and Staphylococcus hominis AUP19 acted as whole-cell biocatalysts for the bioremediation of heat-treated various palm oil mill effluent (Fibriana et al. 2022). This work demonstrated that the blended oily waste medium was appropriate for the low-cost generation of yeast-bacterium CBLs and their prospective uses in solvent-free biodiesel production and wastewater bioremediation. This may significantly contribute to affordable and sustainable biofuel generation and waste treatment.

Nanocatalysts have helped overcome the limitations of conventional enzymes. Cellulase immobilisation on nanostructured supports provides greater surface area, improved enzyme loading, and a stable environment to act efficiently. Different nanostructured supports for cellulases like carbon nanotubes, polymer nanoparticles (NPs), nanohydrogels, nanofibres, and silica NPs have been discussed in detail in an extensive review by Sulman et al. (2022). This review discusses cellulase immobilisation techniques, the shape of nanostructured supports, multienzyme systems, and variables that influence enzyme activity to convert cellulosic biowaste to fermentable sugars as efficiently as possible (Sulman et al. 2022). These nano-cellulases can be used in several applications, and several research groups have studied their role in bioremediation. Ariaeenejad et al. have investigated the role of a nano-biocatalyst, model laccase enzyme (PersiLac1), for the removal of malachite green (MG) and congo red (CR) from water (Ariaeenejad et al. 2022). Through the use of a synergistic adsorption-degradation strategy, this nano-biocatalyst enabled the dye molecules to be readily adsorbed to the nano-carrier and then efficiently destroyed by the enzyme.

Laccase from *Trametes versicolor* was aggregated by the three-phase partitioning technique followed by glutaraldehyde cross-linking, resulting in insolubilised laccase as cross-linked enzyme aggregates (CLEAs). This aggregate was effective in the removal of trace organic contaminants (TrOCs) from municipal wastewater (George et al. 2022). In another report, laccase from *Trametes versicolor* has been immobilized over a magnetic amino-functionalized metal–organic framework Fe_3O_4-NH_2@MIL-101(Cr) (Amari et al. 2021). It effectively degraded Reactive Black 5 (RB) and Alizarin Red S (AR) dyes in water. This laccase system was effective in the total removal of AR and 81% of the Reactive Black 5 dye. In another recent research work, pomegranate peels (PPs) were employed to manufacture profitable, green, and high-quality activated carbon (Al-sareji et al. 2023). This may be effective in eliminating a variety of emerging contaminants from water and wastewater, including diclofenac, amoxicillin, carbamazepine, and ciprofloxacin. Commercial laccase from *Trametes versicolor* was immobilised on Cu/Zn ZIF to act on dye wastewater (Yang et al. 2023). The immobilised enzyme exhibited effective treatment of dye wastewater and could be used in multiple cycles. In another report, laccase from *Weissella viridescens* LB37 was effective in the degradation of Direct Blue 15, Evans Blue, Reactive Black 5, and Acid Red 37 azo dyes when immobilised on magnetic chitosan nanoparticles

(Nadaroglu et al. 2019). A recent study has also reported the efficacy of immobilised laccase. To achieve the recyclability of the laccase, dopamine (DA)-functionalised sodium alginate (SA)/polyethylene glycol (PEG) composite hydrogels were used to manufacture immobilised laccase. As the laccase's protective carrier, the SA/PEG composite hydrogels demonstrated exceptional catalytic stability across a range of interfering conditions (Su et al. 2024).

Research group led by Pounsamy used a protease-immobilized nanoporous carbon catalyst to produce protein hydrolysates by breaking down proteins present in protein-rich tannery wastewater (Pounsamy et al. 2024). The protease that was isolated from *Lysinibacillus fusiformis* MR1, showed 385 U/mL of proteolytic activity. The ideal conditions for immobilization were 2 hours, pH 6-8, and 30-40 °C. Different techniques such as FTIR, TGA, DSC, and XRD analyses were used to test immobilization. HPLC-free amino acid analysis validated the generation of protein hydrolysates, while electroanalytical methods confirmed their redox characteristics. Protein hydrolysates were extracted using a Fe_3O_4-impregnated nanoporous carbon catalyst (Fe_3O_4-NPCC).

In Silico Interventions in Bioremediation

As a science devoted to the computer analysis of large quantities of biological data, bioinformatics combines biology with information technology. The bioinformatics field consists of the storage and search of data, and the development of statistical tools and algorithms to analyse and determine relationships among biological datasets, including sequences of macromolecules and their structures, expression patterns, and different biochemical pathways. In the field of bioinformatics, bioremediation is the new glittering frontier. The use of simple bioinformatics searches that combine a variety of data sources is more accurate and faster than conventional methods for identifying potential bioremediation targets (Suresh et al. 2008). It is possible to screen pollutants for their susceptibility to degradation by enzymes that have already been characterised using the various *in-silico* tools listed in Table 3. Among these tools, protein-ligand docking plays a crucial role in bioremediation by providing insights into the interaction between bioremediation enzymes and their target pollutants. By simulating the binding process, we can identify potential binding sites on the enzyme and optimise the design of bioremediation systems (Rangel et al. 2022). These enzymes can then be used to design synthetic enzymes that are capable of degrading the pollutants. The synthetic enzymes can then be used in bioremediation processes to clean up the environment. A bioinformatics approach combining genomics with proteomics can help in addressing the molecular mechanisms involved in degradation (Fulekar 2007, 2009, Fulekar and Sharma 2008, Malla et al. 2018).

Proteomics-based Bioremediation

As a result of the generation of large and complex genome sequencing datasets, proteomics analysis is a key post-genomic component, involving the analysis of

Table 3: List of *in-silico* tools that assist in the implementation of bioremediation

Tools (Server/ Database)	Description	URL
	Tools for homology modelling	
Swiss Model	A fully automated protein structure homology-modelling server, accessible through the Expasy web server, aiming to make protein modelling accessible to all life science researchers throughout the world	https://swissmodel.expasy.org/
Alpha Fold	Using the primary amino acid sequence and aligned sequences of homologues, AlphaFold predicts the 3D coordinates of all heavy atoms in a given protein	https://alphafold.ebi.ac.uk/
MODELLER	It can also be used to predict the 3D structure of a protein from its amino acid sequence	https://salilab.org/modeller/
I-TASSER	I-Tasser is a hierarchical approach to protein structure prediction and structure-based function annotation	https://zhanggroup.org/I-TASSER/
	Molecular docking tools	
AutoDock	AutoDock Vina is an open-source program for doing molecular docking. It is based on the Vina algorithm	https://vina.scripps.edu/
CB Dock	The CB-Dock2 server integrates cavity detection, docking, and homologous template fitting into the CB-Dock server for blind docking of protein-ligands	https://cadd.labshare.cn/cb-dock2/index.php
H Dock	A highly integrated suite of homology search, template-based modelling, structure prediction, macromolecular docking, biological information integration, and job management, which is capable of robust and fast docking of proteins with each other	http://hdock.phys.hust.edu.cn/
Z Dock	An easy-to-use docking server for protein complexes and symmetric multimers based on the rigid-body docking programs.	https://zdock.wenglab.org/
HADDOCK	It is an information-driven flexible docking approach for modelling biomolecular complexes. It utilizes biochemical and/or biophysical interaction data, including chemical shift perturbation data from NMR titration experiments	https://rascar.science.uu.nl/haddock2.4/
GNINA 1.0	Protein-ligand docking tool based on convolutional neural networks (CNNs) as a scoring function	https://github.com/gnina/gnina

(*Contd.*)

(*Contd.*)

Tools (Server/ Database)	Description	URL
	Tools for homology modelling	
Biovia Discovery Studio	Visualization tool for viewing, sharing, and analyzing protein and modelling data	https://discover.3ds.com/discovery-studio-visualizer-download
LigPlot$^+$	LigPlot+ is used for the generation of 2D ligand-protein interaction diagrams. It is used to visualize and analyze the interactions between ligands and proteins. It can also be used to generate 3D interaction diagrams	https://www.ebi.ac.uk/thornton-srv/software/LigPlus/
PyRx	A virtual screening tool for computational drug discovery, PyRx provides an efficient method for screening libraries of compounds against potential targets	https://pyrx.sourceforge.io/
	Tools for toxicity measurement	
HazardExpert (CompuDrug)	Predictive software tools for pharmaceutical and medicinal chemistry research	http://akosgmbh.de/mobile/compudrug.htm
OncoLogic	Predictive models and tools for assessing chemicals under the Toxic Substances Control Act (TSCA)	https://www.epa.gov/tsca-screening-tools
ADMET Predictor	SwissADME is a web-based tool that provides free access to physicochemical properties, pharmacokinetics, drug-likeness, and medicinal chemistry	http://www.swissadme.ch
ECOSAR	In order to calculate the acute (short-term) and chronic (long-term or delayed) toxicity of a chemical to aquatic organisms, such as fish, aquatic invertebrates, and aquatic plants, the program uses structured activity relationships (SARs)	https://www.epa.gov/tsca-screening-tools/ecological-structure-activity-relationships-ecosar-predictive-model
ECOTOX	ECOTOX is a comprehensive database that provides aquatic and terrestrial species toxicity data for single chemicals	https://cfpub.epa.gov/ecotox/
	Tools for molecular dynamics simulations	
WebGRO	Molecular dynamics simulations can be performed using WebGro, a fully automated online tool for simulations of macromolecules (proteins) alone or in complex with ligands (small molecules)	https://simlab.uams.edu/

(*Contd.*)

GROMAC	The GROMACS package was designed primarily for the simulation of proteins, lipids, and nucleic acids	https://manual.gromacs.org/current/download.html
Tools for pathway prediction		
EAWAG-PPS	Using biotransformation rules, EAWAG-PPS pathways are predicted by identifying the functional chemical groups in a starting compound	http://eawag-bbd.ethz.ch/

protein profiles. Furthermore, it examines proteomes from the perspective of protein composition, structure, and respective function. In addition to determining changes in protein composition and arrangement, proteomic analyses have also been used to identify proteins that contribute to the physiological reaction of microorganisms when they are contaminated by anthropogenic substances (López and dos Santos Silva 2023). In the bioremediation of polycyclic aromatic hydrocarbons (PAH), proteomes play a significant role (Fulekar and Sharma 2008, Bisht et al. 2015). The removal of PAHs from the environment is extremely important. The removal of PAHs *in situ* as well as *ex situ* can be achieved through the use of natural and genetically engineered microorganisms. It is possible to gain additional insight into bioremediation-related genes and their regulation by analysing physiological changes in an organism during bioremediation using a proteomics approach. From *Mycobacterium* sp. strain PYR-1, an 81-kDa protein resembling catalase–peroxidase was detected by 2-DE as a response to pyrene exposure. When phenanthrene, dibenzothiophene, and pyrene were used to expose *Mycobacterium* sp. strain PYR-1 to 2-DE, six major proteins were significantly induced and overexpressed. The N-terminal and internal peptide sequences were used to identify several putative enzymes that are specific to pyrene. Moreover, *Mycobacterium* sp. strain 6PY1 has been shown to induce two ring-hydroxylating dioxygenases during pyrene catabolism by expressing PDO1 and PDO2 (Fulekar and Sharma 2008). The identification of unknown genes and proteins during the anaerobic degradation of toluene and ethylbenzene has progressed. As a result of the recent combination of transcriptomics and proteomics, new pathways for aerobic and anaerobic biodegradation of toxic wastes have been identified, opening the door for additional identification of signature proteins (Chen et al. 2009).

It has been shown that two-dimensional DE can be updated to accommodate compartmental proteomics by following MudPIT (multidimensional protein identification technology) (Paoletti et al. 2004). The eco-proteomic approach has been modified towards bioremediation using mass spectrometry. In two-dimensional gels, MALDI-TOF-MS has been widely used to determine the proteins. The surface-enhanced laser desorption–ionisation mass spectrometry (SELDI-TOF-MS) method has been demonstrated to be an effective tool for identifying proteins involved in the degradation pathway of polycyclic aromatic hydrocarbons. A direct detection and identification of potential contaminants in water has also been described using liquid chromatography–mass spectrometry (LC-MS) together with bioinformatics. This technique is able to detect and identify the presence and concentration of various pollutants in water, such as pesticides, pharmaceuticals, and heavy metals. It is a comprehensive and reliable method that can also be used to monitor changes in water quality over time (Jacob 2010).

Genomics-based Bioremediation

In genomics, all genes in an organism are understood based on their entire DNA sequence, which is a powerful computer technology (Giani et al. 2020). A significant part of this field is the determination of the entire DNA sequences of organisms as well as fine-scale genetic mapping. Recently, a large number of bacterial genomes relevant to biodegradation have been published, making it possible to examine the complex interactions between cellular reactions on a genomic and proteomic level. To gain a quantitative understanding of how cells function, each gene and protein must be considered within their dynamic context, which implies an integrated consideration of a variety of interactions. It was found that *Pseudomonas aeruginosa* N002 possesses genes for degradation of alkanes, alkenes, crude oil and aromatic hydrocarbons (Oyewusi et al. 2021). This suggests that N002 can degrade a wide range of environmental pollutants. The results of this study could have implications for the remediation of polluted environments.

Using next-generation sequencing (NGS) to understand microorganism physiology and genomic characteristics is an integral part of bioremediation. It has been found that pili genes of *Geobacter* species play a significant role in facilitating the association with iron oxides when growing on insoluble oxides, which indicates insoluble oxides stimulate the expression of these genes (Childers et al. 2002). The energy-dependent mechanism by which *Geobacter* species locate and reduce Fe(III) oxides appears to be different from those in other well-studied organisms, such as Shewanella and *Geothrix* species (Singh et al. 2020, Kashyap et al. 2021). For this reason, it is imperative that the hidden physiological characteristics of Geobacter species be explored in order to enhance the limits of Geobacter species to other characteristics and metal contaminants in degraded groundwater by controlling subsurface conditions. Consequently, genomics has identified new roles for genes involved in the bioremediation of polluted environments.

Databases in the Context of Bioremediation

It has been demonstrated that bioinformatics has novel capabilities in the field of bioremediation, however, the resources that are dedicated to bioremediation processes are still scarce. Several innovative and interesting projects have been undertaken with the goal of organising the large amount of data generated through multi-omics methodologies. These projects aim to streamline bioremediation processes, while reducing costs and time. They have the potential to revolutionise the field and lead to more effective and cost-efficient solutions. One of the prime examples of such projects is the Biodegradation Database (UMBBD) (Gao et al. 2010) and the Biotransformation Pathway Database (the enviPath tool), now updated into the EAWAG-BBD/PPS database (Wicker et al. 2016, Hafner et al. 2024). These two databases provide a comprehensive platform for researchers to explore and study the environmental biotransformation pathways of contaminants (Yadav et al. 2024). They also provide invaluable information that can help inform decision-making and policy development. Another prominent example is the MetaCyc and BioCyc databases maintained by SRI International (Caspi et al. 2016). These databases provide researchers with a comprehensive overview of the metabolic pathways of bacteria, archaea, and fungi. They can also be used to identify potential sites of bioremediation and to develop strategies for mitigating the environmental impact of pollutants. A summary of the various biodegradation databases developed for studies of biodegradation and bioremediation is shown in Table 4. With pathway prediction systems, users can predict the degradation of xenobiotics.

Table 4: List of *in silico* databases that assist in the implementation of bioremediation

Database	Description	URL
	Literature Databases	
PUBMED	These databases provide users with access to papers related to biodegradation and bioremediation	https://pubmed.ncbi.nlm.nih.gov/
Google Scholar		https://scholar.google.com/
	Chemical Databases	
PubChem	Chemical compound database for commercially available compounds	https://pubchem.ncbi.nlm.nih.gov/
ZINC 15	Chemical compound database for commercially available compounds	https://zinc15.docking.org/
ACToR	Chemical toxicity data aggregated from all publicly available sources by the Environmental Protection Agency	https://norecopa.no/search?q=*&fq=cat:%22Toxicology%22&fq=db:%223r%22
CPDat	In this database, more than 48,800 chemicals are mapped to a set of terms used to categorize or describe their use in more than 16,600 consumer products	https://19january2021snapshot.epa.gov/chemical-research/chemical-and-products-database-cpdat_.html
T3DB	An innovative bioinformatics resource that provides detailed information about toxins, as well as a comprehensive list of the targets of the toxins	http://www.t3db.ca/toxins
DSSTox	DSSTox offers a comprehensive resource that can be used to enhance predictive toxicology. One of its most useful features is the mapping of bioassay results and physicochemical properties to chemical names	https://www.epa.gov/comptox-tools
ChEMBL	Database for freely available chemical compound	https://www.ebi.ac.uk/chembl/

(*Contd.*)

(*Contd.*)

Database	Description	URL
Contaminant DB	Contains detailed information on contaminants from a variety of online documents and databases. A total of 54,249 compounds are currently included in the database	https://contaminantdb.ca/
	Biodegradative Databases	
MetaCyc	One of the largest collections of metabolic pathways	http://metacyc.org
RHObase	Database of aromatic ring-hydroxylating oxygenases that provides information on bacterial Rieske-type ring-hydroxylating oxygenases	https://bicresources.jcbose.ac.in/ssaha4/Rhobase/
AromaDeg	Devoted to the aerobic degradation of aromatic compounds	http://aromadeg.siona.helmholtz-hzi.de
BioCyc	This database includes 20,005 Pathway/Genome Databases (PGDBs) for model eukaryotes and thousands of microorganisms	https://biocyc.org/
OxDBase	This database includes both monooxygenases and dioxygenases that are involved in the degradation of various xenobiotics	http://www.imtech.res.in/raghava/oxdbase/
PMBD	A database focusing on the microbial degradation of plastics	http://pmbd.genome-mining.cn/home
EAWAGBBD	This database provides information about the biodegradation of various xenobiotic compounds. It includes metabolic pathways, chemical reactions, bacteria, genes, and enzymes involved in biodegradation	http://eawag-bbd.ethz.ch/
KEGG	This database aims to provide a comprehensive understanding of high-level functions and uses of biological systems, including cells, organisms, and ecosystems, based on molecular-level information, particularly large-scale molecular datasets generated by genome sequencing and other high-throughput experiments	https://www.genome.jp/kegg/

Bioinformatics to Identify Enzymes for Bioremediation

Bioinformatics and Molecular Docking (MolDock) can provide insight into bioremediation mechanisms. This technique determines the most appropriate binding position between molecules while forming molecular complexes with the least energy conformation. The interactions occur between proteins and proteins, proteins and DNA, proteins and RNA, proteins and lipids, proteins and sugars, polymers and ligands, and many others. The substrate is a large metabolite molecule and the smaller is the ligand. Additionally, molecular docking facilitates the enzymatic investigation of certain pollutants and improves their properties. It provides insight into the interaction and communication between pollutants and enzyme molecules and can be used to predict the function of enzyme binding, degradation, or adsorption (Basharat et al. 2017). An enzyme's active site can be determined by the damage to stearic acid residues and the nature of active site pockets. In order to identify proteins and contaminants, thousands of conformational docking poses are evaluated, which aim to blindly bind proteins to unknown active sites. This provides information concerning regulation in the active pocket of the site (Bhatt et al. 2020, Verma 2023).

Water pollution caused by herbicides is one of the most pressing global concerns, as these compounds can adversely affect ecosystems as well as human health (e.g. cardiovascular and respiratory effects) (Nicolopoulou-Stamati et al. 2016). The public has received considerable attention regarding wastewater containing glyphosate in particular. An organophosphorus post-emergence herbicide with an annual production volume of 700,000 tons, glyphosate is extensively used around the world to manage the development of weeds. Due to its widespread application in agriculture in massive quantities (Poiger et al. 2020), it is widely detected in surface water and municipal wastewater. As part of the treatment of wastewater contaminated with glyphosate, a number of methods have been employed, primarily physicochemical and biological processes (Li et al. 2018). As physicochemical processes are time-consuming and costly, recent research has focused on integrating bioinformatics and biotechnology with biological organisms (bacteria and fungi) to reduce the cost of bioremediation.

In the bioremediation of glyphosate, bacterial or fungal laccase plays an essential role (Pande et al. 2022). The laccase enzyme of *Bacillus subtilis* (PDB ID: 3ZDW) has active site residues (ILE12 PRO13 THR15 THR59 GLU61 GLY147 ALA148 ILE173 ILE174 HIS175 ASP176 GLU179 LYS180 ARG181 LEU182 LYS183 LEU184 PRO185 SER186 ASP187 GLU188 TYR189 ASP190 VAL191 PRO192 ARG248 LYS249 ARG251 ARG253 LEU282 ASP301 GLN345 LYS346 ASP347 GLU348 SER349 ARG350). When *Bacillus subtilis* laccase comes in contact with glyphosate, it interacts well with pollutants (glyphosate), with binding energies –5.7 kcal/mol, as shown in (Fig. 4A, B and C). In particular, the lowest binding energy of docked complex suggesting laccase may be involved in glyphosate degradation (Bhatt et al. 2023). Additionally, more recent studies have demonstrated that *Bacillus pseudomycoides*, an environmentally friendly strain of bacterium, can be used to treat wastewater effectively. A MolDock analysis indicated that the binding energy between the enzyme and pollutants such as methylene green, crystal violet, and acid blue was between –6.3 and –6.8 kcal/mol. Using MD simulation, we can assess the stability of the enzyme in relation to pollutants. This suggests that the enzyme can effectively decolourise both individual and mixed dyes. Based on the results from the overall MD simulation and MMPBSA, and binding free energy

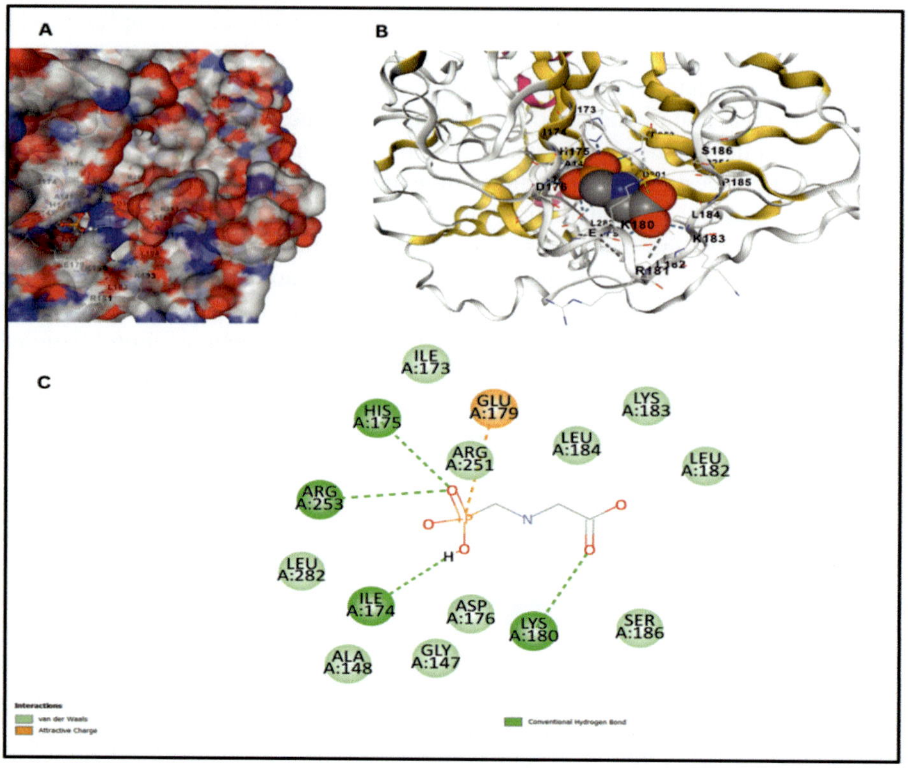

Figure 4: (A) The docked complex of a protein (laccase) and ligand (glyphosate). (B) The molecular interaction between the docked complex and the protein molecules. (C) 2-dimensional structure of the docked complex, reveals presence of vander waals, hydrogen bonds, and attractive charges between *Bacillus subtilis* laccase and glyophosate for biodegradation.

analysis, it was determined that methyl orange is very stable and has excellent binding affinity for laccase. According to the study, laccases from *B. subtilis* were most effective in degrading glyphosate and methyl-orange dyes. As a result of our study (Fig. 4C), laccase from *B. subtilis* is capable of efficiently binding with glyphosate as well as causing its degradation, and similar results have been reported for dye degradation after laccase treatment (Kumar et al. 2023). As a result of this model, it was anticipated that microbial enzymes could be used as a reliable *in silico* predictor for the bioremediation of pollutants in wastewater.

Conclusion and Future Prospects

The presence of pollutants in waterbodies gradually degrades the aquatic ecosystem. Enzyme-mediated treatment of wastewater serves as a sustainable approach, as enzymes can efficiently convert hazardous pollutants into non-toxic forms through enzyme-mediated reaction mechanisms. The enzyme-mediated process is eco-friendly, innocuous, and cost-effective for restoring the biological and physicochemical properties of polluted water. Enhancement of enzyme activity may be carried out by immobilisation on different materials, or molecular tools may be used for improved catalytic activity. Bioinformatics has also been used to combine a variety of data sources more accurately and faster than conventional methods for identifying potential bioremediation targets. It has also become possible to screen pollutants for their susceptibility to degradation by enzymes that have already been characterized by using the various *in-silico* tools. Altogether, enzyme-based wastewater bioremediation is a sustainable, efficient, and practical technique, but it has several challenges associated with it. Immobilization and other enzyme modification approaches have attempted to circumvent these issues. However, further studies on enhancing enzyme activity, as well as isolating new enzymes hold promise for reducing pollutants and improving the water quality for humans and other species. There is still a substantial gap between laboratory-based investigations, field studies, and the actual application of these enzymes at large scale in bioreactors. Future research should concentrate on the practical application of enzymes in contemporary wastewater treatment plants.

References

Aghdam, M.A., Kariminia, H.R. and Safari, S. 2016. Removal of lignin, COD, and color from pulp and paper wastewater using electrocoagulation. Desalination and Water Treatment, 57(21): 9698-9704.

Al-Hussieny, A.A., Imran, S.G. and Jabur, Z.A. 2020. The use of local blue-green algae in the bioremediation of hydrocarbon pollutants in wastewater from oil refineries. Plant Arch, 20: 797-802.

Al-Maqdi, K.A., Elmerhi, N., Athamneh, K., Bilal, M., Alzamly, A., Ashraf, S.S. et al. 2021. Challenges and recent advances in enzyme-mediated wastewater remediation—A review. Nanomaterials, 11(11): 3124. https://www.mdpi.com/2079-4991/11/11/3124.

Al-sareji, O.J., Meiczinger, M., Al-Juboori, R.A., Grmasha, R.A., Andredaki, M., Somogyi, V. et al. 2023. Efficient removal of pharmaceutical contaminants from water and wastewater using immobilized laccase on activated carbon derived from pomegranate peels. Scientific Reports, 13(1): 11933. doi:10.1038/s41598-023-38821-3.

Amari, A., Alzahrani, F.M., Alsaiari, N.S., Katubi, K.M., Rebah, F.B. and Tahoon, M.A. 2021. Magnetic metal organic framework immobilized laccase for wastewater decolorization. Processes, 9(5): 774. https://www.mdpi.com/2227-9717/9/5/774.

An, X., Zhong, B., Chen, G., An, W., Xia, X., Li, H. et al. 2021. Evaluation of bioremediation and detoxification potentiality for papermaking black liquor by a new isolated thermophilic and alkali-tolerant Serratia sp. AXJ-M. Journal of Hazardous Materials, 406: 124285. doi:https://doi.org/10.1016/j.jhazmat.2020.124285.

An, X., Zong, Z., Zhang, Q., Li, Z., Zhong, M., Long, H. et al. 2022. Novel thermo-alkali-stable cellulase-producing Serratia sp. AXJ-M cooperates with Arthrobacter sp. AXJ-M1 to improve degradation of cellulose in papermaking black liquor. Journal of Hazardous Materials, 421: 126811. doi:https://doi.org/10.1016/j.jhazmat.2021.126811.

Ariaeenejad, S., Motamedi, E. and Salekdeh, G.H. 2022. Highly efficient removal of dyes from wastewater using nanocellulose from quinoa husk as a carrier for immobilization of laccase. Bioresource Technology, 349: 126833. doi:https://doi.org/10.1016/j.biortech.2022.126833.

Bhandari, S., Poudel, D.K., Marahatha, R., Dawadi, S., Khadayat, K., Phuyal, S. et al. 2021. Microbial enzymes used in bioremediation. Journal of Chemistry, 2021(1): 8849512.

Bhatia, S.K., Vivek, N., Kumar, V., Chandel, N., Thakur, M., Kumar, D. et al. 2021. Molecular biology interventions for activity improvement and production of industrial enzymes. Bioresource Technology, 324: 124596.

Costa, S., Summa, D., Zappaterra, F., Blo, R. and Tamburini, E. 2021. Aspergillus oryzae grown on rice hulls used as an additive for pretreatment of starch-containing wastewater from the pulp and paper industry. Fermentation, 7(4): 317. https://www.mdpi.com/2311-5637/7/4/317.

Ding, S., Cargill, A.A., Medintz, I.L. and Claussen, J.C. 2015. Increasing the activity of immobilized enzymes with nanoparticle conjugation. Current Opinion in Biotechnology, 34: 242-250.

Fibriana, F., Upaichit, A. and Cheirsilp, B. 2022. Statistical optimization for cost-effective production of yeast-bacterium cell-bound lipases using blended oily wastes and their potential applications in biodiesel synthesis and wastewater bioremediation. Fermentation, 8(8): 411. https://www.mdpi.com/2311-5637/8/8/411.

George, J., Rajendran, D.S., Venkataraman, S., Rathankumar, A.K., Saikia, K., Muthusamy, S. et al. 2022. Insolubilization of *Tramates versicolor* laccase as cross-linked enzyme aggregates for the remediation of trace organic contaminants from municipal wastewater. Environmental Research, 209: 112882. doi:https://doi.org/10.1016/j.envres.2022.112882.

Haq, I., Kumar, S., Kumari, V., Singh, S.K. and Raj, A. 2016. Evaluation of bioremediation potentiality of ligninolytic Serratia liquefaciens for detoxification of pulp and paper mill effluent. Journal of Hazardous Materials, 305: 190-199. doi:https://doi.org/10.1016/j.jhazmat.2015.11.046.

Hazeena, S.H., Nair Salini, C., Sindhu, R., Pandey, A. and Binod, P. 2019. Simultaneous saccharification and fermentation of oil palm front for the production of 2,3-butanediol. Bioresource Technology, 278: 145-149. doi:https://doi.org/10.1016/j.biortech.2019.01.042.

Hernandez, K. and Fernandez-Lafuente, R. 2011. Control of protein immobilization: Coupling immobilization and site-directed mutagenesis to improve biocatalyst or biosensor performance. Enzyme and Microbial Technology, 48(2): 107-122.

Homaei, A.A., Sariri, R., Vianello, F. and Stevanato, R. 2013. Enzyme immobilization: An update. Journal of Chemical Biology, 6: 185-205.

Joseph, A., Joji, J.G., Prince, N.M., Rajendran, R. and Nainamalai, D.M. 2021. Domestic wastewater treatment using garbage enzyme. Proceedings of the International Conference on Systems, Energy & Environment (ICSEE).

Kang, X., Li, C., Ding, W., Ma, Y., Gao, S., Zhou, X. et al. 2023. Optimization of operating conditions in the biological enzymes for efficient waste activated sludge dewatering. Process Safety and Environmental Protection, 170: 545-552.

Koutsandreas, T., Ladoukakis, E., Pilalis, E., Zarafeta, D., Kolisis, F.N., Skretas, G. et al. 2019. ANASTASIA: An automated metagenomic analysis pipeline for novel enzyme discovery exploiting next generation sequencing data. Frontiers in Genetics, 10: 469.

Kumar, P. 2023. Degradation of xenobiotics by bacteria and fungi: An overview. Biopesticides International, 19(2).

Kumar, A., Mishra, T. and Kulshreshtha, A. 2023. Binding interaction of laccases from Bacillus Subtilis after industrial dyes exposure: Molecular docking and molecular dynamics simulation studies. Indian Journal of Biochemistry & Biophysics, 60: 320–330. https://doi.org/10.56042/ijbb.v60i4.69787.

Kumar, V.V., Venkataraman, S., Kumar, P.S., George, J., Rajendran, D.S., Shaji, A. et al. 2022. Laccase production by Pleurotus ostreatus using cassava waste and its application in remediation of phenolic and polycyclic aromatic hydrocarbon-contaminated lignocellulosic biorefinery wastewater. Environmental Pollution, 309: 119729. doi:https://doi.org/10.1016/j.envpol.2022.119729.

Kumari, A., Kaur, R. and Kaur, R. 2019. A review on fate and remediation techniques of oil spills. Int. J. Res. Pharm. Sci., 10(1): 111-116.

Lorenz, P. and Eck, J. 2005. Metagenomics and industrial applications. Nature Reviews Microbiology, 3(6): 510-516.

Massima Mouele, E.S., Tijani, J.O., Badmus, K.O., Pereao, O., Babajide, O., Zhang, C. et al. 2021. Removal of pharmaceutical residues from water and wastewater using dielectric barrier discharge methods—A review. International Journal of Environmental Research and Public Health, 18(4): 1683.

Mathur, P., Sanyal, D. and Dey, P. 2021. Optimization of growth conditions for enhancing the production of microbial laccase and its application in treating antibiotic contamination in wastewater. 3 Biotech, 11(2): 81. doi:10.1007/s13205-020-02627-1.

Meiczinger, M., Varga, B., Wolmarans, L., Hajba, L. and Somogyi, V. 2022. Stability improvement of laccase for micropollutant removal of pharmaceutical origins from municipal wastewater. Clean Technologies and Environmental Policy, 24(10): 3213-3223. doi:10.1007/s10098-022-02336-8.

Nadaroglu, H., Mosber, G., Gungor, A.A., Adıguzel, G. and Adiguzel, A. 2019. Biodegradation of some azo dyes from wastewater with laccase from Weissella viridescens LB37 immobilized on magnetic chitosan nanoparticles. Journal of Water Process Engineering, 31: 100866. doi:https://doi.org/10.1016/j.jwpe.2019.100866.

Narayanan, M., Ali, S.S. and El-Sheekh, M. 2023. A comprehensive review on the potential of microbial enzymes in multipollutant bioremediation: Mechanisms, challenges, and future prospects. Journal of Environmental Management, 334: 117532.

Nimkande, V.D. and Bafana, A. 2022. A review on the utility of microbial lipases in wastewater treatment. Journal of Water Process Engineering, 46: 102591. doi:https://doi.org/10.1016/j.jwpe.2022.102591.

Ortellado, L.E., Villalba, L.L., Zapata, P.D. and Fonseca, M.I. 2022. Mycoremediation of wastewater by fungal lipases. *In*: D. Udayanga, P. Bhatt, D. Manamgoda and J.M. Saez [eds.]. Mycoremediation Protocols, 213-219. New York, NY: Springer US. doi:10.1007/978-1-0716-2006-9_18.

Panda, M.K., Sahu, M.K. and Tayung, K. 2013. Isolation and characterization of a thermophilic Bacillus sp. with protease activity isolated from hot spring of Tarabalo, Odisha, India. Iranian Journal of Microbiology, 5(2): 159.

Pang, H., Pan, X., Li, L., He, J., Zheng, Y., Qu, F. et al. 2020. An innovative alkaline protease-based pretreatment approach for enhanced short-chain fatty acids production via a short-term anaerobic fermentation of waste activated sludge. Bioresource Technology, 312: 123397.

Pang, H., Jiao, Q., He, J., Zhang, Z., Wang, L., Yan, Z., et al. 2022. Enhanced short-chain fatty acids production through a short-term anaerobic fermentation of waste activated sludge: sSynergistic pretreatment of alkali and alkaline hydrolase blend. Journal of Cleaner Production., 342: 130954.

Patel, A.B., Shaikh, S., Jain, K.R., Desai, C. and Madamwar, D. 2020. Polycyclic aromatic hydrocarbons: Sources, toxicity, and remediation approaches. Frontiers in Microbiology, 11: 562813.

Peilin, G., Meng, C., Lichao, Z., Yuejun, S., Minghao, M. and Lingyun, W. 371 2019. Study on water ecological restoration technology of river. IOP Conference Series: Earth and Environmental Science. IOP Publishing, p. 032025 3.

Pinto, É.S.M., Dorn, M. and Feltes, B.C. 2020. The tale of a versatile enzyme: Alpha-amylase evolution, structure, and potential biotechnological applications for the bioremediation of n-alkanes. Chemosphere, 250: 126202. doi:https://doi.org/10.1016/j.chemosphere.2020.126202

Posso Mendoza, H., Pérez Salinas, R., Tarón Dunoyer, A., Tatis, C.C., Morgado-Gamero, W.B., Castillo Ramírez, M. et al. 2020. Evaluation of Enzymatic Extract with Lipase Activity of Yarrowia Lipolytica. An Application of Data Mining for the Food Industry Wastewater Treatment. Cham: Springer International Publishing. pp. 304-313.

Pounsamy, M., Somasundaram, S., Palanivel, S., Balasubramani, R., Chang, S.W., Nguyen, D.D. et al. 2019. A novel protease-immobilized carbon catalyst for the effective fragmentation of proteins in high-TDS wastewater generated in tanneries: Spectral and electrochemical studies. Environmental Research, 172: 408-419. doi:https://doi.org/10.1016/j.envres.2019.01.062.

Pounsamy, M., Rajendran, G., Ramanujam, A., Priyadharsini, V. and Ganesan, S. 2024. Sustainable valorization of protein-rich tannery saline wastewater: Protein hydrolysate synthesis via protease and Fe_3O_4 porous carbon catalyst. Journal of Water Process Engineering, 60: 105149. doi:https://doi.org/10.1016/j.jwpe.2024.105149.

Prayogo, F. A., Budiharjo, A., Kusumaningrum, H.P., Wijanarka, W., Suprihadi, A. and Nurhayati, N. 2020. Metagenomic applications in exploration and development of novel enzymes from nature: A review. Journal of Genetic Engineering and Biotechnology, 18(1): 39. doi:https://doi.org/10.1186/s43141-020-00043-9.

Qin, X., Qian, J., Yao, G., Zhuang, Y., Zhang, S. and Chu, J. 2011. GAP promoter library for fine-tuning of gene expression in Pichia pastoris. Applied and Environmental Microbiology, 77(11): 3600-3608.

Rathod, M.G. and Pathak, A.P. 2018. Efficient decolorization of textile dyes by alkaline protease producing bacterial consortia. Indian J. Mar. Sci., 47(7): 1468-1477.

Samal, K., Bandyopadhyay, R. and Dash, R.R. 2022. Biological treatment of contaminants of emerging concern in wastewater: A review. Journal of Hazardous, Toxic, and Radioactive Waste, 26(2): 04022002.

Şen, A., Kargar, K., Akgün, E. and Pınar, M. Ç. 2020. Codon optimization: A mathematical programing approach. Bioinformatics, 36(13): 4012-4020.

Shakya, M., Lo, C.-C. and Chain, P. S. 2019. Advances and challenges in metatranscriptomic analysis. Frontiers in Genetics, 10: 904.

Sonune, N. and Garode, A. 2018. Isolation, characterization and identification of extracellular enzyme producer Bacillus licheniformis from municipal wastewater and evaluation of their biodegradability. Biotechnology Research and Innovation, 2(1): 37-44. doi:https://doi.org/10.1016/j.biori.2018.03.001.

Su, X., Yang, X., Long, H., Li, Y., Sun, W., Mo, T. et al. 2024. Construction of immobilized laccase hydrogels via sodium alginate-dopamine/polyethylene glycol and its efficient degradation of dyeing wastewater. International Journal of Biological Macromolecules, 134929.

Sulman, A.M., Matveeva, V.G. and Bronstein, L.M. 2022. Cellulase immobilization on nanostructured supports for biomass waste processing. Nanomaterials, 12(21): 3796. https://www.mdpi.com/2079-4991/12/21/3796.

Tongco, J.V., Kim, S., Oh, B.-R., Heo, S.-Y., Lee, J. and Hwang, S. 2020. Enhancement of hydrolysis and biogas production of primary sludge by use of mixtures of protease and lipase. Biotechnology and Bioprocess Engineering, 25: 132-140.

Unuofin, J.O., Okoh, A.I. and Nwodo, U.U. 2019. Aptitude of oxidative enzymes for treatment of wastewater pollutants: A laccase perspective. Molecules, 24(11): 2064. https://www.mdpi.com/1420-3049/24/11/2064.

Vincent, C.S., Nwuche, C.O., Mayel, M.H. and Eze, S.O.O. 2022. Production, partial purification, and characterization of lipase from Aspergillus niger and its application in treatment of vegetable oil effluent. Biomolecular Engineering, 7(1): 1-7.

Vocciante, M., Finocchi, A., De Folly D' Auris, A., Conte, A., Tonziello, J., Pola, A. et al. 2019. Enhanced oil spill remediation by adsorption with interlinked multilayered graphene. Materials, 12(14): 2231.

Waseem, A., Ali, S., Khan, Q.F., Khalid, S.W., Shah, T.A., Salamatullah, A.M., et al. 2024. Exploring chitosan-immobilized Rhizopus oligosporus lipase for olive-mill wastewater treatment. International Journal of Environmental Science and Technology, 21(14): 9097-9110. doi:10.1007/s13762-024-05808-0.

Xu, Q., Wu, B. and Chai, X. 2022. In situ remediation technology for heavy metal contaminated sediment: A review. International Journal of Environmental Research and Public Health, 19(24): 16767.

Xu, X., Pose-Boirazian, T., Eibes, G., McCoubrey, L.E., Martínez-Costas, J., Gaisford, S. et al. 2022. A customizable 3D printed device for enzymatic removal of drugs in water. Water Research, 208: 117861.

Yang, X., Zhao, J., Cavaco-Paulo, A., Su, J. and Wang, H. 2023. Encapsulated laccase in bimetallic Cu/Zn ZIFs as stable and reusable biocatalyst for decolorization of dye wastewater. International Journal of Biological Macromolecules, 233: 123410. doi:https://doi.org/10.1016/j.ijbiomac.2023.123410.

Yao, L.W., Ahmed Khan, F.S., Mubarak, N.M., Karri, R.R., Khalid, M., Walvekar, R. et al. 2022. Insight into immobilization efficiency of Lipase enzyme as a biocatalyst on the graphene oxide for adsorption of Azo dyes from industrial wastewater effluent. Journal of Molecular Liquids, 354: 118849. doi:https://doi.org/10.1016/j.molliq.2022.118849.

Younes, I. and Rinaudo, M. 2015. Chitin and chitosan preparation from marine sources: Structure, properties and applications. Marine Drugs, 13(3): 1133-1174.

Zhang, Y., Peng, D., Luo, Y., Huang, D., Guo, X. and Zhu, L. 2020. Cellulase modified waste biomass to remove sulfamethazine from aqueous solutions. Science of The Total Environment. 731: 138806. doi:https://doi.org/10.1016/j.scitotenv.2020.138806.

Index

For Product Safety Concerns and Information please contact our EU representative GPSR@taylorandfrancis.com Taylor & Francis Verlag GmbH, Kaufingerstraße 24, 80331 München, Germany